Hierarchical Optimization and Mathematical Physics

Applied Optimization

Volume 37

Series Editors:

Panos M. Pardalos
University of Florida, U.S.A.

Donald Hearn
University of Florida, U.S.A.

The titles published in this series are listed at the end of this volume.

Hierarchical Optimization and Mathematical Physics

by

Vladimir Tsurkov

Computing Center,
Russian Academy of Sciences,
Moscow, Russia

SPRINGER-SCIENCE+BUSINESS MEDIA, B.V.

A C.I.P. Catalogue record for this book is available from the Library of Congress.

ISBN 978-1-4613-7112-0 ISBN 978-1-4615-4667-2 (eBook)
DOI 10.1007/978-1-4615-4667-2

Printed on acid-free paper

Contents

Preface

This book should be considered as an introduction to a special class of hierarchical systems of optimal control, where subsystems are described by partial differential equations of various types. Optimization is carried out by means of a two-level scheme, where the center optimizes coordination for the upper level and subsystems find the optimal solutions for independent local problems. The main algorithm is a method of iterative aggregation. The coordinator solves the problem with macrovariables, whose number is less than the number of initial variables. This problem is often very simple. On the lower level, we have the usual optimal control problems of mathematical physics, which are far simpler than the initial statements. Thus, the decomposition (or reduction to problems of less dimensions) is obtained. The algorithm constructs a sequence of so-called disaggregated solutions that are feasible for the main problem and converge to its optimal solution under certain assumptions (e.g., under strict convexity of the input functions).

Thus, we bridge the gap between two disciplines: optimization theory of large-scale systems and mathematical physics. The first motivation was a special model of branch planning, where the final product obeys a preset assortment relation. The ratio coefficient is maximized. Constraints are given in the form of linear inequalities with block diagonal structure of the part of a matrix that corresponds to subsystems. The central coordinator assembles the final production from the components produced by the subsystems. Therefore, the binding constraints of the initial matrix are specific: their submatrices are diagonal. This structure suggests a special decomposition algorithm, where variables from various blocks are aggregated. Here, all the difficulties related to the large number of dimensions, i.e., the large amount of subsystems and components they produce, are reduced to a simple aggregated problem of the upper level, which consists in finding the minimal element of a large-dimension matrix.

Substantiation of the decomposition scheme is based on the duality principles of linear programming. Local monotonicity with respect to the functional of the iterative process is important. This scheme of iterative aggregation is generalized to a wide class of hierarchical problems, for which duality principles hold. It is a question of block separable problems of mathematical programming, calculus of variations, and of optimal control. We go on to consider systems with subsystems described by partial differential equations and consider models of distribution of energy resources, propagation of heat, oscillation damping, etc. The approach is generalized for nonseparable problems: for example, the so-called systems with cross-connection, when, e.g., there is a heat exchange between subsystems.

The format of the book is as follows:

Necessary knowledge from the theory of extremal problems is given in Chapter 1. This is mostly related to duality theory and parametric programming, which are used to justify the decomposition method. Here, the main model of branch planning is described, and the iterative algorithm is constructed. Moreover, we study the properties of the problem's solution with aggregated variables of the upper level.

In Chapter 2, the scheme of iterative aggregation is generalized to block separable problems of linear, quadratic, convex mathematical programming and classical calculus of variations. Particular attention is given to the criterion of optimality of disaggregated solutions (condition of termination of the iterative process) and local monotonicity with respect to the functional. The main mathematical techniques are the duality theorems and Kuhn-Tucker theory. The necessary conditions for optimality in the form of Euler equations for the classical calculus of variations are used in the justification.

In Chapter 3, we state hierarchical problems of optimal control, where subsystems are described by ordinary and partial differential equations. Illustrative examples are given, where all intermediary problems in the iterative process are solved analytically. Particular emphasis is placed on block linear quadratic problems of optimal control. Here, we propose a reduction method for the systems of linear algebraic equations that are finally obtained after application of the Pontryagin maximum principle.

In Chapter 4, we demonstrate the efficiency of the decomposition using iterative aggregation. At first, this is done for nonlinear block statements, for which the direct application of the maximum principle leads to intractable problems. Here, model hierarchical control problems are stated, where the subsystems are described by equations of mathematical physics. Their physical meaning relates to the optimal distribution of resources over subsystems. Use of Fourier series leads either to efficient decomposition or to a reduction in the systems of algebraic equations. The results of the numerical computations testify to the fast convergency and efficiency of the decomposition algorithm.

To relate the algorithm of iterative aggregation to other approaches as well as to areas of application, we give in the Appendix a brief survey of methods in hierarchical optimization. These are schemes based on Dantzig-Wolfe and Kornai-Liptak decomposition, the method of Lagrangian function, parametric decomposition, other approaches to iterative aggregation, etc.

The reader is assumed to be acquainted with mathematical programming, optimal control, and with mathematical physics. References to additional literature are given in comments to the chapters.

In the book, two numbers are used to identify each formulas and theorems. The first number denotes the number of the section of this chapter, the second is the number of the formula or theorem. This should be specially noted when a reference to a theorem or a formula from another chapter is made. In this book, we use common mathematical symbols and notations. Sometimes, their meaning is additionally cleared from the text. In particular, by $[1:N]$, we denote the set of sequential integers from 1 to N inclusive.

Chapter 1

The Main Model and Constructions of the Decomposition Method

First, we consider the foundations of the theory of extremal problems. Particular emphasis is placed on the methods that are used in the construction of the decomposition algorithm based on the iterative aggregation of variables. These methods include the principles of duality, Kuhn-Tucker theorems, and marginal values. The statements are given without proofs. The corresponding justification can be found in the cited literature.

Below, we study a special model of branch planning. The year plan of the branch is formed on the basis of demand for its production. Here, a part of the assignment is fixed rigidly or bounded from below by the external demand, and the other part is at the disposal of the planning body. Maximization of the free nomenclature with a given assignment ratio is taken as the optimization criterion. We obtain an optimization problem of linear programming with block diagonal structure of the constraint part. The binding constraints and criterion have a specificity caused by a particular planning model. This specificity leads to the idea of aggregated variables in the construction of the decomposition method for the solution of the block problem. All difficulties related to the high dimension of the main problem include intermediary aggregated problems, which are reduced to the search for a minimal element of matrices with large number of rows. The decomposition method is an iterative process that forms a sequence of solutions feasible for the main problem. An optimality criterion for these solutions (termination condition of the process) is established. We prove monotonicity with respect to the functional. Results of numerical computations are given. Then, we consider modifications of the model of branch planning and the development of the decomposition method. We study models that are close to the main one from the standpoint of the application of the iterative algorithm. Special emphasis is placed on aggregated problems. The simplicity of solutions of these problems testifies to the efficiency of the decomposition method. We introduce linear binding constraints of general form, and the aggregated problem is reduced to the search for minimal elements of a finite set. In another case, the aggregated problem and the dual problem allow analytical treatment, and one does not have the nonuniqueness of the dual estimates. We study the main model with a more general criterion. Finally, some parameters are considered as random variables. Here, the stochastic programming statement is involved. The corresponding aggregated problem is reduced to the one-dimensional maximization of the concave piecewise linear function. This characterizes the efficiency of the iterative aggregation.

§1. Necessary Knowledge from the Theory of Extremal Problems

We consider the following problem:

$$f(x) \to \max, \tag{1.1}$$

$$g_k(x) \leq 0, \qquad k \in [1:m], \tag{1.2}$$

$$x_s \geq 0, \qquad s \in [1:n]. \tag{1.3}$$

Here, x is the point of the n-dimensional Euclidean space R^n. Real-valued functions $f(x)$, $g_k(x)$, $k \in [1:m]$ are defined on the nonnegative octant $R^n_+\{x = (x_1, \ldots, x_n) | x_s \geq 0\}$.

The point x is called *feasible* if it satisfies (1.2), (1.3). *Extremal (optimal) solution* of problem (1.1)–(1.3) is the point x^* that is feasible, and for which the inequalities $f(x^*) \geq f(x)$ hold for all feasible x.

The notions of convex and concave functions are important. The function $f(x)$ given on the nonnegative octant $\mathbf{R}^n_+$ is called *convex*, if, for any two points x_1 and x_2 from $\mathbf{R}^n_+$ and any real τ, $0 \leq \tau \leq 1$, the inequality $f(\tau x_2 + (1-\tau)x_1) \leq \tau f(x_2) + (1-\tau) f(x_1)$ holds. A function $f(x)$ is called *concave* if $-f(x)$ is convex. The sum of convex functions, multiplied by a nonnegative number, is a convex function. Convex functions given on a closed set are continuous at each internal point of the set. For a continuously differentiable convex function with gradient (grad $f(x)$) and any two points x_1 and x_2, the following inequality holds:

$$f(x_2) \geq f(x_1) + (\text{ grad } f(x_1), (x_2 - x_1)). \tag{1.4}$$

The corresponding inequality for the concave function has the following form:

$$f(x_2) \leq f(x_1) - (\text{ grad } f(x_1), (x_2 - x_1)). \tag{1.5}$$

A linear function is both convex and concave. The quadratic function $f(x) = (x, Dx) + (c, x)$ is convex if and only if the matrix D is nonnegative definite, i.e. $(x, Dx) \geq 0$ for any $x \in \mathbf{R}^n$.

The linear programming problem

$$(c, x) \to \max,$$

$$Ax \leq b,$$

$$x \geq 0$$

is a particular case of the problem (1.1)-(1.3). Here, c, b are vectors of dimensions m, n, respectively, A is an $m \times n$-matrix.

Consider the statement of extremal problems in the infinite-dimensional case. Let B be a Banach space. The set $K \subset B$ is called *the convex cone* if, for all $x_1, x_2 \in K$ and any two nonnegative numbers τ_1, τ_2, the relation $\tau_1 x_1 + \tau_2 x_2 \in K$ holds. The convex cone introduces partial order in B. Namely, $x_2 \geq x_1$ if $x_2 - x_1 \in K$. The cone K is called *closed* if it contains its limit points.

Let operator $T(x)$ take the Banach space B_1 to the Banach space B_2, which is partially ordered by the closed convex cone K. The operator $T(x)$ is called *convex* if

$$T(\tau x_2 + (1-\tau)x_1) \leq \tau T(x_2) + (1-\tau)T(x_1)$$

$$\forall x_1, x_2 \in B_1, \qquad 0 \leq \tau \leq 1.$$

The following inequality that generalizes (1.4) is valid for the convex operator $T(x)$ with Frechet derivative $T'(x)$:

$$T(x_2) \geq T(x_1) + T'(x_1)(x_2 - x_1). \tag{1.6}$$

The functional $\varphi(x)$ takes B_1 to $\mathbf{R}^1$.

Let the space B_1 be partially ordered by the closed convex cone K. Then, the generalization of the problem (1.1)-(1.3) to the normalized spaces has the form:

$$\varphi(x) \to \max,$$

$$T(x) \leq 0(K_2), \tag{1.7}$$

$$x \geq 0(K_1).$$

We return to the initial problem (1.1)–(1.3). Let the function $f(x)$ have the following form:

$$f(x) = \sum_{j=1}^{J} f_j(x_j). \tag{1.8}$$

Here, for each fixed $j \in [1 : J]$, the point $x_j = \{x_j^1, \dots, x_j^i, \dots x_j^I\}$ belongs to the I-dimensional Euclidean space $\mathbf{R}_j^I$. Let the restrictions of the problem split up into two groups, the local ones:

$$\begin{gathered} g_j^k(x_j) \le 0, \qquad j \in [1 : J], \qquad k \in [1 : K_j], \\ x_j \in \mathbf{R}_{j+}^I, \qquad j \in [1 : J] \end{gathered} \tag{1.9}$$

and the general ones:

$$g^l(x) = \sum_{j=1}^{J} \tilde{g}_j^l(x_j) \le 0, \qquad l \in [1 : L]. \tag{1.10}$$

Problem (1.8)–(1.10) is called a *block separable* problem of mathematical programming. Relations (1.9) are *block* constraints, relations (1.10) are *binding* constraints. The solution x of the problem (1.8)–(1.11) is sought in the Cartesian product of spaces $\mathbf{R}_{j+}^I$, $j \in [1 : J]$. The problem has IJ variables. The goal of the decomposition is the reduction of the initial problem to problems of lesser dimensions.

In linear cases, a part of matrices of conditions of the problem (1.8)–(1.10) has the block diagonal structure:

$$\sum_{j=1}^{J} (c_j, x_j) \to \max, \quad A_j x_j \le b_j, \quad x_j \ge 0, \qquad j \in [1 : J], \quad \sum_{j=1}^{J} A_j^0 x_j \le b^0.$$

Here, vectors c_j, b_j, b^0 have dimensions I, K_j, L respectively; matrices A_j, A_j^0 have dimensions $K_j \times I$, $L \times I$, respectively, where $j \in [1 : J]$. The Dantzig-Wolfe and Kornai-Liptak decomposition principles were formulated for the linear problems of this type.

In some cases, the block diagonal structure of problem conditions is connected with the specificity of the considered model. In other cases, special techniques for isolating blocks in matrices are used.

We study the questions of the duality theory and Kuhn-Tucker theory.

Consider the problem of linear programming

$$\sum_{s=1}^{n} c_s x_s \to \max, \tag{1.11}$$

$$\sum_{s=1}^{n} a_{ks} x_s \leq b_k, \qquad k \in [1:m], \tag{1.12}$$

$$x_s \geq 0, \qquad s \in [1:n]. \tag{1.13}$$

The problem (1.11)–(1.13) is tightly connected with the dual problem

$$\sum_{k=1}^{m} b_k \lambda_k \to \min, \tag{1.14}$$

$$\sum_{k=1}^{m} a_{ks} \lambda_k \geq c_s, \qquad s \in [1:n], \tag{1.15}$$

$$\lambda_k \geq 0, \qquad k \in [1:m]. \tag{1.16}$$

The problem (1.11)–(1.13) and (1.14)–(1.16) are called *the conjugated pair*. The following statement of the duality theory holds:

Theorem 1.1. *If x and λ are feasible solutions of the problems (1.11)–(1.13) and (1.14)–(1.16), respectively, then the inequality*

$$\sum_{s=1}^{n} c_s x_s \leq \sum_{k=1}^{m} b_k \lambda_k.$$

holds.

Theorem 1.2. *If, for the feasible solutions x and λ of the problems (1.11)–(1.13) and (1.14)–(1.16), respectively, the equality*

$$\sum_{s=1}^{n} c_s x_s = \sum_{k=1}^{m} b_k \lambda_k \tag{1.17}$$

holds, then x and λ are optimal solutions of the corresponding problems.

Theorem 1.3. *If one of the problems of the conjugated pairs (1.11)–(1.13) and (1.14)–(1.16) has a solution, then the other problem is also solvable. Here, for any optimal solutions x^* and λ^* of the problems (1.11)–(1.13) and (1.14)–(1.16), the equality (1.17) holds.*

Theorem 1.3 is called *the first duality theorem* of linear programming. Consider conditions of conjugated pairs. Inequalities $x_s \geq 0$ in (1.13) and $\sum_{k=1}^{m} a_{ks}\lambda_k \geq c_s$ in (1.15) for the fixed index $s \in [1:n]$ as well as inequalities $\sum_{s=1}^{n} a_{ks}x_s \leq b_k$ in (1.12) and $\lambda_k \geq 0$ in (1.16) for the fixed index $k \in [1:m]$ are called *the pairs of dual conditions* of the conjugated problems.

The condition of the problem (1.11)–(1.13) or the problem (1.14)–(1.16) is called *fixed* if, for any optimal solution of the corresponding pair, it holds as a precise equality. A condition is called *free* if it holds as a strict inequality for at least one optimal solution of the corresponding problem.

Theorem 1.4. *If the conjugated problems (1.11)–(1.13) and (1.14)–(1.16) are solvable, then, in each pair of its dual conditions, one condition is free and the other is fixed.*

Theorem 1.4 is called *the second duality theorem* of linear programming.

Consider the statements of the duality theory for nonlinear problems of mathematical programming. We return to the problem (1.1)–(1.3) under the assumption that the functions $-f(x)$, $g_k(x)$, $k \in [1:m]$ are convex. Compose the Lagrange function (1.1)–(1.3):

$$L(x, \lambda) = f(x) - \sum_{k=1}^{m} \lambda_k g_k(x).$$

Here, $\lambda = (\lambda_1, \ldots, \lambda_m)$ is the vector of Lagrange multipliers (dual variables). Let $\lambda_k \geq 0$, $k \in [1:m]$ be fixed. Consider the function

$$H(\lambda) = \max_{x \in \mathbf{R}_+^n} \left[f(x) - \sum_{k=1}^{m} \lambda_k g_k(x) \right].$$

For $\lambda \in \mathbf{R}_+^m$, the problem $H(\lambda) \to \min$ is called *dual* to the problem (1.1)–(1.3).

The following duality theorems are valid:

Theorem 1.5. *Let $\lambda \in R_+^m$ and x be a feasible solution to the problem (1.1)–(1.3). Then, the inequality*

$$f(x) \leq H(\lambda)$$

holds.

Further, we will use *the Slater condition,* , which is formulated in the following way: there exists a point $x_1 \in \mathbf{R}^n_+$ such that $g_k(x_1) < 0$, $k \in [1:m]$.

Theorem 1.6 If x^* is an extremum solution to the problem (1.1)–(1.3), and the Slater condition holds, then

$$\min_{\lambda \in \mathbf{R}^m_+} H(\lambda) = f(x^*).$$

Theorem 1.7. (Kuhn-Tucker). *Let x^* be an extremum solution to the problem (1.1)–(1.3), and the Slater condition holds. Then, there exists vector $\lambda^* \in \mathbf{R}^m_+$ such that*

$$L(x^*, \lambda^*) = \max_{x \in \mathbf{R}^n_+} L(x, \lambda^*) = H(\lambda^*), \tag{1.18}$$

$$\lambda^*_k g_k(x^*) = 0, \qquad k \in [1:m]. \tag{1.19}$$

Conditions (1.18) and (1.19) are sufficient conditions for the feasible point x^* to be a solution of the problem (1.1)–(1.3). Theorem 1.7 is equivalent to the following statement:

Theorem 1.8 (Kuhn-Tucker theorem on the saddle point). *For a point x^* to be an extremum solution to the problem (1.1)–(1.3), it is necessary that there exists vector $\lambda^* \in \mathbf{R}^m_+$ such that equalities*

$$\max_{x \in \mathbf{R}^n_+} L(x, \lambda^*) = L(x^*, \lambda^*) = \min_{\lambda \in R^m_+} L(x^*, \lambda)$$

hold.

Components of the vector λ^* are called *optimal Lagrange multipliers.*

Consider the statements of dual problems in particular cases. In the matrix form, the problem of quadratic programming is written by the following form:

$$(c, x) + (x, Dx) \to \max,$$

$$Ax \leq b, \tag{1.20}$$

$$x \geq 0.$$

The matrix D is considered to be symmetric, and the corresponding quadratic form is considered to be positive definite. The problem dual to (1.20) is

$$
\begin{aligned}
&-(x, Dx) + (b, \lambda) \rightarrow \min, \\
&-2Dx + A^T\lambda \geq c, \\
&x \geq 0, \qquad \lambda \geq 0.
\end{aligned}
\tag{1.21}
$$

We present the dual problem for the problem (1.1)–(1.3) when the functions $-f(x)$, $g_k(x)$, $k \in [1:m]$ are convex and continuously differentiable:

$$
\begin{aligned}
&\left[f(x) - \sum_{k=1}^{m} \lambda_k g_k(x) \right] \rightarrow \min, \\
&\frac{\partial f(x)}{\partial x_s} - \sum_{k=1}^{m} \lambda_k \frac{\partial g_k(x)}{\partial x_s} = 0, \qquad x_s > 0, \\
&\frac{\partial f(x)}{\partial x_s} - \sum_{k=1}^{m} \lambda_k \frac{\partial g_k(x)}{\partial x_s} \leq 0, \qquad x_s = 0, \qquad s \in [1:n], \\
&\lambda_k \geq 0, \qquad k \in [1:m].
\end{aligned}
\tag{1.22}
$$

Now, we study duality for problem (1.7) written in terms of normalized spaces. Let $T'(x)$ be the Frechet derivative of the operator $T(x)$ at the point x. By definition, $T'(x)$ is a restricted linear operator that acts from B_1 into B_2. Let the spaces B_1^* and B_2^* conjugated for B_1 and B_2 be partially ordered by conjugated cones K_1^+ and K_2^+. If K is a convex cone in B, then, by definition, we have

$$
K^+ = \{\xi | [\xi, y] \geq 0 \text{ for all } y \in K\}.
$$

Here, $[\xi, y]$ denotes the value of the linear functional ξ for the element y. It is known that if K is a convex cone in B, then K^+ is the convex closed cone in B^*. It is assumed that there exists linear operator $T'^*(x)$ acting from B_2^* to B_1^*, which is a conjugated operator for the Frechet derivative $T'(x)$, i.e., the relation $[\eta, T'(x)y] = [T'^*(x)\eta, y]$ holds for all $y \in B_1$ and $\eta \in B_2^*$. Frechet derivative of the functional $\varphi(x)$ can be considered as the element $\omega(x)$ of the conjugated space B_1^*, hence the equality $[\varphi'(x), y] = [\omega(x), y]$ holds for all $y \in B_1$.

In what follows, we consider convex functional $-\varphi(x)$ and operator $T(x)$. For the problem (1.7), we write down the Lagrange functional: $L(x,\lambda) = \varphi(x) - [\lambda, T(x)]$, $\lambda \in B_2^*$.

The problem dual to (1.7) has the following form:

$$\begin{gathered}
\varphi(x) - [\lambda, T(x)] - [\omega(x), x] + [\lambda, T'(x)x] \to \min, \\
\left[T'^*(x), \lambda\right] \geq \omega(x)(K_1^+), \\
\lambda \geq 0(K_2^+), \qquad x \geq 0(K_1).
\end{gathered} \tag{1.23}$$

The Slater condition for infinite-dimensional problems is generalized in the following way: there is a strictly positive functional e in the space B_2, and there exists a feasible element $\beta > 0$ such that $T(x_1) \leq -\beta e(K_2)$. We formulate the Kuhn-Tucker theorems under the assumption of closedness of a certain cone related to the operator $F(x) = T'^*(x)$.

Theorem 1.9. *Let x^* be extremal solution to problem (1.7), and the generalized Slater condition hold. Let the cone K_F*

$$\begin{aligned}
&K_F = \{(\mu, r) | \mu = \sigma - F_\eta,\ r = d + (z^*, \eta), \\
&\quad \sigma \geq 0(K_1^+),\ \eta \geq 0(K_2^+),\ d \geq 0\}, F = F(x^*), \qquad z^* = [T'(x^*), x^*] - T(x^*)
\end{aligned}$$

be closed in the space $B_1^* \times \mathbf{R}^1$. Then, there exists a solution λ^* of problem (1.23) such that conditions of optimality

$$[\lambda^*, T(x^*)] = 0, \qquad [\lambda^*, T'(x^*)x^*] - [\omega(x^*), x^*] = 0 \tag{1.24}$$

hold.

Conditions (2.14) can be written in the form

$$\left[\lambda^*, \frac{\partial L(x^*, \lambda^*)}{\partial \lambda}\right] = 0, \qquad \left[x^*, \frac{\partial L(x^*, \lambda^*)}{\partial x}\right] = 0. \tag{1.25}$$

Theorem 1.10 (about the saddle point). *Let conditions of Theorem 1.9 be satisfied, and element x^* be extremum for problem (1.7). Then, there exists an element $\lambda^* \geq 0$ such that the pair (x^*, λ^*) is the saddle point of the Lagrange functional, i.e., inequalities*

$$L(x, \lambda^*) \leq L(x^*, \lambda^*) \leq L(x^*, \lambda)$$

hold for all $x \geq 0(K_1)$ *and* $\lambda \geq 0(K_2^+)$.

Now, we study the parametric programming and marginal values. Let us consider the problem of linear programming (1.11)–(1.13) and suppose that its coefficients depend on the parameter ρ in the following way: $a_{ks}(\rho) = a_{ks} + \rho\delta_{ks}$, $c_s(\rho) = c_s + \rho\delta_{0s}$, $b_k(\rho) = b_k + \rho\delta_{k0}$. Then, the maximal value of the functional (1.11) is a function of ρ. We denote this function by $\Psi(\rho)$. Assume that the problem (1.11)–(1.13) is solvable in a right neighborhood of the point $\rho = 0$. We are interested in the derivative

$$\Psi'(0) = \lim_{\rho \to +0} [\Psi(\rho) - \Psi(0)]/\rho$$

or, which is the same, the derivative Ψ'_Δ given by the direction $\Delta = \{\delta_{ks}, \delta_{0s}, \delta_{k0}\}$. The derivative $\Psi'(0)$ is called the *marginal value.*

We introduce the function

$$\Gamma(x, \lambda) = \sum_{s=1}^{n} \delta_{0s} x_s + \sum_{k=1}^{m} \delta_{k0} \lambda_k - \sum_{k=1}^{m} \sum_{s=1}^{n} \delta_{ks} \lambda_k x_s,$$

which is given by the direction Δ.

Let $M_x(0)$ and $M_\lambda(0)$ be sets of solutions of the pair of conjugated problems for $\rho = 0$.

Theorem 1.11. *Let $M_x(0)$ and $M_\lambda(0)$ be nonempty bounded sets. Then, the derivative Ψ'_Δ along an arbitrary direction exists, and the following relation holds:*

$$\Psi'(0) = \Psi'_\Delta = \max_{x \in M_x(0)} \min_{\lambda \in M_\lambda(0)} \Gamma(x, \lambda) = \max_{\lambda \in M_\lambda(0)} \min_{x \in M_x(0)} \Gamma(x, \lambda) = \Gamma(x^*, \lambda^*),$$

where (x^*, λ^*) is the saddle point of the functional $\Gamma(x, \lambda)$ under condition $x \in M_x(0)$ and $\lambda \in M_x(0)$.

Theorem 1.11 is deduced from the following statements about the properties of solutions to the problems of parametric linear programming:

Theorem 1.12. *If $M_x(0)$ and $M_\lambda(0)$ are nonempty bounded sets, then, for any Δ, the problem (1.11)–(1.13) is solvable for all values ρ from the right neighborhood of zero.*

By $R(z, M)$, denote the difference from the point z and the closed set M, which is equal to the minimum of distances between the point z and the points of $y \in M$.

Theorem 1.13. *Let nonempty sets $M_x(0)$ and $M_\lambda(0)$ of solutions of conjugated problems be bounded. Then, $R(x(\rho), M_x(0)) \to 0$, $R(\lambda(\rho), M_\lambda(0)) \to 0$ for $\rho \to 0$, where $x(\rho)$, $\lambda(\rho)$ are solutions of conjugated problems for the parameter ρ.*

Theorem 1.13 is a theorem about the stability of the solution of the problem of parametric linear programming.

Now, we study the questions of stability for finite-dimensional convex programming. Let $\Phi(\rho)$ be a point-set mapping, which takes each point ρ of the metric space N to a set $\Phi(\rho)$ of the finite-dimensional Euclidean space. The mapping $\Phi(\rho)$ is said to be *semicontinuous from above at the point* $\rho_0 \in N$, if, for any $\varepsilon > 0$, there exists $\delta > 0$ such that, for $||\rho - \rho_0|| < \delta$, the set $\Phi(\rho)$ is contained in the ε-neighborhood of the set $\Phi(\rho_0)$. Let the function $\chi(x, \lambda, \rho)$ be defined and continuous on the set $P_x \times P_\lambda \times N$, concave with respect to x for all values of λ, ρ from $P_\lambda \times N$ and convex with respect to λ for all values x, ρ from $P_x \times N$. Here, P_x and P_λ are concave closed sets in Euclidean spaces.

We introduce the following notations:

$$\nu(x, \rho) = \min_{\lambda \in P_\lambda} \chi(x, \lambda, \rho), \qquad \pi(\lambda, \rho) = \max_{x \in P_x} \chi(x, \lambda, \rho).$$

Consider the parametric family of problem pairs:

$$\nu(x, \rho) \to \max, \qquad x \in P_x; \tag{1.26}$$

$$\pi(\lambda, \rho) \to \min, \qquad \lambda \in P_\lambda. \tag{1.27}$$

Let

$$v_1(\rho) = \max_{x \in P_x} \nu(x, \rho), \qquad v_2(\rho) = \min_{\lambda \in P_\lambda} \pi(\lambda, \rho).$$

We introduce the sets of solutions of the problems (1.26) and (1.27) depending on the parameter:

$$M_x(\rho) = \{x | \nu(x, \rho) = v_1(\rho),\ x \in P_x\},$$

$$M_\lambda(\rho) = \{\lambda | \pi(\lambda, \rho) = v_2(\rho),\ \lambda \in P_\lambda\}.$$

The following theorem about the stability of the saddle point holds:

Theorem 1.14. *Let, for a point $\rho_0 \in N$, the sets $M_x(\rho_0)$ and $M_\lambda(\rho_0)$ be nonempty and bounded. Then, there exists a neighborhood $\tau(\rho_0) \subset N$ of the* *point ρ_0 such that, for all $\rho \in \tau(\rho_0)$, the sets $M_x(\rho)$ and $M_\lambda(\rho)$ are nonempty, the inequality*

$$v_1(\rho) = v_2(\rho)$$

holds, and the point-set mappings $\rho \in \tau(\rho_0)$ in $M_x(\rho)$, $M_\lambda(\rho)$ are semicontinuous at the point ρ_0.

The proof of Theorem 1.14 uses a variant of the main theorem of antagonistic games by J. von Neumann, which is used below. This theorem is as follows:

Theorem 1.15. *If P_x and P_λ are convex compact sets, the function $\chi(x, \lambda)$ is concave with respect to x for any fixed $x \in P_x$ and continuous with respect to x, λ, then the function $\chi(x, \lambda)$ has at least one saddle point with respect to P_x and P_λ, whereas*

$$\max_{x\in P_x,\ \lambda\in P_\lambda}\min \chi(x,\lambda) = \min_{\lambda\in P_\lambda,\ x\in P_x}\max \chi(x,\lambda) = \chi(x_0,\lambda_0),$$

where (x_0, λ_0) is an arbitrary saddle point of the function $\chi(x, \lambda)$ with respect to the sets P_x and P_λ.

The formula for the marginal value of the saddle point is the consequence of Theorem 1.14.

Let $N = \mathbf{R}^1$ and $\rho_0 = 0$. Under the assumptions of Theorem 1.13, for some ρ_1, the sets $M_x(\rho)$ and $M_\lambda(\rho)$ are nonempty for $0 \le \rho \le \rho_1$, and the relation

$$\mu(\rho) = \max_{x\in P_x,\ \lambda\in P_\lambda}\min \chi(x,\lambda,\rho) = \min_{\lambda\in P_\lambda,\ x\in P_x}\max \chi(x,\lambda,\rho) = \chi(x_\rho,\lambda_\rho,\rho)$$

holds, where $x_\rho \in M_x(\rho)$, $\lambda_\rho \in M_\lambda(\rho)$.

Theorem 1.16 (on the marginal value of the saddle point). *Let the assumptions of Theorem 1.14 hold, the function $\chi(x, \lambda, \rho)$ be differentiable with respect to ρ at the point $\rho = +0$ for all (x, λ) from a neighborhood Ω of the set $M_x(0) \times M_\lambda(0)$. Let derivative $\chi'_\rho(x, \lambda, 0)$ satisfy the condition*

$$[\chi(x,\lambda,\rho) - \chi(x,\lambda,0)]/\rho \to \chi'_\rho(x,\lambda,0)$$

as $\rho \to +0$ uniformly with respect to $(x, \lambda) \in \Omega$. Then, there exists the derivative $\mu'(0)$ of the function $\mu(\rho)$ at the point $\rho = 0$, whereas

$$\mu'(0) = \max_{x\in M_x(0),\ \lambda\in M_\lambda(0)}\min \chi'_\rho(x,\lambda,0) = \min_{\lambda\in M_\lambda(0),\ x\in M_x(0)}\max \chi'(x,\lambda,0).$$

The formula for the marginal value in problems of convex programming is deduced from Theorem 1.15.

Consider the family of problems

$$
\begin{gathered}
f(x,\rho) \to \max, \\
g_k(x,\rho) \leq 0, \qquad k \in [1:m], \\
x \in \mathbf{R}^n_+,
\end{gathered}
\tag{1.28}
$$

where $-f(x,\rho)$, $g_k(x,\rho)$, $k \in [1:m]$ are functions convex with respect to x for $\rho \in [0,\rho_1]$. Let $w(\rho)$ be the extremum value of the functional of problem (1.28) as the function of ρ. We write the Lagrange function as follows:

$$
L(x,\lambda,\rho) = f(x,\rho) - \sum_{k=1}^{m} \lambda_k g_k(x,\rho).
$$

Then, the problem dual to (1.28) has the form

$$
\begin{gathered}
\max_{x\in\mathbf{R}^n_+} L(x,\lambda,\rho) \to \min, \\
\lambda \in \mathbf{R}^m_+.
\end{gathered}
\tag{1.29}
$$

Theorem 1.17. *Let the problems (1.28) and (1.29) be solved for $\rho = 0$, and the sets of its solutions $M_x(0)$ and $M_\lambda(0)$ be bounded. Let the functions $g_0(x,\rho) = f(x,\rho)$, $g_k(x,\rho)$, $k \in [1:m]$ be differentiable at the point $\rho = +0$ for all x from a neighborhood $\bar{M}$ of the set $M_x(0)$, and the derivative $\partial g_k(x,0)/\partial\rho$, $k \in [0:m]$ satisfy the condition of uniform convergence with respect to $x \in \bar{M}$:*

$$
[g_k(x,\rho) - g_k(x,0)]/\rho \to \partial g_k(x,0)/\partial\rho
$$

for $\rho \to +0$. Then, the function $w(\rho)$ has the right derivative $w'(0)$ at the point $\rho = 0$, and the relation

$$
w'(0) = \max_{x\in M_x(0)}\min_{\lambda\in M_\lambda(0)} \partial L(x,\lambda,0)/\partial\rho = \min_{\lambda\in M_\lambda(0)}\max_{x\in M_x(0)} \partial L(x,\lambda,0)/\partial\rho
$$

holds.

§2. Branch Model and Description of the Algorithm

A branch is given by the following sets: $\bar{J} = [1 : J]$ is a set of numbers of the branch plants, $\bar{M} = [1 : M]$ is the set of numbers of the final products of the branch, $\bar{I} = [1 : I]$ is the set of numbers in the nomenclature of component parts at the plants of the branch, K_j is the set of numbers of equipment groups that are at the plants j, $I_j \subset \bar{I}$ is the set of numbers of components produced at the plant j. It is assumed that the technological chain for the production of a given component part at the plant is rigidly fixed. Then, the production capacity of each plant is described within the framework of the model that uses noninterchangeable equipment groups, namely, by the following inequalities:

$$\sum_{i \in I_j} T_j^{ik} x_j^i \le \Phi_j^k, \qquad k \in K_j, \qquad x_j^i \ge 0, \qquad i \in I_j. \tag{2.1}$$

Here, x_j^i is the annual amount of component i produced at the plant j, Φ_j^k is the annual time resource for the operation of the equipment from the k-th group at the plant j; T_j^{ik} are time expenditures for the equipment from the k-th group needed for the production of a unit of component i.

As follows from (2.1), the values x_j^i are defined at each plant j only for $i \in I_j$. In what follows, it is convenient to define them for all $i \in I$ assuming $x_j^i = 0$ for $i \in \bar{I} \backslash I_j$. Note that, by the definition, for each plant with the number j, we have $T_j^{ik} \neq \infty$, $\Phi_j^k \neq 0$, $k \in K_j$, $i \in I_j$.

The relation between the production of the component parts and the final products in the industrial branch is described by the following:

$$\sum_{j=1}^{J} (x_j^i + w_j^i) = \sum_{m=1}^{M} e^{im} Y^m (1 + \delta^m). \tag{2.2}$$

Here, Y^m is the annual final production of item m, w_j^i is the resource of the component parts i at the plant j at the beginning of the planning period, e^{im} is the number of the component part i used in the production of the final product m (completing coefficient), δ^m is the normative reserve of the product m for the next planning period (in units).

Further, we assume that the set of numbers of the final products is partitioned into two disjoint subset

$$\bar{M} = \bar{M}_1 \cup \bar{M}_2, \qquad \bar{M}_1 = [1 : M_1], \qquad \bar{M}_2 = [M_1 + 1 : M].$$

The production from the set $\bar{M}_1$ is most important for the branch from the viewpoint of demand. Its amount is either fixed or given by the lower bounds:

$$Y^m \geq \bar{Y}^m, \qquad m \in [1 : M_1]. \tag{2.3}$$

This nomenclature of the final products is called *obligatory*. It is assumed that there exists vector $\{\bar{Y}^{M_1+1}, \dots, \bar{Y}^M\}$ such that the vector $\{\bar{Y}^1, \dots, \bar{Y}^{M_1}, \bar{Y}^{M_1+1}, \dots \bar{Y}^M\}$ is an admissible plan, i.e., it satisfies (2.1), (2.2) for some x_j^i. As far as the production from the set $\bar{M}_2$ is concerned, the decision about its amount is to some extent defined by the planning body of the branch. This nomenclature is called *free*.

Let the planning body, with due regard to one or another consideration, assigned the vector of free nomenclature

$$\{\bar{Y}^{M_1+1}, \dots, \bar{Y}^M\}. \tag{2.4}$$

Then, two situations are possible.

1. The vector $\{\bar{Y}^1, \dots, \bar{Y}^M\}$ is an unfeasible plan. In other words, there are no values x_j^i such that relations (2.1) and (2.2) are satisfied. In this case, the problem of "cutting off" components of vector (2.4) arises.

2. The vector $\{\bar{Y}^1, \dots, \bar{Y}^M\}$ is a feasible plan. In this case, generally speaking, there exist an infinite number of ways of distributing component parts among plants. Moreover, a certain amount of industrial capacity can remain unloaded.

Thus, in both cases, the planning body of the branch must update the vector (2.4). Further, we propose to interpret the vector (2.4) as a notation about the desire structure of the free nomenclature of the final product of the branch. Finally, the plan is sought in the form

$$Y^{M_1+1} = \theta \bar{Y}^{M_1+1}, \dots, Y^M = \theta \bar{Y}^M. \tag{2.5}$$

Components of the vector (2.4) are called *proportions* or *assortment relations*. The amount θ is called *the value of assortment sets*. Thus, if an assortment relation of the free production (2.4) is assigned, then we come to the following optimization problem: maximize θ under conditions (2.1)–(2.2). The solution of this problem gives the vector $\{Y^{M_1+1}, \dots, Y^M\}$. Analyzing this plan, expert planners can change assignment relation by intuition. Thereafter, the problem must be solved once again. This procedure can be included in the scheme of interaction of the expert with computer. However, in what follows, we will be interested in the optimization problem for the fixed vector of free nomenclature. Further, for simplicity sake, we consider that components of the vector of obligatory nomenclature are rigidly fixed.

We rewrite (2.2) in accordance with the partition of the vector of final production:

$$\sum_{j=1}^{J}(x_j^i + w_j^i) = \sum_{m=1}^{M_1} e^{im}Y^m(1+\delta^m) + \sum_{m=M_1+1}^{M} e^{im}Y^m(1+\delta^m), \qquad i \in [1:I].$$

With the use of (2.5), the latter equation is rewritten in the form

$$\sum_{j=1}^{J} x_j^i + W^i = a^i\theta + H^i, \qquad i \in [1:I].$$

Here, the value $W^i = \sum_{j=1}^{J} w_j^i$ is the total amount of resource of the item i in the branch at the beginning of the planning period. The component

$$H^i = \sum_{m=1}^{M_1} e^{im}\bar{Y}^m(1+\delta^m)$$

is the total amount of the item i, necessary for the final production of the obligatory nomenclature. The value

$$a^i = \sum_{m=M_1+1}^{M} e^{im}\bar{Y}^m(1+\delta^m)$$

is the i-th component of the assortment relation of the production of the component parts. We introduce notation $V^i = W^i - H^i$, $i \in [1:I]$ as well as $b_j^{ik} = T_j^{ik}/\Phi_j^k$, $j \in [1:j]$, $k \in K_j$, $j \in I_j$. Finally, the optimization problem takes the form

$$\theta \longrightarrow \max, \tag{2.6}$$

$$\sum_{i \in I} b_j^{ik} x_j^i \le 1, \qquad k \in K_j, \tag{2.7}$$

$$x_j^i \ge 0, \qquad i \in I_j, \qquad x_j^i = 0, \qquad i \in \bar{I}\backslash I_j, \qquad j \in [1:J], \tag{2.8}$$

$$a^j\theta - \sum_{j=1}^{J} x_j^i = V^i, \qquad i \in [1:I], \tag{2.9}$$

$$\theta \geq 0. \tag{2.10}$$

The problem (2.6)–(2.10) is the problem of linear programming with equalities and inequalities. To solve it, we can use standard methods, such as the simplex method. However, in practice, this is fraught with great difficulties due to the large dimension of practical problems of branch planning. At the same time, it is easily seen that this problem has block structure: constraints (2.7), (2.8) are related to separate plants, and they are blockwise. Constraints (2.9) are branch (binding) constraints. Therefore, it is reasonable to use decomposition algorithms. We should stress that the block problem (2.6)–(2.10) has a specific structure. First, θ does not enter the block conditions. Second, the binding constraints have special form, and only one variable θ enters the criterion. The above considerations are taken into account in the construction of the decomposition algorithm for solution of the stated problem.

The occurrence of sums $\sum_{j=1}^{J} x_j^i$ in equations (2.9) leads to the idea of introducing new variables, complete productions with respect to all plants, i.e.,:

$$X^i = \sum_{j=1}^{J} x_j^i, \qquad i \in [1 : I]. \tag{2.11}$$

The values X^i will be called *the aggregated variables.* We introducc also a share production of the item i at the plant j:

$$\alpha_j^i = x_j^i / X^i, \qquad j \in [1 : J], i \in [1 : I].$$

The conditions

$$\sum_{j=1}^{J} \alpha_j^i = 1, \qquad i \in [1 : I], \tag{2.12}$$
$$\alpha_j^i \geq 0, \qquad i \in I_j, \qquad \alpha_j^i = 0, \qquad i \in \bar{I} \backslash I_j, \qquad j \in [1 : J]$$

are obviously satisfied.

The values α_j^i are called *aggregation weights.*

Let all aggregation weights that satisfy conditions (2.12) be given. Then, substituting variables $x_j^i = \alpha_j^i X^i$, $j \in [1 : J]$, $i \in [1 : I]$ in (2.6)–(2.8) and introducing notation

$$B_j^{ik} = b_j^{ik} \alpha_j^i, \qquad j \in [1 : J], \qquad k \in K_j, \qquad i \in I_j, \tag{2.13}$$

we come to the problem

$$\theta \to \max, \tag{2.14}$$

$$\sum_{i\in I}^{I} B_j^{ik} X^i \leq 1, \qquad j \in [1:J], \qquad k \in K_j, \tag{2.15}$$

$$a^i\theta - X^i = V^i, \tag{2.16}$$

$$X^i \geq 0, \qquad i \in [1:I], \tag{2.17}$$

$$\theta \geq 0. \tag{2.18}$$

The problem (2.14)–(2.18) is called *the aggregated problem (the macroproblem).*

In what follows, we need dual problems for the main problem and for the macroproblem. The problem dual to (2.6)–(2.10) takes the form

$$\varphi = \sum_{j=1}^{J} \sum_{k\in K_j} \xi_j^k + \sum_{i=1}^{I} V^i \mu^i \to \min, \tag{2.19}$$

$$\sum_{k\in K_j} b_j^{ik} \xi_j^k - \mu^i \geq 0, \; j \in [1:J], \; i \in I_j, \tag{2.20}$$

$$\sum_{i=1}^{I} \alpha^i \mu^i \geq 1, \tag{2.21}$$

$$\xi_j^k \geq 0, \qquad j \in [1:J], k \in K_j. \tag{2.22}$$

The dual problem for the aggregated problem has the following form:

$$\psi = \sum_{j=1}^{J} \sum_{k\in K_j} \eta_j^k + \sum_{i=1}^{I} V^i \lambda^i \to \min, \tag{2.23}$$

$$\sum_{j=1}^{J} \sum_{k\in K_j} B_j^{ik} \eta_j^k - \lambda^i \geq 0, \qquad i \in [1:I], \tag{2.24}$$

$$\sum_{i=1}^{I} \alpha^i \lambda^i \geq 1, \tag{2.25}$$

$$\eta_j^k \geq 0, \qquad j \in [1:J], \qquad k \in K_j. \tag{2.26}$$

Optimal solutions of conjugated pairs of the problems (2.14)–(2.18) and (2.23)–(2.26) will be denoted by $\circ$ over the corresponding letters.

Let $\overset{\circ}{\lambda}{}^i$ denote the unique optimal solutions of problem (2.23)–(2.26). For every fixed index j, we formulate plant (local) problems:

$$h_j = \sum_{i \in I_j} \overset{\circ}{\lambda}{}^i x_j^i \to \max, \tag{2.27}$$

$$\sum_{i \in I} b_j^{ik} x_j^i \leq 1, \qquad k \in K_j, \tag{2.28}$$

$$x_j^i \geq 0, \qquad i \in I, \qquad x_j^i = 0, \qquad i \in \bar{I} \backslash I_j. \tag{2.29}$$

For each $j \in [1 : J]$, the dual problems for (2.27)–(2.29) are written in the form

$$\chi_j = \sum_{k \in K_j} \xi_j^k \to \min, \tag{2.30}$$

$$\sum_{k \in K_j} b_j^{ik} \xi_j^k - \overset{\circ}{\lambda}{}^i \geq 0, \qquad i \in I_j, \tag{2.31}$$

$$\xi_j^k \geq 0, \qquad k \in K_j. \tag{2.32}$$

The iterative process is constructed in the following way. For some fixed weights α_j^i that satisfy conditions (2.12), the aggregated problems (2.14)–(2.18) are solved. As we show below, under some assumptions, the problem is solved analytically. The dual estimates $\overset{\circ}{\lambda}{}^i$, $i \in [1 : I]$ form functionals of block problems (2.27)–(2.29). Let $\hat{x}_j^i$, $j \in [1 : J]$, $i \in [1 : I]$ be optimal solutions of these problems. We introduce the variables

$$\overset{\circ}{x}{}_j^i = \alpha_j^i \overset{\circ}{X}{}^i, \qquad j \in [1 : J], \qquad i \in [1 : I],$$

where $\overset{\circ}{X}{}^i$, $i \in [1 : I]$ are optimal solutions of the aggregated problem (2.14)–(2.18). The values $\overset{\circ}{x}{}_j^i$ are called *disaggregated solutions.* New aggregation weights are defined in the form of a function α_j^i of variables ρ_j according to the relation

$$\alpha_j^i(\rho_j) = \begin{cases} \left[\overset{\circ}{x}{}_j^i + \rho_j(\hat{x}_j^i - \overset{\circ}{x}{}_j^i)\right] \left[\sum_{j=1}^{J} \left[\overset{\circ}{x}{}_j^i + \rho_j(\hat{x}_j^i - \overset{\circ}{x}{}_j^i)\right]\right]^{-1}, & i \in I_j, \\ 0 \text{ for } i \in \bar{I} \backslash I_j, \end{cases} \tag{2.33}$$

where $0 \leq \rho_j \leq 1$, $j \in [1 : J]$. Note that conditions (2.12) hold for the weights $\alpha_j^i(\rho_j)$ expressed according to (2.33).

If we treat the aggregated problems (2.14)–(2.18) with the weights (2.33), then the optimal value of the functional is the function of parameters ρ_j. This function will be denoted by $\overset{\circ}{\theta}(\rho_j)$. The problem of maximization of the function $\overset{\circ}{\theta}$ on the unit cube $0 \le \rho_j \le 1$, $j \in [1 : J]$ arises. Let maximum be attained for some $\overset{\circ}{\rho}_j$, $j \in [1 : J]$. Then, the weights for the next step are defined by formula (2.33) for $\rho_j = \overset{\circ}{\rho}_j$.

The algorithm forms the sequence of disaggregated solutions $\overset{\circ}{x}{}^i_j$, which together with $\overset{\circ}{\theta}$ are feasible for the main problem (1.16)–(1.20). We will formulate an optimality criterion of the solution $\{\overset{\circ}{x}{}^i_j, \overset{\circ}{\theta}\}$ for main problem (1.16)–(1.20) (the termination condition for the iterative process) and prove strict monotonicity of the functional values that correspond to the disaggregated solutions $\overset{\circ}{x}{}^i_j$ and $\overset{\circ}{\theta}$.

Below, we assume that the iterative process begins from the aggregation weights that give strict positive value $\overset{\circ}{\theta}$ for the aggregated problem. Since this number increases at each variation, the strict inequality

$$\overset{\circ}{\theta} > 0. \tag{2.34}$$

holds for all macroproblems.

We also assume that the production of each item is strictly positive, i.e.,

$$\overset{\circ}{X}{}^i > 0, \qquad i \in [1 : I]. \tag{2.35}$$

Condition (2.35) is satisfied if $V^i \le 0$, $i \in [1 : I]$ (as a rule, the latter inequalities are valid in practice).

§3. Optimality Criterion and the Aggregated Problem

Assume that, for some weights α^i_j satisfying (2.12), optimal solution $\{\overset{\circ}{X}{}^i, \overset{\circ}{\theta}\}$ of the aggregated problem (2.13)–(2.26) was obtained, and $\overset{\circ}{x}{}^i_j$ is the corresponding disaggregated solution. Let $\overset{\circ}{\lambda}{}^i$ be the unique optimal solution to the dual problem (2.23)–(2.26).

We introduce the notation

$$\hat{h}_j = \sum_{i \in I_j} \overset{\circ}{\lambda}{}^i \hat{x}^i_j, \tag{3.1}$$

$$\overset{\circ}{h}_j = \sum_{i \in I_j} \overset{\circ}{\lambda}{}^i \overset{\circ}{x}{}^i_j. \tag{3.2}$$

The values $\hat{h}_j$ and $\overset{\circ}{h}_j$ are the values of functionals of the local problems (2.27)–(2.29) on the optimal and disaggregated solutions, respectively. We set also $\hat{h} = \sum_{j=1}^{J} \hat{h}_j$, $\overset{\circ}{h} = \sum_{j=1}^{J} \overset{\circ}{h}_j$.

Lemma 3.1. *The following statements hold:*
(a) $\hat{h}_j \geq \overset{\circ}{h}_j$ *for all* $j \in [1:J]$;
(b) $\hat{h} = \overset{\circ}{h}$ *if* $\hat{h}_j = \overset{\circ}{h}_j$ *for all* $j \in [1:J]$.

Proof. The disaggregated solution $\overset{\circ}{x}{}^i_j$ is feasible for local problems. It satisfies conditions (2.29) due to (2.17) and (2.12). This solution satisfies also (2.28) due to the following chains of inequalities:

$$\sum_{i \in I_j} b_j^{ik} \overset{\circ}{x}{}^i_j = \sum_{i \in I_j} b_j^{ik} \alpha_j^i \overset{\circ}{X}{}^i = \sum_{i=1}^{I} B_j^{ik} \overset{\circ}{X}{}^i \leq 1,$$

$$j \in [1:J], \qquad k \in K_j.$$

Since $\overset{\circ}{x}{}^i_j$ are feasible solutions of plant problems, and $\hat{x}^i_j$ are their optimal solutions, the statement (a) of the lemma is proved. Further, due to the definitions $\hat{h}$, $\overset{\circ}{h}$ and statement (a) of the lemma, we establish the validity of the statement (b). The lemma is proved.

The following gives the optimality conditions of the disaggregated solution.

Theorem 3.1. *Let, for the optimal solutions* $\{\overset{\circ}{X}{}^i, \overset{\circ}{\theta}\}$ *and* $\overset{\circ}{\lambda}{}^i$, $i \in [1:I]$ *of the primal and dual macroproblems and the optimal solutions* $\hat{x}^i_j$, $j \in [1:J]$, $i \in I_j$ *of the local problems, the following identities hold:*

$$\hat{h}_j = \overset{\circ}{h}_j, \qquad j \in [1:J]. \tag{3.3}$$

Then, the disaggregated solution $\{\overset{\circ}{x}{}^i_j, \overset{\circ}{\theta}\}$, *where* $\overset{\circ}{x}{}^i_j = \alpha^i_j \overset{\circ}{X}{}^i$, $j \in [1:J]$, $i \in I$, *is the optimal plan of the main problem (2.6)–(2.10).*

Proof. Since $\overset{\circ}{\theta} > 0$ is valid by assumption (2.24), due to the second duality Theorem 1.4, condition (2.25) holds as the strict equality:

$$\sum_{i=1}^{I} a^i \overset{\circ}{\lambda}{}^i = 1.$$

By multiplying equalities (2.16) written for the optimal solution by $\mathring{\lambda}^i$ and summing over i, we obtain

$$\mathring{\theta}\sum_{i=1}^{I} a^i\mathring{\lambda}^i - \sum_{i=1}^{I}\mathring{\lambda}^i\mathring{X}^i = \sum_{i=1}^{I}\mathring{\lambda}^i V^i.$$

The latter equations give

$$\mathring{\theta}\sum_{i=1}^{I}\mathring{\lambda}^i(\mathring{X}^i + V^i). \tag{3.4}$$

Consider the disaggregated solution $\{\mathring{x}^i_j, \mathring{\theta}\}$. It is easily tested that this plan is feasible for the main problem (2.6)–(2.10), where the functional value for this solution is $\mathring{\theta}$. We have

$$\mathring{h} = \sum_{j=1}^{J}\mathring{h}_j = \sum_{j=1}^{J}\sum_{i\in I_j}\mathring{\lambda}^i\mathring{x}^i_j = \sum_{j=1}^{J}\sum_{i\in I_j}\mathring{\lambda}^i\alpha^i_j\mathring{X}^i = \sum_{i=1}^{I}\mathring{\lambda}^i\mathring{X}^i\sum_{j=1}^{J}\alpha^i_j = \sum_{i=1}^{I}\mathring{\lambda}^i\mathring{X}^i.$$

Comparing the latter solution with (3.4), we obtain

$$\mathring{\theta} = \mathring{h} + \sum_{i=1}^{I}\mathring{\lambda}^i V^i. \tag{3.5}$$

Further, we consider optimal solutions $\hat{\xi}^k_j$, $j \in [1:J]$, $k \in K_j$ of the dual local problems (2.30)–(2.32). It is easily tested that the set $\{\hat{\xi}^k_j, \mathring{\lambda}^i\}$, $j \in [1:J]$, $k \in K_j$, $i \in [1:I]$ is a feasible plan for the dual problem (2.19)-(2.22). For this problem, functional φ takes the value

$$\hat{\varphi} = \sum_{j=1}^{J}\sum_{k\in K_j}\hat{\xi}^k_j + \sum_{i=1}^{I}\mathring{\lambda}^i V^i. \tag{3.6}$$

By the first duality theorem, the following equalities hold for the local problems:

$$\sum_{k\in K_j}\hat{\xi}^k_j = \sum_{i\in I_j}\mathring{\lambda}^i\hat{x}^k_j = \hat{h}_j, \qquad j \in [1:J].$$

Substituting the latter relations into (3.6), we obtain

$$\hat{\varphi} = \hat{h} + \sum_{i=1}^{I}\mathring{\lambda}^i V^i. \tag{3.7}$$

Thus, we have feasible solutions of conjugated problems (2.6)–(2.10) and (2.19)–(2.22), whose functional values are expressed through (3.5) and (3.7). According to Theorem 1.2, the equality

$$\overset{\circ}{\theta} = \hat{\varphi} \tag{3.8}$$

ensures optimality of solution $\{\overset{\circ}{x}{}^i_j, \overset{\circ}{\theta}\}$ for the problem (2.6)–(2.10). However, relations (3.5), (3.7) imply that identity (3.8) is equivalent to $\hat{h} = \overset{\circ}{h}$ or, due to Lemma 3.1, $\hat{h}_j = \overset{\circ}{h}_j$, $j \in [1 : J]$. Theorem 3.1 is proved.

Relations (3.3) express the termination condition of the iterative process.

The following statement is established analogously:

Theorem 3.2. *Let the main problem (2.6)–(2.10) has a solution and, for some α^i_j, the disaggregated plan $\overset{\circ}{x}{}^i_j$ be obtained, such that the solution $\{\overset{\circ}{x}{}^i_j, \overset{\circ}{\theta}\}$ is not optimal for the problem (2.6)–(2.10). Then, indices j exist such that $\hat{h}_j > \overset{\circ}{h}_j$ and, therefore,*

$$\hat{h} > \overset{\circ}{h}\,. \tag{3.9}$$

Below, we consider the macroproblem (2.14)–(2.18) in more detail.

The main point of the decomposition method is the multiple solving of the aggregated problem (2.14)–(2.18) and the problem (2.23)–(2.26) dual to it. At each step of the iterative process, in the maximization of the function $\overset{\circ}{\theta}(\rho_j)$ on the unit cube, the macroproblem is solved for some parameter values ρ_j, $j \in [1 : J]$. Therefore, it is very important to obtain the properties of a solution to the aggregated problem.

We express the values X^i through θ according to (2.16):

$$X^i = a^i\theta - V^i, \qquad i \in [1 : I]. \tag{3.10}$$

Substituting (3.10) in (2.15), we obtain

$$\theta \sum_{i=1}^{I} B^{ik}_j a_i - \sum_{i=1}^{I} B^{ik}_j V^i \leq 1, \qquad j \in [1 : J], k \in K_j. \tag{3.11}$$

From inequalities (3.11), we obtain an expression for the optimal value of the variable:

$$\overset{\circ}{\theta} = \min_{\substack{j \in [1:J] \\ k \in K_j}} \left(1 + \sum_{i=1}^{I} B^{ik}_j V^i\right) \Big/ \left(\sum_{i=1}^{I} B^{ik}_j a^i\right). \tag{3.12}$$

The optimal values $\overset{\circ}{X}{}^i$, $i \in [1:I]$ are found as functions of $\overset{\circ}{\theta}$ by relation (3.10).

Consider the dual problem (2.23)-(2.26). Due to (2.35) and the second duality Theorem 1.4, relations (2.24) hold for the optimal solution as equalities, i.e.,

$$\overset{\circ}{\lambda}{}^i = \sum_{j=1}^{J} \sum_{k \in K_j} B_j^{ik} \overset{\circ}{\eta}{}_j^k. \tag{3.13}$$

Substituting (3.13) into (2.23) and (2.25), we obtain the following problem:

$$\psi = \sum_{j=1}^{J} \sum_{k \in K_j} \overset{\circ}{\eta}{}_j^k \left(1 + \sum_{i=1}^{I} B_j^{ik} V^i\right) \to \min, \tag{3.14}$$

$$\sum_{j=1}^{J} \sum_{k \in K_j} \overset{\circ}{\eta}{}_j^k \left(\sum_{i=1}^{I} B_j^{ik} a^i\right) = 1. \tag{3.15}$$

Let the minimum in (3.12) be attained for the unique pair of indices (j^*, k^*), then we have

$$\overset{\circ}{\theta} = \left(1 + \sum_{i=1}^{I} B_{j^*}^{ik^*} V^i\right) \Big/ \left(\sum_{i=1}^{I} B_{j^*}^{ik^*} a^i\right). \tag{3.16}$$

In the considered case, the problem (3.14), (3.15) has the unique solution:

$$\begin{aligned} \overset{\circ}{\eta}{}_{j^*}^{k^*} &= 1 \Big/ \left(\sum_{i=1}^{I} B_{j^*}^{ik^*} a^i\right), \\ \overset{\circ}{\eta}{}_j^k &= 0, \qquad (j,k) \neq (j^*, k^*). \end{aligned} \tag{3.17}$$

According to (3.13), (3.17), the values $\overset{\circ}{\lambda}{}^i$ are equal to

$$\overset{\circ}{\lambda}{}^i = B_{j^*}^{ik^*} \Big/ \left(\sum_{i=1}^{I} B_{j^*}^{ik^*} a^i\right), \qquad i \in [1:I] \tag{3.18}$$

and are found uniquely. This case will be called *simple*.

Let the minimum in (3.12) be attained for a set of indices (j, k), which we denote by R. Then, (3.12) can be written in the form

$$\overset{\circ}{\theta} = \left(1 + \sum_{i=1}^{I} B_j^{ik} V^i\right) \Big/ \left(\sum_{i=1}^{I} B_j^{ik} a^i\right), \qquad (j,k) \in R.$$

For the values $\overset{\circ}{\eta}{}_j^k$, we have the following problem:

$$\sum_{(j,k)\in R} \overset{\circ}{\eta}{}_j^k \left(1+\sum_{i=1}^{I} B_j^{ik} V^i\right) \to \min, \tag{3.19}$$

$$\begin{gathered}\sum_{(j,k)\in R} \overset{\circ}{\eta}{}_j^k \left(\sum_{i=1}^{I} B_j^{ik} a^i\right) = 1, \\ \overset{\circ}{\eta}{}_j^k = 0, \qquad (j,k) \notin R.\end{gathered} \tag{3.20}$$

The problem (3.19), (3.20) has a set of solutions, and therefore, an additional question arises about the definition of the values $\overset{\circ}{\lambda}{}^i$ that form the functionals of the block problems. This case will be called *degenerate.*

We should outline that, in both cases, finding an optimal solution of the aggregated problem is reduced to the determining a minimum element of a higher-dimension matrix. Thus, all difficulties related to higher-dimensions are reduced to exhaustive search.

Further, we study the degenerate case. The following reasoning shows that an optimal solution of the initial problem (2.6)–(2.10) corresponds necessarily to an aggregated problem that is degenerate in the above sense. In fact, let $\{\overset{*}{x}{}_j^i, \overset{*}{\theta}\}$ be an optimal plan for the problem (2.6)–(2.10), and $\overset{*}{\alpha}{}_j^i$ be the corresponding aggregation weights. Consider the macroproblem with these weights. We can state that, for each j, there exists an index $k_j^* \in K_j$ such that the corresponding constraint of the problem (2.6)–(2.10) and macroproblem (2.14)–(2.18) holds as an equality. In fact, assuming the converse, for a fixed j, we can add $a^i\varepsilon$ to all x_j^i, $i \in I_j$, where $\varepsilon > 0$. Thus, due to the smallness of ε, relations (2.7) and (2.15) that correspond to all k will hold as strict inequalities, and relations (2.9) and (2.16) will also hold, however, with the value of the functional equal to $\overset{*}{\theta} + \varepsilon$.

Consider the problem (3.19) and (3.20) under degeneracy. The set of its solution belongs to the polyhedron with the vertices

$$\overset{\circ}{\eta}{}_{j^0}^{k^0} = 1\Big/\left(\sum_{i=1}^{I} B_j^{ik^0} a^i\right), \qquad (j^0, k^0) \in R. \tag{3.21}$$

We enumerate sequentially all the elements of the set R and denote the obtained set of numbers by $\bar{S} = [1:S]$. We substitute (3.21) into (3.13) and obtain the following expression:

$$\overset{\circ}{\lambda}{}^i_s = B^{ik}_j \Big/ \left(\sum_{i=1}^{I} B^{ik}_j a^i \right), \qquad (j,k) \in R, \qquad s \in \bar{S}. \tag{3.22}$$

Here, we assume a one-to-one correspondence between the elements of the sets R and $\bar{S}$. Thus, all solutions of the dual macroproblems can be represented in the form

$$\overset{\circ}{\lambda}{}^i = \sum_{s=1}^{S} \pi_s \overset{\circ}{\lambda}{}^i_s, \tag{3.23}$$

where the values π_s belong to the set

$$\Pi = \left\{ \pi = (\pi_1, \ldots, \pi_S) \middle| \sum_{s=1}^{S} \pi_s = 1, \quad \pi_s \geq 0, \quad s \in [1:S] \right\}. \tag{3.24}$$

By M^j_x, we denote the sets of feasible solutions of block problems for each $j \in [1:J]$ or, in other words, the sets of values x^i_j, $i \in I_j$ that satisfy the constraints (2.28), (2.29) for fixed $j \in [1:J]$. Consider the following maximin problem:

$$\max_{\substack{x^i_j \in M^j_x \\ j \in [1:J]}} \min_{\pi \in \Pi} (h - \overset{\circ}{h}) = \max_{\substack{x^i_j \in M^j_x \\ j \in [1:J]}} \min_{\pi \in \Pi} \sum_{j=1}^{J} \sum_{i \in I_j} \sum_{s=1}^{S} \pi_s \overset{\circ}{\lambda}{}^i_s (x^i_j - \overset{\circ}{x}{}^i_j). \tag{3.25}$$

If M^j_x, $j \in [1:J]$ are bounded sets, then, by Theorem 1.14, the problem (3.2) has a saddle point, which we denote by $(\bar{x}^i_j, \bar{\pi}_s)$. Here, for each j, the values $\bar{x}^i_j$, $i \in I_j$ are optimal solutions of the block problems with functionals that contain $\overset{\circ}{\lambda}{}^i = \sum_{s=1}^{S} \bar{\pi}_s \overset{\circ}{\lambda}{}^i_s$.

Denote by $\bar{h}$ the value of h for the saddle point $\bar{x}^i_j$. We come to the following optimality criterion for the disaggregated solution $\overset{\circ}{x}{}^i_j$ in the degenerate case:

Theorem 3.3. *In the degenerate case, the satisfaction of the equation*

$$\bar{h} = \overset{\circ}{h} \tag{3.26}$$

is a sufficient condition for the optimality of the disaggregated solution $\{\overset{\circ}{x}{}^i_j, \overset{\circ}{\theta}\}$ *for the main problem (2.6)–(2.10). If the problem (2.6)–(2.10) has a solution, and the disaggregated plan* $\{\overset{\circ}{x}{}^i_j, \overset{\circ}{\theta}\}$ *is not optimal for it, then the equality sign* $=$ *is replaced by* $>$.

Note that, in the case of degeneracy, the components of the saddle point $\bar{x}^i_j$ are taken as the values $\hat{x}^i_j$ in the formula for the new aggregation weights (2.33).

§4. Local Monotonicity with Respect to the Functional and Numerical Computation

The main result of this section is the proof that the value of the functional grows at each step of the iterative process. This is obtained by study of the differential properties of the function $\overset{\circ}{\theta}(\rho_j)$ introduced in the second section. We assume that the main problem (2.6)–(2.10) is solvable in general.

First, consider the nondegenerate case. The following statement holds:

Theorem 4.1. *Let, for some weights* α^i_j, *an optimal solution for the aggregated problems (2.14)–(2.18)* $\{\overset{\circ}{X}{}^i, \overset{\circ}{\theta}\}$ *be obtained, and assume the dual problem (2.23)–(2.26) has a unique solution. Let the disaggregated solution* $\{\overset{\circ}{x}{}^i, \overset{\circ}{\theta}\}$, *where* $\overset{\circ}{x}{}^i_j = a^i_j \overset{\circ}{X}{}^i$, *be not optimal for the main problem (2.6)–(2.10). Then, there exists a solution feasible for (2.6)–(2.10) with the value of functional strictly greater than* $\overset{\circ}{\theta}$.

Proof. The value of the functional of the main problem for the disaggregated solution $\{\overset{\circ}{x}{}^i_j, \overset{\circ}{\theta}\}$ is equal to $\overset{\circ}{\theta}$. Consider the function $\alpha^i_j(\rho_j)$ in a neighborhood of $\rho_j = 0$, $j \in [1 : J]$. Then, as already mentioned, the optimal solution $\overset{\circ}{\theta}$ is a function of ρ_j, which was denoted by $\overset{\circ}{\theta}(\rho_j)$. There exists a neighborhood of the point $\rho_j = 0$, $j \in [1 : J]$, where the basis is retained, i.e., minimum in (3.12) is attained for the pair of indices (j^*, k^*). For simplicity, we take $\rho_j = \rho$, $j \in [1 : J]$ and consider macroproblems as parametrically dependent on ρ.

According to (3.16) and (2.33), we can write an explicit statement for the optimal solution $\overset{\circ}{\theta}$ as a function of ρ in a neighborhood of point $\rho = 0$:

$$\overset{\circ}{\theta}(\rho) = \left(1 + \sum_{i \in I_{j^*}} b^{ik^*}_{j^*} \alpha^i_{j^*}(\rho) V^i\right) \Big/ \left(\sum_{i \in I_{j^*}} b^{ik^*}_{j^*} \alpha^i_{j^*}(\rho) a^i\right). \tag{4.1}$$

We formulate an additional statement about the derivative $(\overset{\circ}{\theta}(0))'$ of the function $\overset{\circ}{\theta}(\rho)$ in the point $\rho = 0$.

Lemma 4.1. *The derivative* $(\overset{\circ}{\theta}(0))'$ *is expressed by the formula*

$$(\overset{\circ}{\theta}(0))' = (\hat{h} - \overset{\circ}{h}) + \left(1 - \sum_{i \in I_{j*}} b_{j*}^{ik^*} x_{j*}^i\right) \overset{\circ}{\eta}_{j*}^{k^*}. \tag{4.2}$$

We present two proofs of the lemma. The first proof is based on the direct differentiation of the right-hand side of (4.1), the second one uses Theorem 1.11 about the marginal value.

First Proof of Lemma 4.1. For simplicity of transformations, we assume $V^i = 0$, $i \in [1 : I]$. Note that, for $V^i \neq 0$, the proof is almost the same. Differentiating the right-hand side in (4.1), we obtain

$$(\overset{\circ}{\theta}(0))' = -\left(\sum_{i \in I_{j*}} b_{j*}^{ik^*} (\alpha_{j*}^i(0))' a^i\right) \Big/ \left(\sum_{i \in I_{j*}} b_{j*}^{ik^*} \alpha_{j*}^i(0) a^i\right)^2 . \tag{4.3}$$

According to (2.33), the derivatives $(\alpha_{j*}^i(0))'$, $i \in I_{j*}$ take the form

$$(\alpha_{j*}^i(0))' = \left(\hat{x}_{j*}^i - \alpha_j^{i^*} \sum_{p=1}^{J} \hat{x}_p^i\right) (\overset{\circ}{X}{}^i)^{-1}. \tag{4.4}$$

Substituting (4.4) in the right-hand side of (4.3), we have

$$(\overset{\circ}{\theta}(0))' = -\sum_{i \in I_{j*}} b_{j*}^{ik^*} \hat{x}_{j*}^i a^i (\overset{\circ}{X}{}^i)^{-1} \Big/ \left(\sum_{i \in I_{j*}} b_{j*}^{ik^*} \alpha_{j*}^i a^i\right) +$$
$$+ \sum_{i \in I_{j*}} b_{j*}^{ik^*} \alpha_{j*}^i (\overset{\circ}{X}{}^i)^{-1} \Big/ \left(\sum_{i \in I_{j*}} b_{j*}^{ik^*} \alpha_{j*}^i a^i\right)^2 .$$

For $V^i = 0$, $i \in [1 : I]$, we have $a^i \overset{\circ}{\theta} = \overset{\circ}{X}{}^i$, $i \in [1 : I]$.. Then, taking into account expressions for $\overset{\circ}{\eta}_{j*}^{k^*}$ and $\overset{\circ}{\lambda}_i$ as well as the definition $\hat{h}$, the latter relation gives

$$(\overset{\circ}{\theta}(0))' = \hat{h} - \left(\sum_{i \in I_{j*}} b_{j*}^{ik^*} \hat{x}_j^i\right) \overset{\circ}{\eta}_{j*}^{k^*}. \tag{4.5}$$

Furthermore, due to (3.4), we have $\mathring{\theta} - \mathring{h} = 0$. However, due to (3.16), (4.7), for $V^i = 0$, $i \in [1 : I]$, $\mathring{\theta}$ is equal to $\mathring{\eta}^{k^*}_{j^*}$. Substituting the term $\mathring{\theta} - \mathring{\eta}^{k^*}_{j^*}$ into the right-hand side of (4.5), we obtain (4.2), which proves the lemma.

Note that if we reject the assumption $\rho_j = \rho$, $j \in [1 : J]$, we need to consider partial derivatives $\partial \mathring{\theta}(0)/\partial \rho_j$ of the function $\mathring{\theta}(\rho_j)$ for $\rho_j = 0$, $j \in [1 : J]$. In similar way, we obtain the relation

$$\partial \mathring{\theta}(0)/\partial \rho_j = \begin{cases} (\hat{h}_j - \mathring{h}_j) \text{ for } j \neq j^*, \\ (\hat{h}_{j^*} - \mathring{h}_{j^*}) + \left(1 - \sum\limits_{i \in I_{j^*}} b^{ik^*}_{j^*} \hat{x}^{k^*}_{j^*}\right) \mathring{\eta}^{k^*}_{j^*} \text{ for } j = j^*. \end{cases}$$

Second proof of Lemma 4.1.

Taking into account (4.4), we consider coefficients B^{ik}_j of the aggregated problem (2.14)–(2.18), which depend on parameter ρ in the following way:

$$B^{ik}_j(\rho) = b^{ik}_j \alpha^i_j(\rho) = b^{ik}_j \left(\alpha^i_j + \left(\hat{x}^i_j - \alpha^i_j \sum_{p=1}^{j} \hat{x}^i_p \right) (\mathring{X}^i)^{-1} \rho \right). \tag{4.6}$$

Using Theorem 1.11 about marginal values, we obtain an expression for the derivative:

$$(\breve{\theta}(0))' = - \sum_{i \in I_{j^*}} b^{ik^*}_{j^*} \left(\hat{x}^i_j - \alpha^i_j \sum_{p=1}^{j} \hat{x}^i_p \right) (\mathring{X}^i)^{-1} \mathring{X}^i \mathring{\eta}^{k^*}_{j^*}. \tag{4.7}$$

Here, we take into account the assumption about the uniqueness of the solution of the conjugated problems (2.14)–(2.18) and (2.23)–(2.26) for $\rho = 0$. After cancellation in (4.7), using expressions for $\mathring{\eta}^{k^*}_{j^*}$ and $\mathring{\lambda}^i$ as well as definition (2.13), we obtain (4.5). By the first duality theorem, the equality

$$\breve{\eta}^{k^*}_{j^*} + \sum_{i=1}^{I} \mathring{\lambda}^i V^i - \mathring{\theta} = 0$$

holds for the primal and dual macroproblems.

By (3.5), the value $\sum\limits_{i=1}^{i} \mathring{\lambda}^i V^i - \mathring{\theta}$ is equal to $-\mathring{h}$. By adding to (4.7) or, which is the same, to (4.5), the term $\mathring{\eta}^{k^*}_{j^*} - \mathring{h}$ equal to zero, we obtain (4.2). The lemma is proved.

We continue the proof of Theorem 4.1. Since, by the assumption, the disaggregated solution $\{\overset{\circ}{x}{}^i_j, \overset{\circ}{\theta}\}$ is not optimal for the main problem (2.6)–(2.10), then, according to Theorem 3.2, we have the inequality $\hat{h} > \overset{\circ}{h}$. Then, the first term of the right-hand side (4.2) is strictly greater than zero. The second term is greater or equal zero due to (2.26), (2.28). Thus, in a neighborhood of point $\rho = 0$, the estimate $\overset{\circ}{\theta}(\rho) > \overset{\circ}{\theta}(0)$ holds, which proves Theorem 4.1.

We establish local monotonicity with respect to the functional in the case of degeneracy.

Theorem 4.2. *Let, for some weights α^i_j, the macroproblem (2.14)–(2.18) be degenerate. If the disaggregated solution $\{\overset{\circ}{x}{}^i_j, \overset{\circ}{\theta}\}$ is nonoptimal for the main problem (2.6)–(2.10), then there exists a solution with a functional value strictly greater than $\overset{\circ}{\theta}$.*

Proof. As before, we consider the parametric family of aggregated problems (2.14)–(2.18) that depends on ρ_j and let the weights α^i_j be expressed in form (2.33), and $\rho_j = \rho$, $j \in [1 : J]$. Recall that, under degeneracy, the components $\bar{x}^i_j$ of the saddle point are taken in (2.33) instead of $\hat{x}^i_j$.

The following lemma holds for the derivative $(\overset{\circ}{\theta}(0))'\}$ of the function $\overset{\circ}{\theta}(\rho)$ at the point $\rho = 0$ under degeneracy:

Lemma 4.2. *The derivative $(\overset{\circ}{\theta}(0))'$ satisfies the inequality*

$$(\overset{\circ}{\theta}(0))' \geq (\bar{h} - \overset{\circ}{h}). \tag{4.8}$$

Here, $\bar{h}$ is the value h of the function at the saddle point $(\bar{x}^i_j, \bar{\lambda}^i)$ of the problem (3.25).

Proof of lemma 4.2. Consider the change of parameters of the macroproblem (2.14)–(2.18) when coefficients B^{ik}_j take the form (4.6). We use Theorem 1.11 from Chapter 1 about the marginal value, taking account of the uniqueness of the optimal solution $\overset{\circ}{X}{}^i$ to the primal problem (2.14)–(2.18) and nonuniqueness of variables $\overset{\circ}{\eta}{}^k_j$ of the dual problem (2.23)–(2.26) for $\rho = 0$. We have the following:

$$(\overset{\circ}{\theta}(0))' = \min_{\eta^k_j \in \Omega_\eta} \left[- \sum_{(j,k)\in R} \sum_{i \in I_j} b^{ik}_j \left(\bar{x}^i_j - \alpha^i_j \sum_{p=1}^{J} \bar{x}^i_p \right) (\overset{\circ}{X}{}^i)^{-1} (\overset{\circ}{x}{}^i) \overset{\circ}{\eta}{}^k_j \right] =$$

$$= \min_{\overset{\circ}{\eta}{}_j^k \in \Omega_\eta} \left[- \sum_{(j,k)\in R} \sum_{i\in I_j} b_j^{ik} \bar{x}_j^i \overset{\circ}{\eta}{}_j^k + \sum_{(j,k)\in R} \sum_{i\in I_j} B_j^{ik} \left(\sum_{p=1}^{J} \bar{x}_p^i \right) \overset{\circ}{\eta}{}_j^k \right] \quad (4.9)$$

Here, Ω_η denotes the set of optimal solutions $\overset{\circ}{\eta}{}_j^k$ of the dual problem (2.23)–(2.26). This set is a polyhedron with vertices given by (3.21), so that, for each point $\overset{\circ}{\eta}{}_j^k \in \Omega_\eta$, the expression $\overset{\circ}{\eta}{}_j^k = \sum_{s=1}^{S} \pi_s \overset{\circ}{\eta}{}_s^k$ holds, where the values π_s belong to the set Π, and components $\overset{\circ}{\eta}{}_s^k$, $s \in [1:S]$ are equivalent to $\overset{\circ}{\eta}{}_j^k$, $(j,k) \in R$ because of introduced renumeration.

Let the minimum in (4.9) be attained at a point $\tilde{\eta}_j^k$, which corresponds to $\tilde{\pi}_s$, $s \in [1:S]$. Then, using (3.21)–(3.23), the second term in (4.9) at the minimum point will take the following form:

$$\tilde{h} = \sum_{j=1}^{J} \sum_{i\in I_j} \sum_{s=1}^{S} \tilde{\pi}_s \overset{\circ}{\lambda}{}_s^i \bar{x}_j^i.$$

Due to the properties of the saddle point (the relation of Theorem 1.9), we have the inequality

$$\tilde{h} \geq \bar{h}. \quad (4.10)$$

As applied to the conjugated problems (2.14)–(2.18) and (2.23)–(2.26), using (3.5) and (3.23), the first duality Theorem 1.3 gives the equality

$$\sum_{(j,k)\in R} \tilde{\eta}_j^k - \overset{\circ}{h} = 0. \quad (4.11)$$

Adding the left-hand side (4.11) to (4.9) and taking into account the definition of $\tilde{h}$, we obtain

$$(\overset{\circ}{\theta}(0))' = (\tilde{h} - \overset{\circ}{h}) + \sum_{(j,k)\in R} \left(1 - \sum_{i\in I_j} b_j^{ik} \bar{x}_j^i \right) \tilde{\eta}_j^k. \quad (4.12)$$

Since the second term is greater then or equal to zero due to (2.26), (2.28), relations (4.10), (4.12) imply estimate (4.8). The lemma is proved.

We continue the proof of Theorem 4.2. Since the disaggregated solution $\{\overset{\circ}{x}{}_j^i, \overset{\circ}{\theta}\}$ is nonoptimal for the main problem (2.6)–(2.10), then the inequality $\bar{h} > \overset{\circ}{h}$ of the Theorem 3.3 is not valid. Taking into account the latter relation and (4.8), we establish the validity of the estimate $\mathring{\theta}(\rho) > \mathring{\theta}(0)$ in a neighborhood of the point $\rho = 0$. Theorem 4.2 is proved.

Two series of numerical computation were carried out for the model using the method described. In the first series, we took block problems of lower dimension. As indicated earlier, one of the difficulties in solving of the iterative algorithm is the nonuniqueness of solutions of dual macroproblems. It was stressed that the indicated degeneracy becomes essential when the initial problem approaches the optimal solution. However, in the proposed numerical example, the degeneration is not taken into account. In other words, optimal dual variables $\mathring{\lambda}^i$ were computed starting from the fact that the minimum in (3.12) is attained on a single pair of indices, even when approaching the optimum of the main problem. Consider the following example:

$$\theta \to \max,$$

$$x_1^1 + 2x_1^2 \leq 1,$$

$$2x_1^1 + x_1^2 \leq 1,$$

$$x_2^1 + 3x_2^2 \leq 1,$$

$$3x_2^1 + x_2^2 \leq 1,$$

$$\theta - x_1^1 - x_2^1 = 1,$$

$$\theta - x_1^2 - x_2^2 = 1,$$

$$\theta, x_1^1, x_1^2, x_2^1, x_2^2 \geq 0.$$

The initial weights are taken in the form: $\alpha_1^{1(0)} = 0,5$, $\alpha_1^{2(0)} = 0,8$, $\alpha_2^{1(0)} = 0,5$, $\alpha_2^{2(0)} = 0,2$. The optimal solution of the macroproblem for these weights is $\mathring{\theta}^{(0)} = 1,476$, $\mathring{X}^{1(0)} = 0,476$, $\mathring{X}^{2(0)} = 0,476$. Components of the matrix of the right-side (3.12) are $\{1,476, 1,555, 1,558, 1909\}$. The disaggregated solution is $\mathring{x}_1^{1(0)} = 0,238$, $\mathring{x}_1^{2(0)} = 0,381$, $\mathring{x}_2^{1(0)} = 0,238$, $\mathring{x}_2^{2(0)} = 0,095$. The block problems give the solutions $\hat{x}_1^{1(0)} = 0$, $\hat{x}_1^{2(0)} = 0,5$, $\hat{x}_2^{1(0)} = 0$, $\hat{x}_2^{2(0)} = 0,333$.

The value $\Delta h^{(0)} = h^{(0)} - \mathring{h}^{(0)}$ is equal to 0,126.

The next iteration gives $\mathring{\rho}^{(1)} = 0,25$, $\mathring{\theta}^{(1)} = 1,526$, $\mathring{X}^{1(1)} = 0,583$. $\mathring{X}^{2(1)} = 0,583$. The weights are equal to $\alpha_1^{1(1)} = 0,5$, $\alpha_1^{2(1)} = 0,6$, $\alpha_2^{1(1)} = 0,5$, $\alpha_2^{2(1)} = 0,4$. Components of the matrix in (3.12) are $\{1,588, 1,625, 1526, 1588\}$.

The disaggregated solution is $\overset{\circ}{x}{}_1^{1(1)} = 0,263$, $\overset{\circ}{x}{}_1^{2(1)} = 0,316$, $\overset{\circ}{x}{}_2^{1(1)} = 0,263$, $\overset{\circ}{x}{}_2^{2(1)} = 0,211$. Optimal solutions of the main problems are $\hat{x}_1^{1(1)} = 0,333$, $\hat{x}_1^{2(1)} = 0,333$, $\hat{x}_2^{1(1)} = 0,25$, $\hat{x}_2^{2(1)} = 0,25$, $\Delta h^{(1)} = 0,05$.

At the second iteration, we obtain a precise optimal solution of the main problem. We have the following: $\overset{\circ}{\rho}{}^{(2)} = 0,9$, $\overset{\circ}{\theta}{}^{(2)} = 1,583$, $\overset{\circ}{X}{}^{1(2)} = 0,583$, $\overset{\circ}{X}{}^{2(2)} = 0,583$.

The weights are $\alpha_1^{1(2)} = 0,57$, $\alpha_1^{2(2)} = 0,57$, $\alpha_2^{1(2)} = 0,428$, $\alpha_2^{2(2)} = 0,428$.

The disaggregated solution $\overset{\circ}{x}{}_1^{1(2)} = 0,33$, $\overset{\circ}{x}{}_1^{2(2)} = 0,33$, $\overset{\circ}{x}{}_2^{1(2)} = 0,25$, $\overset{\circ}{x}{}_2^{2(2)} = 0,25$ coincides with the unique optimal solution of the main problem. The matrix (3.12) has the form $\{1,583; 1,583; 1,583; 1,583\}$, i.e., the minimum in the right-hand side of (3.12) is attained for four elements, corresponding to the degeneracy. Nevertheless, optimal dual estimates $\overset{\circ}{\lambda}{}^i$ are computed by the nondegenerate dual macroproblem. Finally, $\Delta h^{(2)} = 0,004$, $\hat{x}_1^{1(2)} = 0$, $\hat{x}_1^{2(2)} = 0,5$, $\hat{x}_2^{1(2)} = 0,25$, $\hat{x}_2^{2(2)} = 0,25$.

In other small problems of the series, the nondegeneracy appeared even far from optimum. In this case, we took an optimal solution $\overset{\circ}{\lambda}{}^i$ of the dual macroproblem, so that the monotonicity of the functional in the iterative process was not violated.

In the second series, computation for the problem of dimension $J = 20$, $I = 50$, $K_j = 23$, $j \in [1 : 20]$ were carried out. The local problems were solved by simplex method. To find a saddle point of a maximin problem, we applied procedure based on the method of Brown. In the search of the maximum of the function $\overset{\circ}{\theta}(\rho_j)$ on the unit cube, it was assumed that $\rho_j = \rho$, $j \in [1 : J]$. For the one-dimensional extremum search, the golden section method was used. For fairly remote initial aggregation weights, approximately 10 iterations were used to obtain a solution that practically agreed with respect to the functional. Here, not all possibilities for accelerating the iterative algorithm were used.

§5. Modification of the Main Model

We add arbitrary binding constraints of the form

$$\sum_{j=1}^{J} \sum_{i \in I_j} \bar{b}_j^{il} x_j^i \leq 1, \qquad l \in [1 : L] \tag{5.1}$$

to the conditions of the problem (2.6)–(2.10).

The economic meaning of the written general branch conditions is the following. Assume that the production of the i-th item at the jth plant uses some types of resources $i \in [1 : L]$ that are at the disposal of the branch, while the general amount is constrained. Then, the value $\bar{b}_j^{il}$ is equal to the ratio of the l-th resource necessary for the production of the i-th item at the j-th plant. Adding (5.1) to (2.6), we obtain a branch problem that takes into account general binding constraints.

We describe iterative decomposition method using variable aggregation as applied to the problem (2.6)–(2.10), (5.1). Consider the dual problem for (2.6)–(2.10), (5.1):

$$\begin{aligned}
&\varphi = \sum_{j=1}^{J} \sum_{k \in K_j} \xi_j^k + \sum_{i=1}^{I} V^i \mu^i + \sum_{l=1}^{L} \omega^l \to \min, \\
&\sum_{k \in K_j} b_j^{ik} \xi_j^k - \mu^i + \sum_{l=1}^{L} \bar{b}_j^{il} \omega^l \geq 0, \qquad j \in [1 : J], \qquad i \in I_j, \\
&\sum_{i=1}^{I} a^i \mu^i \geq 1, \\
&\xi_j^k \geq 0, \qquad j \in [1 : J], k \in K_j, \qquad \omega^i \geq 0, \qquad l \in [1 : L].
\end{aligned} \tag{5.2}$$

Here, the dual variables ω^l that correspond to (5.1) occurred. The macroproblem for (2.6)–(2.10), (5.1) has the form

$$\begin{aligned}
&\theta \to \max, \\
&\sum_{i=1}^{I} B_j^{ik} X^i \leq 1, \qquad j \in [1 : J], \qquad k \in K_j, \\
&a^i \theta - X^i = V^i, \\
&X^i \geq 0, \qquad \theta \geq 0, \qquad i \in [1 : I], \\
&\sum_{i=1}^{I} B^{il} X^i \leq 1, \qquad l \in [1 : L].
\end{aligned} \tag{5.3}$$

Here, as before, we introduce the aggregated variables

$$X^i = \sum_{j=1}^{J} x_j^i, \qquad i \in [1 : I]$$

and the aggregation weights $\alpha_j^i = x_j^i / X^i$. Moreover, along with $B_j^{ik} = b_j^{ik}\alpha_j^i$, we introduce the notation

$$B^{il} = \sum_{j=1}^{J} \bar{b}_j^{il}\alpha_j^i, \qquad i \in [1:I], \qquad l \in [1:L].$$

The dual problem for (5.3) has the form

$$\begin{aligned}
&\psi = \sum_{j=1}^{J}\sum_{k\in K_j} \eta_j^k + \sum_{i=1}^{I} V^i\lambda^i + \sum_{l=1}^{L}\delta^l \to \min,\\
&\sum_{j=1}^{J}\sum_{k\in K_j} B_j^{ik}\eta_j^k - \lambda^i + \sum_{l=1}^{L} B^{il}\delta^l \geq 0, \qquad i \in [1:I],\\
&\sum_{i=1}^{I} a^i\lambda^i \geq 1,\\
&\eta_j^k \geq 0, \qquad j \in [1:J], \qquad k \in K_j, \qquad \delta^l \geq 0, \qquad l \in [1:L].
\end{aligned} \tag{5.4}$$

Variables δ^l correspond to the new constraints of macroproblem (5.3).

We formulate problems for separate blocks. Let $\{\mathring{\lambda}^i, \mathring{\delta}^l\}$ be unique dual estimates (optimal solutions of the problem (5.4)). For each $j \in [1:J]$, we have

$$\begin{aligned}
&h_j = \sum_{i\in I_j}\left(\mathring{\lambda}^i - \sum_{l=1}^{L}\bar{b}_j^{il}\mathring{\delta}^l\right)x_j^i \to \max,\\
&\sum_{i\in I_j} b_j^{ik}x_j^i \leq 1, \qquad k \in K_j,\\
&x_j^i \geq 0, \qquad i \in I_j, \qquad x_j^i = 0, \qquad i \in I\backslash I_j.
\end{aligned} \tag{5.5}$$

In (5.5), the dual problems are written in the form

$$\begin{aligned}
&\chi_j = \sum_{k\in K_j}\zeta_j^k \to \min,\\
&\sum_{k\in K_j} b_j^{ik}\zeta_j^k \geq \mathring{\lambda}^i - \sum_{l=1}^{L}\bar{b}_j^{il}\mathring{\delta}^l, \qquad i \in I_j,\\
&\zeta_j^k \geq 0, \qquad k \in K_j.
\end{aligned}$$

The algorithm is constructed along the scheme described above.

Let, for some aggregation weights α_j^i, the solution $\{\overset{\circ}{X}{}^i, \overset{\circ}{\theta}\}$ of the macroproblem be obtained, where the dual problem (5.4) has a single solution $\{\overset{\circ}{\lambda}{}^i, \overset{\circ}{\delta}{}^l\}$. Let these dual estimates correspond to optimal solutions of block problems (5.5). We establish a sufficient condition for the optimality of the disaggregated solution $\{\overset{\circ}{x}{}_j^i, \overset{\circ}{\theta}\}$, where $\overset{\circ}{x}{}_j^i = \alpha_j^i \overset{\circ}{X}{}^i$, for the problem (2.6)–(2.10), (5.1).

Theorem 5.1. *A sufficient condition for the optimality of solution* $\{\overset{\circ}{x}{}_j^i, \overset{\circ}{\theta}\}$ *of the problem (2.6)–(2.10), (5.1) is satisfaction of the following equality:*

$$\sum_{j=1}^{J} \sum_{i \in I_j} \left(\overset{\circ}{\lambda}{}^i - \sum_{l=1}^{L} \bar{b}_j^{il} \overset{\circ}{\delta}{}^l \right) \hat{x}_j^i + \sum_{l=1}^{L} \overset{\circ}{\delta}{}^l - \sum_{j=1}^{J} \sum_{i \in I_j} \overset{\circ}{\lambda}{}^i \overset{\circ}{x}{}_j^i = 0. \tag{5.6}$$

If the problem (2.6)–(2.10), (5.1) is solvable, and the disaggregated solution $\{\overset{\circ}{x}{}_j^i, \overset{\circ}{\theta}\}$ *is not optimal for it, then the symbol* $=$ *in the relation (5.6) is replaced by* $>$.

The proof of Theorem 5.1 is carried out by analogy to the proof of Theorem 3.1.

Consider the aggregated problem (6.5). Expressing X^i as $a^i\theta - V^i$ and substituting it in the inequalities, we obtain

$$\theta \left(\sum_{i=1}^{I} B_j^{ik} a^i \right) \le 1 + \sum_{i=1}^{I} B_j^{ik} V^i, \qquad j \in [1:J], \qquad k \in K_j,$$
$$\theta \left(\sum_{i=1}^{I} B^{il} a^i \right) \le 1 + \sum_{i=1}^{I} B^{il} V^i, \qquad l \in [1:L],$$

From the last equations, we deduce the formula for the optimal solution $\overset{\circ}{\theta}$ of the problem (5.3):

$$\overset{\circ}{\theta} = \min \left[\min_{\substack{j \in [1:j] \\ k \in K_j}} \left(1 + \sum_{i=1}^{I} B_j^{ik} V^i \right) \Big/ \left(\sum_{i=1}^{I} B_j^{ik} a^i \right), \right.$$
$$\left. \min_{l \in [1:L]} \left(1 + \sum_{i=1}^{I} B^{il} V^i \right) \Big/ \left(\sum_{i=1}^{I} B^{il} a^i \right) \right]. \tag{5.7}$$

Here, as before, it is assumed that the strict inequalities $\overset{\circ}{\theta} > 0$, $X^i > 0$, $i \in [1:I]$ hold.

Three possibilities arise in the solution of the conjugated macroproblems (5.3) and (5.4).

Case A. Let minimum in (5.7) be attained for a unique pair of indices (j^*, k^*). Then,

$$\overset{\circ}{\theta} = \left(1 + \sum_{i=1}^{I} B_{j^*}^{ik^*} V^i\right) \Big/ \left(\sum_{i=1}^{I} B_{j^*}^{ik^*} a^i\right). \tag{5.8}$$

The optimal solution of the problem (5.4) takes the form

$$\begin{aligned}
&\overset{\circ}{\eta}{}_{j^*}^{k^*} = \left(1 + \sum_{i=1}^{I} B_{j^*}^{ik^*} V^i\right) \Big/ \left(\sum_{i=1}^{I} B_{j^*}^{ik^*} a^i\right), \\
&\overset{\circ}{\eta}{}_j^k = 0 \ \text{ for } \ (j,k) \neq (j^*, k^*), \\
&\overset{\circ}{\lambda}{}^i = B_{j^*}^{ik^*} \left(1 + \sum_{i=1}^{I} B_{j^*}^{ik^*} V^i\right) \Big/ \left(\sum_{i=1}^{I} B_{j^*}^{ik^*} a^i\right), \\
&\overset{\circ}{\delta}{}^l = 0, \qquad l \in [1:L].
\end{aligned} \tag{5.9}$$

Case B. Let minimum in (5.7) be attained for a unique index $l^* \in [1:L]$.. We have

$$\overset{\circ}{\theta} = \left(1 + \sum_{i=1}^{I} B^{il^*} V^i\right) \Big/ \left(\sum_{i=1}^{I} B^{il^*} a^i\right). \tag{5.10}$$

The formulas for the optimal dual variables have the following form:

$$\begin{aligned}
&\overset{\circ}{\eta}{}_j^k = 0, \qquad j \in [1:J], \qquad k \in K_j, \\
&\overset{\circ}{\delta}{}^{l^*} = \left(1 + \sum_{i=1}^{I} B^{il^*} V^i\right) \Big/ \left(\sum_{i=1}^{I} B^{il^*} a^i\right), \qquad \overset{\circ}{\delta}{}^l = 0 \text{ for } l \neq l^*, \\
&\overset{\circ}{\lambda}{}^i = B^{il^*} \left(1 + \sum_{i=1}^{I} B^{il^*} V^i\right) \Big/ \left(\sum_{i=1}^{I} B^{il^*} a^i\right).
\end{aligned} \tag{5.11}$$

Case C. Let the minimum in (5.7) be attained for a set R_1 of the indicc pairs (j,k) and the set R_2 of indices l. Then, we have

$$\begin{aligned}
&\overset{\circ}{\eta}{}_j^k = 0 \text{ for } (j,k) \in R_1, \qquad \overset{\circ}{\delta}{}^l = 0 \text{ for} l \in R_2, \\
&\overset{\circ}{\lambda}{}^i = \sum_{(j,k) \in R_1} B_j^{ik} \overset{\circ}{\eta}{}_j^k + \sum_{l \in R_2} B^{il} \overset{\circ}{\delta}{}^l.
\end{aligned} \tag{5.12}$$

The dual estimates $\overset{\circ}{\eta}{}_j^k$ for $(j,k) \in R_1$ and $\overset{\circ}{\delta}{}^l$ for $l \in R_2$ are nonunique and belong to the set $\Omega_{\eta,\delta}$, which is given by the constraints

$$\sum_{i=1}^{I}\left(\sum_{(j,k)\in R_1} B_j^{ik} a^i \overset{\circ}{\eta}{}_j^k + \sum_{l\in R_2} B^{il} a^i \overset{\circ}{\delta}{}^l\right) = 1, \qquad (5.13)$$
$$\overset{\circ}{\eta}{}_j^k \geq 0, \qquad \overset{\circ}{\delta}{}^l \geq 0.$$

The following statement establishes local monotonicity with respect to the functional for the iterative process:

Theorem 5.2. *Let, for some weights α_j^i, an optimal solution $\{\overset{\circ}{X}{}^i, \overset{\circ}{\theta}\}$ of the macroproblem (5.3) be obtained, and the corresponding disaggregated solution $\{\overset{\circ}{x}{}_j^i, \overset{\circ}{\theta}\}$, where $\overset{\circ}{x}{}_j^i = \alpha_j^i, \overset{\circ}{X}{}^i$, is not optimal for the problem (2.6)–(2.10), (5.1). Then, there exists a solution feasible for (2.6)–(2.10), (5.1) with the functional strictly greater than the value $\overset{\circ}{\theta}$.*

For cases A and B, the statement of Theorem 5.2 follows from the Taylor expansion in the neighborhood of the point $\rho = 0$ of optimal values $\overset{\circ}{\theta}(\rho)$ from (5.8), (5.10) because of (5.9), (5.11). Here, $\alpha_j^i(\rho_j)$ are taken in the form (2.33) and $\rho_j = \rho$, $j \in [1 : J]$. The indicated expansion has the form

$$\overset{\circ}{\theta}(\rho) = \overset{\circ}{\theta} + (\sigma_1 + \sigma_2)\rho + o(\rho), \qquad (5.14)$$

where, in the case A, one has

$$\sigma_1 = \sum_{j=1}^{J}\sum_{i\in I_j} \overset{\circ}{\lambda}{}^i \hat{x}_j^i - \sum_{j=1}^{J}\sum_{i\in I_j} \overset{\circ}{\lambda}{}^i \overset{\circ}{x}{}_j^i > 0,$$
$$\sigma_2 = \overset{\circ}{\eta}{}_{j^*}^{k^*}\left(1 - \sum_{i\in I_{j^*}} b_{j^*}^{ik^*} x_{j^*}^i\right) \geq 0,$$

in the case B, one has

$$\sigma_1 = \sum_{j=1}^{J}\sum_{i\in I_j}\left(\overset{\circ}{\lambda}{}^i - \bar{b}_j^{il^*}\overset{\circ}{\delta}{}^{l^*}\right)\hat{x}_j^i + \overset{\circ}{\delta}{}^{l^*} - \sum_{j=1}^{J}\sum_{i\in I_j} \overset{\circ}{\lambda}{}^i \overset{\circ}{x}{}_j^i > 0,$$
$$\sigma_2 = 0$$

Here, the values $\overset{\circ}{\lambda}{}^i$ are expressed through (5.9) and (5.11), respectively.

In the degenerate case C, the maximin problem is considered: find

$$\max_{x_j^i \in M_x^j} \min_{\{\overset{\circ}{\eta}_j^k, \overset{\circ}{\delta}^l\} \in \Omega_{\eta,\delta}} P(x_j^i, \overset{\circ}{\eta}_j^k, \overset{\circ}{\delta}^l), \tag{5.15}$$

where

$$P(x_j^i, \overset{\circ}{\eta}_j^k, \overset{\circ}{\delta}^l) = \sum_{j=1}^{J} \sum_{i \in I_j} \left(\overset{\circ}{\lambda}^i - \sum_{l \in R_2} \bar{b}_j^{il} \overset{\circ}{\delta}^l \right) x_j^i + \sum_{l \in R_2} \overset{\circ}{\delta}^l,$$

and M_x^j are bounded sets given by conditions (2.2), (2.8) for each j, and the values $\overset{\circ}{\lambda}^i$ are expressed through $\overset{\circ}{\eta}_j^k$, $\overset{\circ}{\delta}^l$ according to (5.12).

Let $\bar{x}_j^i$, $\bar{\eta}_j^k$, $\bar{\delta}^l$ be components of the saddle point of the problem (5.15). We formulate a sufficient condition for the optimality of the disaggregated solution $\{\overset{\circ}{x}_j^i, \overset{\circ}{\theta}\}$ for (2.6)–(2.10), (5.1) in the case of degeneracy.

Theorem 5.3. *A sufficient condition of optimality of $\{\overset{\circ}{x}_j^i, \overset{\circ}{\theta}\}$ for the problem (2.6)–(2.10), (5.1) under the degeneracy is the satisfaction of the equality*

$$P(\bar{x}_j^i, \bar{\eta}_j^k, \bar{\delta}^l) - \overset{\circ}{\theta} = 0. \tag{5.16}$$

If the disaggregated solution $\{\overset{\circ}{x}_j^i, \overset{\circ}{\theta}\}$ is not optimal for (2.6)–(2.10), (5.1), then the sign $=$ in the relation (5.16) is replaced by $>$.

We study the problem of monotonicity with respect to the functional under degeneracy. It turns out that expansion (5.14) takes place, where σ_1 satisfies the inequality

$$\sigma_1 \geq P(\bar{x}_j^i, \bar{\eta}_j^k, \bar{\delta}^l) - \overset{\circ}{\theta} > 0, \tag{5.17}$$

and the value σ_2 is expressed by the formula

$$\sigma_2 = \sum_{(j,k) \in R_1} \tilde{\eta}_j^k \left(1 - \sum_{i \in I_j} b_j^{ik} \bar{x}_j^i \right) \geq 0, \tag{5.18}$$

where $\tilde{\eta}_j^k \geq 0$.

Inequality (5.17) and relation (5.18) are established in the same way as in the proof of Theorem 4.2. Finally, we obtain the estimate $\overset{\circ}{\theta}(\rho) > \overset{\circ}{\theta}$ in a neighborhood of the point $\rho = 0$.

Thus, for the branch model with common binding constraints, the problem is reduced to finding minimal matrix entries. The typical situation is when we can indicate the constraints, either block or binding, that are essential when the optimum of problem (2.6)–(2.10), (5.1) is approached. If these are binding constraints, then starting from some iteration, the block conditions are omitted, and the problem is essentially simplified. If the minimum in (5.7) is attained for indices (j, k), then the binding constraints are omitted, and the problem is reduced to the one considered in §§2–4.

Let the branch be able to sell the free nomenclature of final production and c^m, $m \in [M_1+1 : M]$ be the cost per unit of the final production Y^m. Then, we can consider the problem of maximizing the profit.

The functional has the following form:

$$f = \sum_{m=M_1+1}^{M} c^m Y^m \to \max. \tag{5.19}$$

We write the following constraints:

$$\begin{gathered}
\sum_{i \in I_j} b_j^{ik} x_j^i \le 1, \qquad k \in K_j, \\
x_j^i \ge 0 \text{ for } i \in I_j, \qquad x_j^i = 0 \text{ for } i \in I \backslash I_j, \qquad j \in [1:J], \\
-\sum_{j=1}^{J} x_j^i + \sum_{m=M_1+1}^{M} e^{1m}(1+\delta^m) Y^m = V^i, \qquad i \in [1:I], \\
Y^m \ge 0, \qquad m \in [M_1+1:M].
\end{gathered} \tag{5.20}$$

The meaning of the introduced variables is the same as in §2.

We fix the weights α_j^i and consider the macroproblem for (5.19)–(5.20):

$$g = \sum_{m=M_1+1}^{M} c^m Y^m \to \max, \tag{5.21}$$

$$\sum_{i=1}^{I} B_j^{ik} X^i \le 1, \qquad j \in [1:J], \qquad k \in K_j, \tag{5.22}$$

$$-X^i + \sum_{m=M_1+1}^{M} e^{im}(1+\delta^m) Y^m = V^i, \qquad i \in [1:I], \tag{5.23}$$

$$X^i \ge 0, \qquad i \in [1:I], \qquad Y^m \ge 0, \qquad m \in [M_1+1:M]. \tag{5.24}$$

Expressing X^i through Y^m from (5.23) and substituting it in (5.24), we have

$$
\begin{gathered}
\sum_{m=M_1+1}^{M} c^m Y^m \to \max, \\
\sum_{m=M_1+1}^{M} D_j^{mk} Y^m \leq p_j^k, \qquad j \in [1:J], \qquad k \in K_j, \\
Y^m \geq 0, \qquad m \in [M_1+1:M],
\end{gathered} \tag{5.25}
$$

where the following notation is used:

$$
D_j^{mk} = \sum_{i=1}^{I} e^{im}(1+\delta^m)B_j^{ik}, \qquad p_j^k = 1 + \sum_{i=1}^{I} B_j^{ik} V^i.
$$

As before, it is assumed that $\overset{\circ}{X}{}^i > 0$, $i \in [1:I]$. Recall that the latter inequalities denote that the reserves of components of each type are less than necessary for the final product.

We present the dual problem for (5.21)–(5.24):

$$
\psi = \sum_{j=1}^{J} \sum_{k \in K_j} \eta_j^k + \sum_{i=1}^{I} V^i \lambda^i \to \min, \tag{5.26}
$$

$$
\sum_{j=1}^{J} \sum_{k \in K_j} B_j^{ik} \eta_j^k - \lambda^i = 0, \qquad i \in [1:I], \tag{5.27}
$$

$$
\sum_{i=1}^{I} e^{im}(1+\delta^m)\lambda^i \geq c^m, \qquad m \in [M_1+1:M], \tag{5.28}
$$

$$
\eta_j^k \geq 0, \qquad j \in [1:J], \qquad k \in K_j. \tag{5.29}
$$

We substitute the values λ^i expressed, according to (5.27), through η_j^k into relations (5.28) and the functional (5.26) and obtain finally the problem dual to (5.25):

$$
\begin{gathered}
\psi = \sum_{j=1}^{J} \sum_{k \in K_j} p_j^k \eta_j^k \to \min, \\
\sum_{j=1}^{J} \sum_{k \in K_j} D_j^{mk} \eta_j^k \geq c^m, \qquad m \in [M_1+1:M], \\
\eta_j^k \geq 0, \qquad j \in [1:J], \qquad k \in K_j.
\end{gathered} \tag{5.30}
$$

Thus, the values $\overset{\circ}{X}{}^i$ are defined from (5.23) after solving the primal problem (5.25), and the values $\overset{\circ}{\lambda}{}^i$ are computed according to (5.27) after solving the dual problem (5.30). The indicated conjugated problems (5.25) and (5.30) are solved iteratively. The macroproblem (5.25) has a small number of variables Y^m and a large number of constraints. This problem can be solved by methods that take into account of this situation (for example, by the relaxation method).

The optimality condition formulated in Section 3 is important for the studied problem. Let, for some weights α_j^i, the optimal plan $\{\overset{\circ}{X}{}^i, \overset{\circ}{Y}{}^m\}$ of the macroproblem (5.21)–(5.24) be obtained. Then, the disaggregated solution $\{\overset{\circ}{x}{}_j^i, \overset{\circ}{Y}{}^m\}$, where $\overset{\circ}{x}{}_j^i = \alpha_j^i \overset{\circ}{X}{}^i$, is optimal for the main problem (5.19), (5.20) if the following equations hold:

$$\hat{h}_j = \overset{\circ}{h}_j, \qquad j \in [1:J].$$

The meaning of the values $\hat{h}_j$, $\overset{\circ}{h}_j$ is given in Section 3.

In fact, solution $\{\overset{\circ}{x}{}_j^i, \overset{\circ}{Y}{}^m\}$ is feasible for the problem (5.19)–(5.20). We compute the value of the functional for this plan. We multiply each equality (5.23) by $\overset{\circ}{\lambda}{}^i$ and sum over all $i \in I$:

$$\sum_{i=1}^{I} \left(\sum_{m=M_1+1}^{M} e^{im}(1+\delta^m)\, \overset{\circ}{Y}{}^m \right) \overset{\circ}{\lambda}{}^i = \sum_{i=1}^{I} (\overset{\circ}{X}{}^i \overset{\circ}{\lambda}{}^i + V^i \overset{\circ}{\lambda}{}^i).$$

By changing the summing order in the left-hand side of the latter equation, we obtain

$$\sum_{m=M_1+1}^{M} \left(\sum_{i=1}^{I} e^{im}(1+\delta^m)\, \overset{\circ}{\lambda}{}^i \right) \overset{\circ}{Y}{}^m = \sum_{i=1}^{I} (\overset{\circ}{X}{}^i \overset{\circ}{\lambda}{}^i + V^i \overset{\circ}{\lambda}{}^i).$$

The sum

$$\sum_{i=1}^{I} e^{im}(1+\delta^m)\, \overset{\circ}{\lambda}{}^i$$

is equal to c^m if $\overset{\circ}{Y}{}^m > 0$, therefore, the left-hand side is equal to

$$\sum_{m=M_1+1}^{M} c^m \overset{\circ}{Y}{}^m,$$

which is the optimal value of the macroproblem (5.21)–(5.24). Thus, for the plan $\{\overset{\circ}{x}{}^i_j, \overset{\circ}{Y}{}^m\}$, the functional of the mail problem (5.19), (5.20) is equal to

$$\overset{\circ}{f} = \sum_{i=1}^{I}(\overset{\circ}{X}{}^i \overset{\circ}{\lambda}{}^i + V^i \overset{\circ}{\lambda}{}^i) = \overset{\circ}{h} + \sum_{i=1}^{I} V^i \overset{\circ}{\lambda}{}^i.$$

The further consideration is almost analogous to the proof of Theorem 3.1. We can show that there exist dual variables $\hat{\xi}^k_j$, $\overset{\circ}{\lambda}{}^i$, where $\hat{\xi}^k_j$ are optimal solutions of the dual block problems. Moreover, these variables are feasible for the problem dual to (5.19), (5.20), and the functional for it is equal to

$$\hat{\varphi} = \hat{h} + \sum_{i=1}^{I} V^i \overset{\circ}{\lambda}{}^i.$$

Finally, due to Theorem 1.2, relations $\hat{h} = \overset{\circ}{h}$ or $\hat{h}_j = \overset{\circ}{h}_j$, $j \in [1 : J]$ imply that $\{\overset{\circ}{x}{}^i_j, \overset{\circ}{Y}{}^m\}$ is an optimal solution of the problem (5.19)–(5.20).

Monotonicity with respect to the functional is established by applying the theorem about the marginal value, as in Section 4. It is important to stress that the optimal solution of the problem (5.19), (5.20) does not necessarily correspond to the degenerate macroproblem, although it always holds for the main problem. Under condition $M - M_1 > J$, the absence of degeneracy is possible.

Introducing a procedure to find the saddle point in the degenerate macroproblem makes the decomposition method more complicated. The following version of the model has unique dual estimates.

Assume that, for each equipment group $k \in K_j$, each plant j uses a minimal fund $\hat{\Phi}^k_j \geq 0$. We introduce additional (standby) funds y^k_j, $j \in [1 : J]$, $k \in K_j$, the use of which is connected with certain expenditures. Let $q^k_j > 0$ be the cost of the use of a unit of the standby fund y^k_j. We can consider the planning process, where the total expenditures with respect to the introduced additional funds are bounded by the value Φ, which is at the disposal of the branch. We come to the following problem:

$$\theta \longrightarrow \max, \tag{5.31}$$

$$\sum_{i \in I_j} T^{ik}_j x^i_j - y^k_j = \hat{\Phi}^k_j, \qquad k \in K_j, \tag{5.32}$$

$$x^i_j \geq 0, \quad i \in I_j, \quad x^i_j = 0, \quad i \in I \backslash I_j, \quad y^k_j \geq 0,$$

$$k \in K_j, \qquad j \in [1:J],$$

$$a^i\theta - \sum_{j=1}^{J} x_j^i = V^i, \qquad i \in [1:I], \tag{5.33}$$

$$\sum_{j=1}^{J} \sum_{k \in K_j} q_j^k y_j^k \leq \Phi, \tag{5.34}$$

$$\theta \geq 0. \tag{5.35}$$

Here, inequality (5.34) is obviously the binding constraint. Consider the application of the decomposition method based on the aggregation for the problems (5.31)–(5.35).

By fixing the weights α_j^i, we have the macroproblem

$$\theta \to \max, \tag{5.36}$$

$$\sum_{i=1}^{I} D_j^{ik} X^i - y_i^k = \hat{\Phi}_j^k, \qquad y_j^k \geq 0, \qquad j \in [1:J], \qquad k \in K_j, \tag{5.37}$$

$$a^i\theta - X^i = V^i, \qquad i \in [1:I], \tag{5.38}$$

$$\sum_{j=1}^{J} \sum_{k \in K_j} q_j^k y_j^k \leq \Phi, \tag{5.39}$$

$$\theta \geq 0, \qquad X^i \geq 0, \qquad i \in [1:I], \tag{5.40}$$

where notation $D_j^{ik} = T_j^{ik}\alpha_j^i$ is used. Relations (5.37) and (5.39) give

$$\overset{\circ}{\theta} = \left[\Phi + \sum_{j=1}^{J} \sum_{k \in K_j} \left(\hat{\Phi}_j^k + \right.\right.$$
$$\left.\left. + \sum_{i=1}^{I} D_j^{ik} V^i\right) q_j^k\right] \Big/ \left(\sum_{j=1}^{J} \sum_{k \in K_j} \sum_{i=1}^{I} D_j^{ik} a^i q_j^k\right). \tag{5.41}$$

As before, it is assumed that $\overset{\circ}{\theta} > 0$, $\overset{\circ}{X}{}^i > 0$, $i \in [1:I]$, and the dual problem for (5.36)–(5.40) is considered:

$$\psi = \sum_{j=1}^{J} \sum_{k \in K_j} \hat{\Phi}_j^k \eta_j^k + \Phi\eta + \sum_{i=1}^{I} V^i \lambda^i \to \min, \tag{5.42}$$

$$\sum_{i=1}^{I} a^i \lambda^i = 1, \tag{5.43}$$

$$\sum_{j=1}^{J} \sum_{k \in K_j} D_j^{ik} \eta_j^k - \lambda^i = 0, \qquad i \in [1:I], \tag{5.44}$$

$$-\eta_j^k - q_j^k \eta \geq 0, \qquad j \in [1:J], \qquad k \in K_j. \tag{5.45}$$

$$\eta \geq 0. \tag{5.46}$$

Here, η is the dual variable that corresponds to (5.39).

Assume that (5.45) is satisfied as the equalities $\eta_j^k = q_j^k \eta$. We substitute the latter equations in (5.44), then we substitute the values λ^i expressed through η in (5.43). Thus, we obtain

$$\eta \sum_{j=1}^{J} \sum_{k \in K_j} \sum_{i=1}^{I} D_j^{ik} a^i q_j^k = 1.$$

This gives us

$$\eta = 1/\overset{\circ}{D}, \qquad \eta_j^k = q_j^k / \overset{\circ}{D}, \qquad \lambda^i = \sum_{j=1}^{J} \sum_{k \in K_j} D_j^{ik} q_j^k / \overset{\circ}{D}, \tag{5.47}$$

where $\overset{\circ}{D} = \sum_{j=1}^{J} \sum_{k \in K_j} \sum_{i=1}^{I} D_j^{ik} a^i q_j^k$.

The solution given by (5.47) is feasible for the dual problem (5.42)–(5.46), and the functional ψ for this solution is equal to the right-hand side of (5.41). Therefore, due to Theorem 1.2, the plan is optimal for (5.42)–(5.46). It is important to note that, in the model considered, the dual estimates are defined uniquely. However, it easily shown that the local problems can have unbounded solutions. We introduce the additional assumption

$$x_j^i \leq \bar{\kappa} < \infty, \qquad j \in [1:J], \qquad i \in I_j,$$

where the constant $\bar{\kappa}$ satisfies the inequality

$$\bar{\kappa} > (\max_{i \in [1:I]} (V^i)) + (\max_{i \in [1:I]} (a^i)) \left(\Phi + \sum_{j=1}^{J} \sum_{k \in K_j} \hat{\Phi}_j^k q_j^k \right) \times$$
$$\times \left[(\min_{\substack{j \in [1:J] \\ k \in K_j,\, i \in I_j}} (T_j^{ik})) \min_{\substack{j \in [1:J] \\ k \in K_j}} (q_j^k)) \times \min_{i \in [1:I]} (a^i)) \left(\min_{j \in [1:J]} \left(\sum_{i \in I_j} \alpha_j^i \right) \right) \right]^{-1}. \tag{5.48}$$

Assume also that all minimums in the right-hand side of (5.48) are strictly positive, and $\hat{\Phi}_j^k \geq 0$, $j \in [1:J]$, $k \in K_j$. Then, the relations (5.41), (5.47) remain valid.

Let us note some consequences. Since the relations in (5.45) hold as equalities, then, by the second duality Theorem 1.4, all y_j^k are strictly greater than zero. Since (5.39) is an equality, this means that the general standby fund is totally distributed, and each group gets a nonzero part.

All other constructions of the decomposition algorithm are the same as for the problem (5.31)–(5.35). We consider the local problems, where the sums

$$-\sum_{k \in K_j} \overset{\circ}{\eta}\, q_j^k y_j^k, \qquad j \in [1:J]$$

are added to the functionals. Assume that, for the fixed weights α_j^i, we obtain the solution $\{\overset{\circ}{X}{}^i, \overset{\circ}{\theta}, \overset{\circ}{y}{}_j^k\}$ of the macroproblem, and $\overset{\circ}{x}{}_j^i$ is the disaggregated plan. Furthermore, let $\{\hat{x}_j^i, \hat{y}_j^k\}$ be optimal solutions of the local problems, whose functionals include $\overset{\circ}{\lambda}{}^i$ and $\overset{\circ}{\eta}$ according to (5.47). Then, a sufficient condition of optimality $\{\overset{\circ}{x}{}_j^i, \overset{\circ}{\theta}, \overset{\circ}{y}{}_j^k\}$ for the problem (5.31)–(5.35) is the satisfaction of the equalities

$$\hat{h} = \overset{\circ}{h},$$

where

$$\hat{h} = \sum_{j=1}^{J} \hat{h}_j, \qquad \hat{h}_j = \sum_{i \in I_j} \overset{\circ}{\lambda}{}^i \hat{x}_j^i - \sum_{k \in K_j} \overset{\circ}{\eta}\, q_j^k \hat{y}_i^k,$$

$$\overset{\circ}{h} = \sum_{j=1}^{J} \overset{\circ}{h}_j, \qquad \overset{\circ}{h}_j = \sum_{i \in I_j} \overset{\circ}{\lambda}{}^i \overset{\circ}{x}{}_j^i - \sum_{k \in K_j} \overset{\circ}{\eta}\, q_j^k \overset{\circ}{y}{}_i^k.$$

As in Section 4, we can prove the local monotonicity of the iterative process with respect to the functional.

We note some properties of the function $\overset{\circ}{\theta}(\rho_j)$, which is obtained if one substitutes $T_j^{ik} \alpha_j^i(\rho_j)$ instead of D_j^{ik} in the right-hand side of (5.41), where the weights $\alpha_j^i(\rho_j)$ are expressed through (2.33). The function $\overset{\circ}{\theta}(\rho_j)$ is linear-fractional with respect to variables ρ_j. If we assume that the denominator of this function does not vanish, then the maximum of the function $\overset{\circ}{\theta}(\rho_j)$ is attained on the boundary of the unit cube, and to find it, we can apply algorithms of linear-fractional programming.

§6. Random Parameters in the Branch Model

Assume that some coefficients of the main model are not defined precisely. Various problem statements of stochastic programming exist for this situation. In a number of cases, the analysis of stochastic problems is reduced to the study of their deterministic equivalents. We will consider the so-called two-stage problem of stochastic programming.

First, we study the two-stage statement for the general problem of linear programming. Consider the following problem:

$$\begin{aligned} &(c, x) \to \min, \\ &Ax = b, \\ &A^1 x = b^1, \\ &x \geq 0. \end{aligned} \tag{6.1}$$

Let entries of the matrix $A = A(\omega)$ and vectors $b = b(\omega)$, $c = c(\omega)$ depend on the random parameter that belong to a set Ω_ω of random events. Assume decision x must be made before knowing the value of the random parameter. Then, the two-stage problem is written in the form

$$\min_x M_\omega\{c(\omega), x) + \min_y[(q, y)|By = b(\omega) - A(\omega)x,\ y \geq 0]\},$$

$$A^1 x = b^1,$$

$$x \geq 0,$$

where M_ω denotes (mathematical) expectation, B and y are so-called compensation matrix and vector, and q is a penalty vector.

For our main model, we assume that the components of technological matrices T_j^{ik} and fund vectors Φ_j^k are random. The compensation matrix is assumed to have the simplest form $B = (E, -E)$, where E is the unit matrix. Vectors y and q are partitioned into two parts that correspond to submatrices E, $-E$, which are denoted by y^+, y^- and q^+, q^-. The two-stage problem is reduced to the form

$$\min_x\{(c,x)+M_\omega[\min_y((q^+,y^+)+(q^-,y^-)|y^+-y^-=$$
$$=b(\omega)-A(\omega)x, \qquad y^+\geq 0, \qquad y^-\geq 0]\},$$
$$A^1x=b^1, \tag{6.2}$$
$$x\geq 0.$$

Here, $y^+=b-Ax$ and $y^-=0$ for $b\geq Ax$; $y^+=0$ and $y^-=b-Ax$ for $b<Ax$.

For branch planning, the values y^+ and y^- are interpreted as the idle time and time deficit, the values q^+ and q^- are interpreted as the cost of the idle time and overtime.

In general, the analysis of two-stage problems for the given distributions of random parameters is fairly complicated. We will make another assumption. We assume that there exist only a finite number of realizations of technological matrices and fund vectors $T^{(1)}$, $\Phi^{(1)},\dots,T^{(N)}$, $\Phi^{(N)}$, whose probabilities are $p^{(1)},\dots p^{(N)}$, where $\sum\limits_{n=1}^{N}p^{(n)}=1$. We use the notation

$$[y]^+=\max(0,y), \qquad [y]^-=\max(0,-y).$$

Then, problem (6.2) for the main model of branch planning is reduced to the following two equivalent deterministic problems:

$$f=\theta-\sum_{n=1}^{N}p^{(n)}\sum_{j=1}^{J}\sum_{k\in K_j}\left(q_j^{+k}[y_j^{k(n)}]^+ + q_j^{-k}[y_j^{k(n)}]^-\right)\to\max,$$
$$\sum_{i\in I_j}T_j^{ik(n)}x_j^i+y_j^{k(n)}=\Phi_j^{k(n)}, \qquad n\in[1:N],$$
$$j\in[1:J], \qquad k\in K_j, \tag{6.3}$$
$$x_j^i\geq 0 \text{ for } i\in I_j, \qquad x_j^i=0 \text{ for } i\in\bar{I}\backslash I_j, \qquad j\in[1:J],$$
$$a^i\theta-\sum_{j=1}^{J}x_j^i=V^i, \qquad i\in[1:I], \qquad \theta\geq 0$$

or

$$f = \theta - \sum_{n=1}^{N} p^{(n)} \sum_{j=1}^{J} \sum_{k \in K_j} \left(q_j^{+k} y_j^{+k(n)} + q_j^{-k} y_j^{-k(n)} \right) \to \max,$$

$$\sum_{i \in I_j} T_j^{ik(n)} x_j^i + y_j^{+k(n)} - y_1^{-k(n)} = \Phi_j^{k(n)}, \qquad n \in [1:N],$$

$$n \in [1:N], \qquad j \in [1:J], \qquad k \in K_j, \tag{6.4}$$

$$x_j^i \geq 0 \text{ for } i \in I_j, \qquad x_j^i = 0 \text{ for } i \in \bar{I} \backslash I_j, \qquad j \in [1:J],$$

$$y_1^{+k(n)} \geq 0, \quad y_j^{-k(n)} \geq 0, \quad n \in [1:N], \quad j \in [1:J], \quad k \in K_j,$$

$$a^i\theta - \sum_{j=1}^{J} x_j^i = V^i, \qquad i \in [1:I], \qquad \theta \geq 0,$$

The problem (6.3) is represented in the first canonical form of the problems of piecewise linear programming. The problem (6.4) is a linear programming problem equivalent to (6.3). Below, we study the solution of problem (6.3) when the iterative algorithm is applied to it. In the cases suitable for the analysis, problem (6.3) is considered in the equivalent form (6.4).

We have the following aggregated problem for (6.3):

$$g = \theta - \sum_{n=1}^{N} p^{(n)} \sum_{j=1}^{J} \sum_{k \in K_j} \left(q_j^{+k} [y_j^{k(n)}]^+ + q_j^{-k} [y_j^{k(n)}]^- \right) \to \max, \tag{6.5}$$

$$\sum_{i=1} D_j^{ik(n)} X^i + y_j^{k(n)} = \Phi_j^{k(n)}, \tag{6.6}$$

$$y_j^{k(n)} \geq 0, \qquad n \in [1:N], \qquad j \in [1:J], \qquad k \in K_j,$$

$$a^i\theta - X^i = V^i, \tag{6.7}$$

$$X^i \geq 0, \qquad i \in [1:I], \qquad \theta \geq 0.$$

Here, we accept the notation $D_j^{ik(n)} = T_j^{ik(n)} \alpha_j^i$.

We express X^i through θ according to (6.7) and, substituting it in (6.6), we find the values $y_j^{k(n)}$ expressed through θ. Then, substituting $y_j^{k(n)}$ in

(8.5), we obtain

$$g(\theta) = \theta - \sum_{n=1}^{N} p^{(n)} \sum_{j=1}^{J} \sum_{k\in K_j} q_j^{+k} \left[\left(\Phi_j^{k(n)} + \sum_{i=1}^{I} D_j^{ik(n)} V^i \right) - \right.$$
$$\left. -\theta \sum_{i=1}^{I} D_j^{ik(n)} a^i \right]^+ + \tag{6.8}$$
$$+ \quad q_j^{-k} \left[\left(\Phi_j^{k(n)} + \sum_{i=1}^{I} D_j^{ik(n)} V^i \right) - \theta \sum_{i=1}^{I} D_j^{ik(n)} a^i \right]^- .$$

Thus, solution of the problem (6.5)–(6.7) is reduced to finding the maximum of the concave piecewise linear function $g(\theta)$, which has break points with coordinates

$$\theta_j^{k(n)} = \left(\Phi_j^{k(n)} + \sum_{i=1}^{I} D_j^{ik(n)} V^i \right) \left(\sum_{i=1}^{I} D_j^{ik(n)} a^i \right)^{-1} .$$

We assume that the maximum of the function $g(\theta)$ is attained for some $\overset{\circ}{\theta}$ and has the following two properties:

Property A. $\overset{\circ}{\theta}$ is strictly greater than zero, and the values $\overset{\circ}{X}{}^i = a^i \overset{\circ}{\theta} - V^i$, $i \in [1:I]$ are also strictly greater than zero.

Property B. The coordinate $\overset{\circ}{\theta}$ is the break point and corresponds to the unique triple of indices, which will be denoted by (n^*, j^*, k^*).

The property B is equivalent to the assumption on nondegeneracy.

Moreover, by K^+ and K^-, we denote the sets of index triples (n,j,k), for which $y_j^{k(n)} > 0$ and $y_j^{k(n)} < 0$, respectively. The following statement establishes the condition of existence of a solution of problem (6.5)–(6.7):

Theorem 8.1. *Satisfaction of the following inequalities is a necessary and sufficient condition for the existence of the maximum of the function $g(\theta)$ with properties A and B:*

$$D^- = \sum_{i=1}^{I} \sum_{(n,j,k)\in K^-} D_j^{ik(n)} a^i q_j^{-k} p^{(n)} - \sum_{i=1}^{I} \sum_{(n,j,k)\in K^+} D_j^{ik(n)} a^i q_j^{+k} p^{(n)} +$$
$$+ q_{j^*}^{k^*(n^*)} p^{(n^*)} \sum_{i=1}^{I} D_{j^*}^{ik^*(n^*)} a^i - 1 \geq 0, \tag{6.9}$$

$$D^+ = \sum_{i=1}^{I} \sum_{(n,j,k)\in K^+} D_j^{ik(n)} a^i q_j^{+k} p^{(n)} - \sum_{i=1}^{I} \sum_{(n,j,k)\in K^-} D_j^{ik(n)} a^i q_j^{-k} p^{(n)} +$$

$$+ q_{j^*}^{k^*(n^*)} p^{(n^*)} \sum_{i=1}^{I} D_{j^*}^{ik^*(n^*)} a^i + 1 \geq 0. \tag{6.10}$$

Proof. Since (6.5)–(6.7) is a problem of piecewise linear programming, a necessary and sufficient condition for its solution is the existence of Lagrange multipliers $\eta_j^{k(n)}$ that correspond to (6.6) and λ^i that correspond to (6.7), which satisfy the following relations:

$$\sum_{i=1} a^i \lambda^i = 1, \tag{6.11}$$

$$\sum_{n=1}^{N} \sum_{j=1}^{J} \sum_{k\in K_j} D_j^{ik(n)} \eta_j^{k(n)} - \lambda^i = 0, \qquad i \in [1:I], \tag{6.12}$$

$$\eta_j^{k(n)} = -q_j^{+k} p^{(n)}, \qquad y_j^{k(n)} > 0, \qquad (n,j,k) \in K^+, \tag{6.13}$$

$$-q_{j^*}^{+k^*} p^{(n^*)} \leq \eta_{j^*}^{k^*(n^*)} \leq q_j^{-k^*p(n^*)}, \qquad y_{j^*}^{k^*(n^*)} = 0, \tag{6.14}$$

$$\eta_j^{k(n)} = q_j^{-k} p^{(n)}, \qquad y_j^{k(n)} < 0, \qquad (n,j,k) \in K^-. \tag{6.15}$$

We express λ^i through (6.12) by replacing $\eta_j^{k(n)}$ from their expressions according to (6.13), (6.15):

$$\lambda^i = \left(- \sum_{(n,j,k)\in K^+} D_j^{ik(n)} q_j^{+k} p^{(n)} + \sum_{(n,j,k)\in K^+} D_j^{ik(n)} q_j^{-k} p^{(n)} \right) +$$

$$+ D_{j^*}^{ik^*(n^*)} \eta_{j^*}^{k^*(n^*)}. \tag{6.16}$$

Then, substituting (6.16) in (6.11), we obtain

$$\eta_{j^*}^{k^*(n^*)} = \left(1 + \sum_{i=1}^{I} \sum_{(n,i,k)\in K^+} D_j^{ik(n)} a^i q_j^{+k} p^{(n)} - \right.$$

$$\left. - \sum_{i=1}^{I} \sum_{(n,j,k)\in K^-} D_j^{ik(n)} a^i q_j^{-k} p^{(n)} \right) \left(\sum_{i=1}^{I} D_{j^*}^{ik^*(n^*)} a^i \right)^{-1}. \tag{6.17}$$

We substitute the latter expression in (6.14), thus arriving at inequalities (6.9), (6.10). The theorem is proved.

We clear up the economical meaning of conditions (6.9), (6.10). Consider the planning when the idle time is not penalized, but the funds should be used completely. In this case, we have $q_j^{+k} = 0$, $y_j^{+k(n)} = 0$, $n \in [1:N]$, $j \in [1:J]$, $k \in [1:K_j]$, and only the condition (6.9), which has the following form:

$$\sum_{n=1}^{N}\sum_{j=1}^{J}\sum_{k\in K_j}\sum_{i\in I_j} T_j^{ik(n)} a_j^i q_j^{-k} p^{(n)} a^i - 1 \geq 0,$$

remains. For the sake of simplicity, we consider that $V^i = 0$, $i \in [1:I]$, then the latter inequality is reduced to the following one:

$$\sum_{j=1}^{J}\sum_{k\in K_j}\sum_{i\in I_j} M_\omega(T_j^{ik(n)})\, \overset{\circ}{x}{}_j^i q_j^{-k} - \overset{\circ}{\theta} \geq 0.$$

Here, $\overset{\circ}{x}{}_j^i$ is the support disaggregated plan. This inequality means that the average expenditures for the production of the final product valued at overtime prices should overweight the cost of the final product.

If we consider planning without a standby fund and penalized idle time, then only condition (6.10) remains. For fixed $j \in [1:J]$, the functionals of these problems have the form

$$h_j = \sum_{i\in I_j} \overset{\circ}{\lambda}{}^i x_j^i - \sum_{n=1}^{N} p^{(n)} \sum_{k\in K_j} (q_j^{+k} y_j^{+k(n)} + q_j^{-k} y_j^{-k(n)}) \to \max,$$

where the values $\overset{\circ}{\lambda}{}^i$ are expressed according to (6.16), (6.17).

Let, for some weights, the optimal solution of the macroproblem (6.5)–(6.7) be obtained: $\{\overset{\circ}{X}{}^i, \overset{\circ}{\theta}, \overset{\circ}{y}{}_j^{+k(n)}, \overset{\circ}{y}{}_j^{-k(n)}\}$, and $\overset{\circ}{x}{}_j^i = \alpha_j^i X^i$ is the corresponding disaggregated plan. Let $\{\hat{x}_j^i, \hat{y}_i^{+k(n)}, \hat{y}_j^{-k(n)}\}$ be optimal solutions of block problems. For each $j \in [1:J]$, we introduce the values

$$\overset{\circ}{h}_j = \sum_{i\in I_j} \overset{\circ}{\lambda}{}^i \overset{\circ}{x}{}_j^i - \sum_{n=1}^{N} p^{(n)} \sum_{k\in K_j} (q_j^{+k} \overset{\circ}{y}{}_j^{+k(n)} + q_j^{-k} \overset{\circ}{y}{}_j^{-k(n)}),$$

$$\hat{h}_j = \sum_{i\in I_j} \overset{\circ}{\lambda}{}^i \hat{x}_j^i - \sum_{n=1}^{N} p^{(n)} \sum_{k\in K_j} (q_j^{+k} \hat{y}_j^{+k(n)} + q_j^{-k} \hat{y}_j^{-k(n)}).$$

As in Theorem 3.1, we can prove that $\hat{h}_j \geq \overset{\circ}{h}_j$ holds always, and the equality for all $j \in [1:J]$ ensures the optimality of the solution $\{\hat{x}_j^i, \hat{y}_i^{+k(n)},$

$\hat{y}_j^{-k(n)}\}$ for problem (6.4). Finally, if the indicated solution is nonoptimal for (6.4), then, for some indices j, strict inequalities $\hat{h}_j > \overset{\circ}{h}_j$ should hold.

The local monotonicity with respect to the functional of the iterative process follows from the expansion of $\overset{\circ}{g}(\overset{\circ}{\theta}(\rho_j))$ into Taylor series. The function $\overset{\circ}{g}(\overset{\circ}{\theta}(\rho_j))$ is expressed from equality (6.8), where $\overset{\circ}{\theta}$ is substituted with account of expansion of weights $\alpha_j^i(\rho_j)$ in the neighborhood $\rho_j = 0$, $j \in [1 : J]$. According to (2.33), this expansion has the following form:

$$a_j^i(\rho_j) = \alpha_j^i + \rho_j(\hat{x}_j^i - \overset{\circ}{x}{}_j^i)(\overset{\circ}{X}{}^i)^{-1} -$$

$$-\alpha_j^i \sum_{p=1}^{J} \rho_p(\hat{x}_p^i - \overset{\circ}{x}{}_p^i)(\overset{\circ}{X}{}^i)^{-1} + o(\rho_j).$$

Finally, we obtain the following expression for the function $\overset{\circ}{g}(\overset{\circ}{\theta}(\rho_j))$:

$$\overset{\circ}{g}(\overset{\circ}{\theta}(\rho_j)) = \overset{\circ}{g}(\overset{\circ}{\theta}(0)) + \sum_{j=1}^{J} \rho_j(\hat{h}_j - \overset{\circ}{h}_j) + \rho_{j^*}\left[D^+\hat{y}_{j^*}^{+k^*(n^*)} + \right.$$

$$\left. + D^-\hat{y}_{j^*}^{-k^*(n^*)}\right] \overset{\circ}{T}{}^{-1} + \sum_{(n,j,k)\in K^+} \rho_j p^{(n)}(q_j^{-k} + q_j^{+k})\hat{y}_j^{-k(n)} +$$

$$+ \sum_{(n,j,k)\in K^-} \rho_j p^{(n)}(q_j^{-k} + q_j^{+k})\hat{y}_i^{+k(n)} + o(\rho_j).$$

In this relation, the values D^+ and D^- are left-hand sides of (6.9) and (6.10). By definition, the constant $\overset{\circ}{T}$ is equal to

$$\sum_{i\in I_{j^*}} T_{j^*}^{ik^*(n^*)} \alpha_{j^*}^i a^i.$$

Therefore, similar to Section 4, the estimate and its corollaries are valid in a neighborhood of a point $\rho_j = 0$, $j \in [1 : J]$.

In the previous analysis, a discrete distribution of random parameters was assumed. Now, we take a continuous distribution of the fund vector. For example, let components satisfy the exponential distribution law:

$$F_j^k(\Phi_j^k) = \begin{cases} 0 & \text{for } \Phi_j^k < \gamma_j^k, \\ 1 - \exp(-(\Phi_j^k - \gamma_j^k)/\bar{\Phi}_j^k) & \text{for } \Phi_j^k \geq \gamma_j^k. \end{cases}$$

Here, $\bar{\Phi}_j^k$ is the expectation of the values Φ_j^k, and γ_j^k are fixed constants.

The two-stage problem (6.1) of the main model is written as follows:

$$
\begin{aligned}
& q = \theta - \sum_{j=1}^{J} \sum_{k \in K_j} q_j^{+k} (\kappa_j^{k(1)} - \kappa_j^{k(2)}) - \\
& -0.5 \sum_{j=1}^{J} \sum_{k \in K_j} \frac{(q_j^{+k} + q_j^{-k})}{\bar{\Phi}_j^k} (\kappa_j^{k(2)})^2 \to \max, \\
& \sum_{i=1}^{I} T_j^{ik} x_j^i + (\kappa_j^{k(1)} - \kappa_j^{k(2)}) = \gamma_j^k, \quad j \in [1:J], \; k \in K_j; \\
& a^i \theta - \sum_{j=1}^{J} x_j^i = V^i, \quad i \in [1:I]; \\
& x_j^i \geq 0, \quad j \in [1:J]; \quad i \in I_j; \qquad x_j^i = 0, \qquad i \in I \backslash I_j \\
& \kappa_j^{k(1)} \geq 0, \quad \kappa_j^{k(2)} \geq 0, \\
& j \in [1:J], \quad k \in [1:K_j]; \quad \theta \geq 0,
\end{aligned}
\tag{6.18}
$$

where the values $\kappa_j^{k(1)}$, $\kappa_j^{k(2)}$ are related to random funds Φ_j^k, and it is assumed that $\kappa_j^{k(2)} \ll \bar{\Phi}_j^k$, $j \in [1:J]$, $k \in K_j$ holds.

The aggregated problem for (6.18) has the following form:

$$g \to \max \tag{6.19}$$

under conditions of problem (6.18), where the aggregated variables for fixed weights are substituted. The dual conditions for (6.19) have the following form:

$$\sum_{i=1}^{I} a^i \lambda^i = 1, \tag{6.20}$$

$$\lambda^i = \sum_{j=1}^{J} \sum_{k \in K_j} T_j^{ik} \alpha_j^i \eta_j^k, \quad i \in [1:I]; \tag{6.21}$$

$$
\begin{aligned}
& \eta_j^k \geq -q_j^{-k}, \quad -\eta_j^k \geq q_j^{+k} - \left(q_j^{+k} + q_j^{-k}\right) \kappa_j^{k(2)} / \bar{\Phi}_j^k, \\
& j \in [1:J], \quad k \in [1:J]
\end{aligned}
\tag{6.22}
$$

and will be used later, where η_j^k, $j \in [1:J]$, $k \in [1:K_j]$, λ^i, $i \in [1:I]$ are the dual variables that correspond to the equalities in (6.19).

Next transformations will be carried out for the sake of simplicity in the case $V^i = 0$, $i \in [1 : I]$, though all considerations are transferred word for word when $V^i \neq 0$, $i \in [1 : I]$. We express the values X^i, $i \in [1 : I]$ through θ from the second equalities in (6.19) and substitute the result into the first equalities of (6.19). After that, the values $(\kappa_j^{k(1)} - \kappa_j^{k(2)})$, $j \in [1 : J]$, $k \in [1 : K_j]$ are expressed from the first equalities of (6.19) and substituted into the cost function of the problem (6.19). Using the condition of this problem $\kappa_j^{k(1)} \geq 0$, $j \in [1 : J]$, $k \in [1 : K_j]$, we obtain

$$\begin{aligned}
&\theta\left[1+\sum_{j=1}^{J}\sum_{k\in K_j}\sum_{i=1}^{I} q_j^{+k} T_j^{ik} \alpha^i a^i\right] - \\
&-\sum_{j=1}^{J}\sum_{k\in K_j}\frac{(q_j^{+k}+q_j^{-k})}{\bar{\Phi}_j^k}(\kappa_j^{k(2)})^2 \to \max, \\
&\kappa_j^{k(2)} - \theta\left(\sum_{i=1}^{I} T_j^{ik}\alpha_j^i a^i\right) \geq -\gamma_j^k, \quad j \in [1:J],\ k \in [1:K_j]; \\
&\theta \geq 0, \quad \kappa_j^{k(2)} \geq 0, \quad j \in [1:J],\ k \in [1:K_j].
\end{aligned} \tag{6.23}$$

We consider the simplest problem

$$\begin{aligned}
&\theta - \frac{1}{2}q\kappa^2 \to \max, \\
&\kappa - \theta a \geq -\gamma,\ \kappa \geq 0,\ \theta \geq 0,
\end{aligned} \tag{6.24}$$

which will be called *the trivial model of the problem (6.23)*. The constant parameters a, γ, q are positive. The solution of (1.7) is constructed by the following. The equality $\kappa = \theta a - \gamma$ is substituted into the cost function. We have the optimal values

$$\overset{\circ}{\theta} = (1 + aq\gamma)/a^2 > 0, \quad \overset{\circ}{\kappa} = 1/aq > 0,$$

which are positive.

We remember the conditions $\kappa_j^{k(2)} \ll \bar{\Phi}_j^k$, $j \in [1 : J]$, $k \in [1 : K_j]$. If we identity approximately the value orders $q_j^{+k} \sim q^+$, $q_j^{-k} \sim q^-$, $\bar{\Phi}_j^k = \Phi$ as well as

$$a_j^k = \sum_{i=1}^{I} T_j^{ik}\alpha_j^i a^i \sim a,$$

then we obtain

$$(1 + q^+ a)\Phi/(q^+ + q^-)a \ll \Phi.$$

The last inequality is valid if $aq^- \gg 1$.

The problem (6.23) is solved by analogy to (6.24). The equalities

$$\kappa_j^{k(2)} = a_j^k \theta - \gamma_j^k, \qquad j \in [1:J],\ k \in [1:K_j]$$

are substituted into the cost function of (6.24). We have

$$\theta \left(1 + \sum_{j=1}^{J} \sum_{k \in K_j} a_j^k q_j^k \gamma_j^k \right) - \frac{\theta^2}{2} \sum_{j=1}^{J} q_j^k (\alpha_j^k)^2 \to \max,$$

where the following notations are used:

$$q_j^k = (q^{+k} + q_j^{-k}) / \bar{\Phi}_j^k D, \quad D = (1 + \sum_{j=1}^{J} \sum_{k \in K_j} \sum_{i=1}^{I} T_j^{ik} \alpha_j^i a^i q_j^{+k}).$$

We obtain the optimal values

$$\overset{\circ}{\theta} = (1 + \sum_{j=1}^{J} \sum_{k \in K_j} q_j^{+k} a_j^k \gamma_j^k),$$
$$\overset{\circ}{\kappa}{}_j^{k(2)} = a_j^k (1 + \sum_{j=1}^{J} \sum_{k \in K_j} q_j^k a_j^k \gamma_j^k) / \sum_{j=1}^{J} \sum_{k \in K_j} (a_j^k)^2 q_j^k - \gamma_j^k. \tag{6.25}$$

It is easy to determine the positivity of the optimal values $\overset{\circ}{\kappa}{}_j^{k(2)}$, $j \in [1:J]$, $k \in [1:K_j]$, which are supposed to be valid at all steps of the iterative decomposition algorithm. Using (6.25), we obtain immediately the formulas for $\overset{\circ}{\lambda}{}^i$, $i \in [1:I]$, $\overset{\circ}{\eta}{}_j^k$, $j \in [1:J]$, $k \in [1:K_j]$, also recalling (6.21) as well as second inequalities of (6.22) that are the strict equalities because of $\overset{\circ}{\kappa}{}_j^{k(2)} > 0$, $j \in [1:J]$, $k \in [1:K_j]$.

The lower level solves the independent problems of the subsystems:

$$\begin{aligned} &\sum_{i=1}^{I} \overset{\circ}{\lambda}{}^i x_j^i - \sum_{k \in K_j} q_j^{+k} (\kappa_j^{k(1)} - \kappa_j^{k(2)}) - \\ &-0.5 \sum_{k \in K_j} \frac{(q_j^{+k} + q_j^{-k})}{\bar{\Phi}_j^k} (\kappa_j^{k(2)})^2 \to \max, \\ &\sum_{i=1}^{I} T_j^{ik} x_j^i + (\kappa_j^{k(1)} - \kappa_j^{k(2)}) = \gamma_j^k, \quad k \in [1:K_j]; \\ &x_j^i \geq 0, \quad i \in [1:I], \quad \kappa_j^{k(1)} \geq 0, \quad \kappa_j^{k(2)} \geq 0, \quad k \in [1:K_j]. \end{aligned} \tag{6.26}$$

Finally, we obtain the optimality criterion

$$\hat{h} = \overset{\circ}{g}\,.$$

If the collection $(\overset{\circ}{\theta}, \overset{\circ}{x}{}^i_j, \overset{\circ}{\kappa}{}^{k(1)}_j, \overset{\circ}{\kappa}{}^{k(2)}_j)$ is not optimal for the main problem (1.1), then the following strict inequality is valid:

$$\hat{h} > \overset{\circ}{g}\,. \tag{6.27}$$

We study the monotonicity with respect to the functional. Consider the Lagrangian function of the aggregated problem (6.19)

$$\begin{aligned}
L(\rho, X^i, \theta, \kappa^{k(1)}_j, \kappa^{k(2)}_j, \lambda, \eta^k_j) &= \theta - \sum_{j=1}^{J}\sum_{k\in K_j} q^{+k}_j(\kappa^{k(1)}_j - \kappa^{k(2)}_j) - \\
&-0.5\sum_{j=1}^{J}\sum_{k\in K_j}(q^{+k}_j + q^{-k}_j)(\kappa^{k(2)}_j)^2\Big/\bar{\Phi}^k_j + \\
&+\sum_{j=1}^{J}\sum_{k\in K_j}\left(\eta^k_j(\gamma^k_j - \sum_{i=1}^{I} T^{ik}_j\alpha^i_j X^i) - \eta^k_j(\kappa^{k(1)}_j - \kappa^{k(2)}_j\right) + \\
&+\sum_{i=1}^{I}\lambda^i(X^i - a^i\theta).
\end{aligned}$$

Using the theorem of the marginal value 1.16, we obtain

$$\begin{aligned}
(\overset{\circ}{g}(0))' = \partial L/\partial\rho\Big|_{\rho=0} &= -\sum_{j=1}^{J}\sum_{k\in K_j}\sum_{i=1}^{I} T^{ik}_j(\hat{x}^i_j - \alpha^i_j\sum_{p=1}^{J}\hat{x}^i_p)\,\overset{\circ}{\eta}{}^k_j = \\
&= \sum_{j=1}^{J}\sum_{k\in K_j}\sum_{i=1}^{I}\sum_{p=1}^{J} T^{ik}_j\alpha^i_j\hat{x}^i_p\,\overset{\circ}{\eta}{}^k_j - \sum_{j=1}^{J}\sum_{k\in K_j}\sum_{i=1}^{I} T^{ik}_j\hat{x}^i_j\,\overset{\circ}{\eta}{}^k_j.
\end{aligned} \tag{6.28}$$

We transform the first term of the right-hand side in (6.28) taking into account (6.21) and use the equalities in the problems of the subsystems (6.26) for the second terms in (6.28). We have

$$\begin{aligned}
(\overset{\circ}{g}(0))' &= \sum_{j=1}^{J}\sum_{i=1}^{I}\overset{\circ}{\lambda}{}^i\hat{x}^i_j - \sum_{j=1}^{J}\sum_{k\in K_j} q^{+k}_j(\hat{\kappa}^{k(1)}_j - \hat{\kappa}^{k(2)}_j) + \\
&\sum_{j=1}^{J}\sum_{k\in K_j}(q^{+k}_j + q^{-k}_j)(\hat{\kappa}^{k(1)}_j - \hat{\kappa}^{k(2)}_j)\,\overset{\circ}{\kappa}{}^{k(2)}_j\Big/\bar{\Phi}^k_j - \\
&\sum_{j=1}^{J}\sum_{k\in K_j}\sum_{i=1}^{I} T^{ik}_j\alpha^i_j\overset{\circ}{X}{}^i\overset{\circ}{\eta}{}^k_j - \sum_{j=1}^{J}\sum_{k\in K_j}(\overset{\circ}{\kappa}{}^{k(1)}_j - \overset{\circ}{\kappa}{}^{k(2)}_j)\,\overset{\circ}{\eta}{}^k_j.
\end{aligned}$$

In the second term of the last relation, we use the second inequalities of (6.22) that hold as exact equalities. The third term is transformed.

We have

$$(\overset{\circ}{g}(0))' = \sum_{j=1}^{J}\sum_{i=1}^{I}\overset{\circ}{\lambda}^i \hat{x}_j^i - \sum_{j=1}^{J}\sum_{k\in K_j} q_j^{+k}(\hat{\kappa}_j^{k(1)} - \hat{x}_j^{k(2)}) -$$

$$-0.5\sum_{j=1}^{J}\sum_{k\in K_j}(q_j^{+k} + q_j^{-k})(\hat{\kappa}_j^{k(2)})^2\Big/\bar{\Phi}_j^k + 0.5\sum_{j=1}^{J}\sum_{k\in K_j}(q_j^{+k} + q_j^{-k})(\hat{\kappa}_j^{k(2)})^2\Big/\bar{\Phi}_j^k -$$

$$-\sum_{j=1}^{J}\sum_{k\in K_j}(q_j^{+k} + q_j^{-k})\hat{\kappa}_j^{k(2)}\overset{\circ}{\kappa}_j^{k(2)}\Big/\bar{\Phi}_j^k + \sum_{j=1}^{J}\sum_{k\in K_j}(q_j^{+k} + q_j^{-k})\hat{\kappa}_j^{k(1)}\overset{\circ}{\kappa}_j^{k(2)}\Big/\bar{\Phi}_j^k -$$

$$-\sum_{j=1}^{J}\sum_{k\in K_j} q_j^{+k}(\overset{\circ}{\kappa}_j^{k(1)} - \overset{\circ}{\kappa}_j^{k(2)}) - \sum_{j=1}^{J}\sum_{k\in K_j}(q_j^{+k} + q_j^{-k})(\overset{\circ}{\kappa}_j^{k(1)} - \overset{\circ}{\kappa}_j^{k(2)})\overset{\circ}{\kappa}_j^{k(2)}\Big/\bar{\Phi}_j^k +$$

$$+0.5\sum_{j=1}^{J}\sum_{k\in K_j}(q_j^{+k} + q_j^{-k})(\overset{\circ}{\kappa}_j^{k(2)})^2\Big/\bar{\Phi}_j^k + 0.5\sum_{j=1}^{J}\sum_{k\in K_j}(q_j^{+k} + q_j^{-k})(\overset{\circ}{\kappa}_j^{k(2)})^2\Big/\bar{\Phi}_j^k$$

and obtain definitively

$$(\overset{\circ}{g}(0))' = (\hat{h} - \overset{\circ}{g}) + 0.5\sum_{j=1}^{J}\sum_{k\in K_j} q_j^{+k}(\hat{\kappa}_j^{k(2)} - \overset{\circ}{\kappa}_j^{k(2)})^2\Big/\bar{\Phi}_j^k +$$
$$+\sum_{j=1}^{J}\sum_{k\in K_j}(q_j^{+k} + q_j^{-k})\hat{\kappa}_j^{k(1)}\overset{\circ}{\kappa}_j^{k(2)}\Big/\bar{\Phi}_j^k. \tag{6.29}$$

The first term in (6.29) is positive in view of (6.27), the last ones are not negative, which proves that the cost function is monotone in the iterative process.

Applicability of the iterative aggregation method for the quadratic programming is established in the next chapter.

Comments and References to Chapter 1

The problem of linear programming and duality are broadly covered, for example, in the book by D.B. Yudin and E.G. Gol'shtein [23]. The generalization in normalized spaces is studied by A.M. Ter-Krikorov [13–15]. Block statements are considered in the book of V.I. Tsurkov [19]. The Dantzig-Wolfe and Kornai-Liptak decomposition principles are formulated in [1] and [7], respectively. The Kuhn-Tucker theory is considered in the book of

J. Hadley [6] and E.G. Gol'shtein [4]. Parametric programming and theorems on marginal values can be found in books by E.G. Gol'shtein and D.B. Yudin [5] and E.G. Gol'shtein [4] as well as in the paper of E.S. Levitin [9]. The main model of the branch and the construction of the method of iterative aggregation for it represented in the paper of I.A. Vatel and Yu. A. Flerov [20]. The local monotonicity with respect to the functional was established by V.I. Tsurkov [16]. The approach of the iterative aggregation of variables entering the same blocks is represented by V.G. Mednitskii [11]. Information about the Brown method can be obtained from game theory, for example, in the book of E.G. Gol'shtein and D.E. Yudin [5]. In particular, the golden section search is represented in the monograph of J.D. Wilde [20]. Modifications of the main branch model are given in the paper by V.I. Tsurkov [17] and J. Silva, Yu. T. Treskov and V.I. Tsurkov [12]. The relaxation method is represented in the works by A.M. Geoffrion [3] and L.S. Lasdon [8]. One can learn more about linear fractional programming in the cited book of E.G. Gol'shtein and D.B. Yudin [5]. Stochastic programming is studied, in particular, by Yu.M. Ermol'ev [2] and D.B. Yudin [22].

References to Chapter 1

[1] Dantzig D.G. and Wolfe P., Decomposition Principle for Linear Programs, *Oper. Res.*, 1960, vol. 8, no. 1, pp. 101-111.

[2] Ermol'ev Yu. M., *Metody stohasticheskogo programmirovanija* (Methods of Stochastic Programming), Moscow, Nauka, 1976.

[3] Geoffrion A.M., *Relaxation and the Dual Method in Mathematical Programming*, Working Paper 135. – Western Management Science Institute, UCLA, 1968.

[4] Gol'shtein E.G., *Teorija dvoistvennosti v mathematicheskom programmirovanii i eyo prilozhenija* (Duality Theory in Mathematical Programming and Its Applications), Moscow: Nauka, 1971.

[5] Gol'shtein E.G. and D.B. Yudin, *Novye napravlenija v lineinom programmirovanii* (New Directions in Linear Programming), Moscow: Sov. Radio, 1966.

[6] Hadley J., *Nelinejnoe i dinamicheskoe programmirovanie* (Nonlinear and Dynamic Programming), Moscow: Mir, 1967.

[7] Kornai I. and Liptak T., *Programming at Two Levels, in: Mathematical Applications in Economical Studies*, vol. 3, Moscow: Mysl., 1965.

[8] Ladson L.S., *Optimization of Large Systems*, Moscow, Nauka, 1975.

[9] Levitin E.S., *O differentsiryemosti po parametry optimal'nogo znachenija parametricheskikh zadach mathematicheskogo programmirovania* (On Differentiability with Respect to Parameter of the Optimal Value of Parametric Problems of Mathematical Programming), Kibernetika, 1976, no. 1, pp. 44–59.

[10] Lions J.-L., *Optimal Controls by Systems Described by Partial Differential Equations*, Moscow: Mir, 1972.

[11] Mednitskii V.G., *Ob optimal'nosti agregirovanija v blochnoy zadache lineinogo programmirovania* (On Optimality of Aggregation in the Block Problem of Linear Programming), in: Mathematical Methods for Slution of Economical Problems, 1972, vol. 3, pp. 3–17.

[12] Silva G. N., Treskov Yu. P. and Tsurkov V. I., The Hierarchical Model of the Branch Planning with Random Coefficients, *Izv. Ross. Akad. Nauk, Teor. Sist. Upr.*, 1998, no. 4, pp. 11-26.

[13] Ter-Krikorov A. M., *Linejnye zadachi optimal'nogo upravleniya s phazovymi ogranicheniyami* (Linear Optimal Control Problems with Phase Constrains), *Zh. Vych. Math. Mat. Fiz.*, 1975, vol.15, no.1, pp. 55–66.

[14] Ter-Krikorov A. M., *Voprosy vypuklogo programmirovaniya v prostranstve, sopryazhennom prostranstve Banakha, i zadachi lineinogo programmirovaniya s phazovymi ogranocheniyami* (Questions of Convex Programming in the Space Conjugated to the Banach Space and Linear Optimal Control Problems with Phase Constrains) Zh. Vych. Math. Mat. Fiz., 1976, vol.16, no.3, pp. 597–604.

[15] Ter-Krikorov A. M., *Optimal'noe upravlenie i matematicheskaja economica* (Optimal Control and Mathematical Economy, Moscow: Nauka, 1977.

[16] Tsurkov V.I., *Agregirovanie v zadache otraslevogo planirovania* (Aggregation in the Problem of Branch Programming, Economics and Mathematical Methods), 1980, vol. XVI, no. 3, pp. 535–544.

[17] Tsurkov V.I., *Razvitie modeli tekushchego otraslevogo planirovanija* (Development of a Model of Current Branch Planning), in: Proc. III All-Union Conf. on Oper. Res., Gorkii, 1978.

[18] Tsurkov V.I., *Dvukhetapnaja zadacha stokhasticheskogo programmirovania blochno tipa* (Two-level Problem of Stochastic Programming of Block Type), Zh. Vych. Mat. Mat. Fiz., 1978, vol. 18, no. 2, pp. 360–369.

[19] Tsurkov V.I., *Dekompozichiya v zadachakh bolshoj razmernosti* (Decomposition in Large-scale Problems), Moscow: Nauka, 1981.

[20] Vatel' I.A. and Flerov Yu.A., *Model' godovogo planirovania v otrasli* (A Model of Annual Planning in a Branch), in: Programming Control Method, Moscow: Vych. Tsentr Akad. Nauk, no.3, 1976.

[21] Wilde D. J., *Metody poiska ekstremuma* (Methods of Search for Extremum), Moscow, Nauka, 1967.

[22] Yudin D.B., *Matematicheskie metody upravlenia v usloviyakh nepolnoj informatsii* (Mathematical Control Methods under Complete Uncertainty), Moscow: Sov. Radio, 1974.

[23] Yudin D. B. and Gol'shtein E. G., *Zadachi i metody lineinogo programmirovania* (Problems and Methods of Linear Programming), Moscow: Sov. Radio, 1964.

Chapter 2

Generalization of the Decomposition Approach to Mathematical Programming and Classical Calculus of Variations

In this chapter, the iterative decomposition method based on the iterative aggregated variables from various blocks is applied to the wide class of extremal problems of hierarchical nature. We consider the use of the approach for the problems of linear and quadratic programming with the block diagonal structure having arbitrary binding constraints. The same is made for finite-dimensional block separable mathematical programming problems of convex type. The constructions of the method are represented in the classical calculus of variations. Here, the main concern is the condition of optimality of the disaggregated solution and the monotonicity with respect to the functional of the iterative process.

§1. Linear Programming

The problem of linear programming with block-diagonal structure of a part of the matrix of constraints is written in the form

$$f = \sum_{j=1}^{J} \sum_{i \in I_j} c_j^i x_j^i \to \max, \tag{1.1}$$

$$\sum_{i \in I_j} b_j^{ik} x_j^i \leq p_j^k, \tag{1.2}$$

$$x_j^i \geq 0 \text{ for } i \in I_j, \qquad x_j^i = 0 \text{ for } i \in \bar{I} \backslash I_j, \qquad j \in [1:J], \tag{1.3}$$

$$\sum_{j=1}^{J} \sum_{i \in I_j} \bar{b}_j^{il} x_j^l \leq p^l, \qquad l \in [1:L]. \tag{1.4}$$

Here, the relations (1.2), (1.3) are local constraints, where the number j, as earlier, corresponds to a fixed block. Relations (1.4) are binding constraints.

We consider the dual problem for the problem (1.1)–(1.4), where the variables ξ_j^k and ν^l correspond to (1.2) and (1.4), respectively:

$$\varphi = \sum_{j=1}^{J} \sum_{k \in K_j} p_j^k \xi_j^k + \sum_{l=1}^{L} p^l \nu^l \to \min, \tag{1.5}$$

$$\begin{aligned} &\sum_{k \in K_j} b_j^{ik} \xi_j^k + \sum_{l=1}^{L} \bar{b}_j^{il} \nu^l \geq c_j^l, \qquad j \in [1:J], \qquad i \in I_j, \\ &\xi_j^k \geq 0, \qquad j \in [1:J], \qquad k \in K_j, \nu^l \geq 0, \qquad l \in [1:L]. \end{aligned} \tag{1.6}$$

As in Chapter 1, we introduce the aggregated variables

$$X^i = \sum_{j=1}^{J} x_j^i, \qquad i \in [1:I],$$

and the aggregation weights are $\alpha_j^i = x_j^i / X^i$. We fix the weights α_j^i and substitute $x_j^i = \alpha_j^i X^i$ in (1.1)–(1.4). We come to the following aggregated problem:

$$g = \sum_{i \in I_j}^{I} C^i X^i \to \max, \tag{1.7}$$

$$\sum_{i=1}^{I} B_j^{ik} X^i \leq p_j^k, \qquad j \in [1:J], \qquad k \in K_j, \tag{1.8}$$

$$\sum_{i=1}^{I} B^{il} X^i \leq p^l, \qquad l \in [1:L], \tag{1.9}$$

$$X^i \geq 0, \qquad i \in [1:I]. \tag{1.10}$$

Here, we introduce the notation

$$B_j^{ik} = b_j^{ik} \alpha_j^i, \qquad B^{il} = \sum_{j=1}^{J} \bar{b}_j^{il} \alpha_j^i, \qquad C^i = \sum_{j=1}^{J} c_j^i \alpha_j^i. \tag{1.11}$$

The problem dual to (1.7)–(1.10) has the form

$$\psi = \sum_{j=1}^{J} \sum_{k \in K_j} p_j^k \eta_j^k + \sum_{l=1}^{L} p^l \delta^l \to \min, \tag{1.12}$$

$$\sum_{j=1}^{J}\sum_{k\in K_j} B_j^{ik}\eta_j^k + \sum_{l=1}^{L} B^{il}\delta^l \geq C^i, \qquad i\in[1:I], \tag{1.13}$$

$$\eta_j^k \geq 0, \qquad j\in[1:J], \qquad k\in K_j, \qquad \delta^l\geq 0, l\in[1:L], \tag{1.14}$$

where the variables η_j^k and δ^l correspond to (1.8) and (1.9), respectively.

Let $\overset{\circ}{\delta}{}^l$ be optimal dual estimates of the problem (1.12)–(1.14). We formulate the problems for separate blocks:

$$h_j = \sum_{i\in I_j}\left(c_j^i - \sum_{l=1}^{L}\bar{b}_j^{il}\overset{\circ}{\delta}{}^l\right)x_j^i \to \max, \tag{1.15}$$

$$\sum_{i\in I_j} b_j^{ik}x_j^i \leq p_j^k, \qquad k\in K_j, \tag{1.16}$$

$$x_j^i \geq 0 \text{ for } i\in I_j, \qquad x_j^i = 0 \text{ for } i\in \bar{I}\backslash I_j. \tag{1.17}$$

The dual problems for the block has the following form:

$$\chi_j = \sum_{k\in K_j} p_j^k\zeta_j^k \to \min, \tag{1.18}$$

$$\sum_{k\in K_j} b_j^{ik}\zeta_j^k \geq c_j^i - \sum_{l=1}^{L}\bar{b}_j^{il}\delta^l, \qquad i\in I_j, \tag{1.19}$$

$$\zeta_j^k \geq 0, \qquad k\in K_j, \tag{1.20}$$

where the dual variables ζ_j^k correspond to (1.16).

The algorithm is constructed according to the method from Section 2 of Chapter 1. Here, we obtain the reduction to a problem of lesser dimension. In fact, the initial problem has $\sum_{j=1}^{J} I_j$ variables. The aggregated problem contains I variables, each local problem has I_j variables, and, finally, the problem of maximization of the function $\overset{\circ}{g}(\rho_j)$ on the unit cube involves J variables.

During the steps of the iteration process, the aggregated problem is solved many times. As a rule, this problem has a small number of variables X^i, $i\in[1:I]$ and a large number of constraints. We establish an optimality criterion for the intermediate disaggregated solution.

Assume that, for some weights α_j^i, we obtained the optimal solution of the macroproblem (1.7)–(1.10)$\overset{\circ}{X}{}^i$, and that $\overset{\circ}{x}{}_j^i = \alpha_j^i\overset{\circ}{X}{}^i$ is the corresponding disaggregated solution. Let $\overset{\circ}{\delta}{}^l$ be the unique optimal solution of problem (1.12)–(1.14), and $\hat{x}_j^i$ are the optimal solutions of the local problems (1.15)–(1.17).

Theorem 1.1. *Satisfaction of the equality*

$$\sum_{j=1}^{J}\sum_{i\in I_j}\left(c_j^i - \sum_{l=1}^{L}\bar{b}_j^{il}\,\overset{\circ}{\delta}{}^l\right)\hat{x}_j^i + \sum_{l=1}^{L}p^l\,\overset{\circ}{\delta}{}^l - \sum_{j=1}^{J}\sum_{i\in I_j}c_j^i\,\overset{\circ}{x}{}_j^i = 0 \tag{1.21}$$

is a sufficient condition $\overset{\circ}{x}{}_j^i$ *for optimality of the solution of problem (1.1)–(1.4). If the problem (1.1)–(1.4) is solvable, and* $\overset{\circ}{x}{}_j^i$ *is nonoptimal for it, then the sign* $=$ *is replaced by* $>$ *in relation (1.21).*

Proof. The disaggregated solution $\overset{\circ}{x}{}_j^i$ is feasible for the problem (1.1)–(1.4). This is shown by direct substitution $\overset{\circ}{x}{}_j^i = \alpha_j^i\,\overset{\circ}{X}{}^i$ in (1.2), (1.4) with account of (1.8), (1.9) and notations (1.11). The value of the functional $\overset{\circ}{f}$ that corresponds to the feasible solution $\overset{\circ}{x}{}_j^i$ for (1.1)–(1.4) is equal to

$$\overset{\circ}{f} = \sum_{j=1}^{J}\sum_{i\in I_j}c_j^i\,\overset{\circ}{x}{}_j^i.$$

The set $\{\hat{\zeta}_j^k, \overset{\circ}{\delta}{}^l\}$, where $\hat{\zeta}_j^k$, $j \in [1:J]$, $k \in K_j$ are optimal solutions of the problems dual to the local problems (1.18)–(1.21), is a feasible solution to the problem dual to (1.1)–(1.4). This follows from the relations (1.19), (1.20), (1.14), (1.6). The value of the functional $\hat{\varphi}$ for this solution is

$$\hat{\varphi} = \sum_{j=1}^{J}\sum_{k\in K_j}p_j^k\hat{\zeta}_j^k + \sum_{l=1}^{L}p^l\,\overset{\circ}{\delta}{}^l.$$

According to Theorem 1.2 of Chapter 1, the equality $\hat{\varphi} = \overset{\circ}{f}$ provides the optimality of the feasible solution $\overset{\circ}{x}{}_j^i$ for the problem (1.1)–(1.4). If $\overset{\circ}{x}{}_j^i$ is nonoptimal for the problem (1.1)–(1.4) under conditions of its solvability, then $\hat{\varphi} > \overset{\circ}{f}$.

Summing up, we obtain the optimality criterion in the form

$$\sum_{j=1}^{J}\sum_{k\in K_j}p_j^k\hat{\zeta}_j^k + \sum_{l=1}^{L}p^l\,\overset{\circ}{\delta}{}^l = \sum_{j=1}^{J}\sum_{i\in I_j}c_j^i\,\overset{\circ}{x}{}_j^i. \tag{1.22}$$

Since $\hat{\zeta}_j^k$, $i \in [1:J]$ are optimal solutions of the problems (1.18)–(1.20), then, applying the first duality Theorem 1.3 of Chapter 1 to the conjugated local problems (1.15)–(1.17) and (1.18)–(1.20), we have

$$\sum_{k\in K_j} \hat{\zeta}_j^k = \sum_{i\in I_j} \left(c_j^i - \sum_{l=1}^{L} \bar{b}_j^{il} \overset{\circ}{\delta}{}^l \right) \hat{x}_j^i, \qquad i \in [1:J].$$

The substitution of these relations in (1.22) proves the equality (1.21). The case of nonoptimal $\overset{\circ}{x}{}_j^i$ is considered similarly.

We prove local monotonicity for the functional of the iterative process on the assumption that all components $\overset{\circ}{X}{}^i$, $i \in [1:I]$ are strictly greater than zero. Let, for some weights α_j^i, optimal solutions $\overset{\circ}{X}{}^i$ of the aggregated problems variables (1.7)–(1.10) and $\{\overset{\circ}{\eta}{}_j^k, \overset{\circ}{\delta}{}^l\}$ of the dual problem (1.12)–(1.14) be obtained. First, we assume that these solutions are unique. We show that a nonoptimal disaggregated solution $\overset{\circ}{x}{}_j^k = \alpha_j^i \overset{\circ}{X}{}^i$ can be improved with respect to the functional.

Theorem 1.2. *For the disaggregated solution $\overset{\circ}{x}{}_j^i$, which is not optimal for the problem (1.1)–(1.4), there exists a feasible solution with the functional value strictly greater than $\overset{\circ}{f} = f(\overset{\circ}{x}{}_j^i)$.*

Proof. The reasoning is carried out according to the proof scheme of Theorem 4.1 from Chapter 1, where

$$(\overset{\circ}{g}(0))' = \sigma_1 + \sigma_2, \qquad \sigma_1 > 0, \quad \sigma_2 > 0 \tag{1.23}$$

is the marginal value $(\overset{\circ}{g}(0))'$ of parametric aggregated problems (1.7)–(1.10), and $\alpha_j^i(\rho_j)$ take form (2.23) Chapter 1 when $\rho_j = \rho$, $j \in [1:J]$.

Let R_1 be the set of index pairs (j,k), for which $\overset{\circ}{\eta}{}_j^k > 0$, and R_2 is the set of indices l, such that $\overset{\circ}{\delta}{}^l > 0$. We use Theorem 1.11 of Chapter 1 on the marginal value as applied to the conjugated problems (1.7)–(1.10) and (1.13)–(1.14). Taking into account notation (1.11) and expressions for arbitrary $(\alpha_j^i(0))'$ at the point $\rho = 0$, we have

$$(\overset{\circ}{g}(0))' = \sum_{j=1}^{J}\sum_{i=1}^{I} c_j^i \left(\hat{x}_j^i - \alpha_j^i \sum_{r=1}^{J} \hat{x}_r^i \right) -$$

$$- \sum_{(j,k)\in R_1} \sum_{i=1}^{I} b_j^{ik} \left(\hat{x}_j^i - \alpha_j^i \sum_{r=1}^{J} \hat{x}_r^i \right) \overset{\circ}{\eta}{}_j^k -$$

$$-\sum_{l\in R_2}\sum_{j=1}^{J}\sum_{i=1}^{I}\bar{b}_j^{il}\left(\hat{x}_j^i-\alpha_j^i\sum_{r=1}^{J}\hat{x}_r^i\right)\overset{\circ}{\delta}{}^l. \qquad (1.24)$$

Taking into account notation (1.11), the right-hand side of (1.24) is written in the form

$$\sum_{j=1}^{J}\sum_{i\in I_j}c_j^i\hat{x}_j^i-\sum_{(j,k)\in R_1}\sum_{i\in I_j}b_j^{ik}\hat{x}_j^i\overset{\circ}{\eta}{}_j^k-\sum_{l\in R_2}\sum_{j=1}^{J}\sum_{i\in I_j}\bar{b}_j^{il}\hat{x}_j^i\overset{\circ}{\delta}{}^l+$$
$$+\sum_{i=1}^{I}\left(\sum_{(j,k)\in R_1}B_j^{ik}\overset{\circ}{\eta}{}_j^k+\sum_{l\in R_2}B^{il}\overset{\circ}{\delta}{}^l-C^i\right)\sum_{r=1}^{J}\hat{x}_r^i.$$

Since $\overset{\circ}{X}{}^i>0$, $i\in[1:I]$, relations (1.13) hold as equalities according to the second duality Theorem 1.4 of Chapter 1. Therefore, the last sum of the written relation is equal to zero. Thus, we have

$$(\overset{\circ}{g}(0))'=\sum_{j=1}^{J}\sum_{i\in I_j}c_j^i\hat{x}_j^i-\sum_{(j,k)\in R_1}\sum_{i\in I_j}b_j^{ik}\hat{x}_j^i\overset{\circ}{\eta}{}_j^k-\sum_{l\in R_2}\sum_{j=1}^{J}\sum_{i\in I_j}\bar{b}_j^{il}\hat{x}_j^i\overset{\circ}{\delta}{}^l. \qquad (1.25)$$

According to the first duality Theorem 1.3 of Chapter 1, the equality

$$\sum_{(j,k)\in R_1}p_j^k\overset{\circ}{\eta}{}_j^k+\sum_{l\in R_2}p^l\overset{\circ}{\delta}{}^l-\sum_{j=1}^{J}\sum_{i\in I_j}c_j^i\overset{\circ}{x}{}_j^i=0$$

holds for conjugated problems (1.7)–(1.10) and (1.12)–(1.14). Summing the left-hand side of the latter equation with (1.25), we finally obtain (1.23), where

$$\sigma_1=\sum_{j=1}^{J}\sum_{i\in I_j}c_j^i\hat{x}_j^i-\sum_{l\in R_2}\sum_{j=1}^{J}\sum_{i\in I_j}b_j^{ik}\hat{x}_j^i\overset{\circ}{\delta}{}^l+\sum_{l\in R_2}p^l\overset{\circ}{\delta}{}^l-\sum_{j=1}^{J}\sum_{i\in I_j}c_j^i\overset{\circ}{x}{}_j^i,$$

$$\sigma_2=\sum_{(j,k)\in R_1}\overset{\circ}{\eta}{}_j^k\left(p_j^k-\sum_{i\in I_j}b_j^{ik}\hat{x}_j^i\right).$$

The value σ_1 coincides with the left-hand side of (1.21), and, due to Theorem 1.1, we have $\sigma_1>0$. Moreover, we have $\sigma_2\geq 0$ according to (1.14), (1.16). Relation (1.23) is established. It implies the estimate $\overset{\circ}{g}(\rho)>\overset{\circ}{g}(0)$ in a neighborhood of the point $\rho=0$, which proves Theorem 1.2.

Consider the degenerate case, where the unique $\overset{\circ}{X}^i$, $i \in [1 : I]$ optimal solution of the problem (1.7)–(1.10) corresponds to nonunique solutions $\{\overset{\circ}{\eta}{}_j^k, \overset{\circ}{\delta}{}^l\}$ of the dual problem (1.12)- (1.14). By $\Omega_{\eta,\delta}$, denote the set of vectors $\{\overset{\circ}{\eta}{}_j^k, \overset{\circ}{\delta}{}^l\}$ – dual estimates under degeneracy – restricted by the assumption. Let Ω_δ be the set of vectors $\overset{\circ}{\delta}{}^l$ such that the vector $\{\overset{\circ}{\eta}{}_j^k, \overset{\circ}{\delta}{}^l\}$ belongs to $\Omega_{\eta,\delta}$. Finally, denote by M_x^l the restricted sets of elements x_j^i, $i \in I_j$ that satisfy conditions (1.2), (1.3) or, which is the same, (1.16), (1.17). We introduce the following maximin problem: find

$$\max_{x_j^i \in M_x^j} \min_{\overset{\circ}{\delta}{}^l \in \Omega_\delta} P(x_j^i, \overset{\circ}{\delta}{}^l), \tag{1.26}$$

where

$$P(x_j^i, \overset{\circ}{\delta}{}^l) = \sum_{j=1}^{J} \sum_{i \in I_j} \left(c_j^i - \sum_{l \in R_2} \bar{b}_j^{il} \overset{\circ}{\delta}{}^l \right) x_j^i + \sum_{l \in R_2} p^l \overset{\circ}{\delta}{}^l.$$

Here, as before, R_2 denotes the set of indices $l \in [1 : L]$, for which $\overset{\circ}{\delta}{}^l > 0$.

Let $\bar{x}_j^i$, $\bar{\delta}^l$ be the saddle point of the problem (1.26). Then, we substitute $\bar{x}_j^i$ for $\hat{x}_j^i$ in $\alpha_j^i(\rho_j)$ from formula (2.33) from Chapter 1. We obtain the following statement similar to Theorem 3.3 from Chapter 1.

Theorem 1.3. *The satisfaction of the equality*

$$P(\bar{x}_j^i, \bar{\delta}^l) - \sum_{j=1}^{J} \sum_{i \in I_j} c_j^i \overset{\circ}{x}{}_j^i = 0 \tag{1.27}$$

is a sufficient optimality condition for the optimality of a disaggregated solution $\overset{\circ}{x}{}_j^i$ *for the problem (1.1)–(1.4) under degeneration. If the problem (1.1)–(1.4) is solvable, and* $\overset{\circ}{x}{}_j^i$ *is not optimal for it, then the sign = is replaced by > in (1.27).*

The proof of Theorem 1.3 is based on the fact that the values $\bar{x}_j^i$, $j \in [1 : J]$, $i \in I_j$ are solutions of local problems (1.15)–(1.17) with functionals with $\bar{\delta}^l$.

We establish local monotonicity by the functional of the iterative process under degeneracy. This fact follows from the inequality

$$(\overset{\circ}{g}(0))' \geq P(\bar{x}_j^i, \bar{\delta}^l) - \sum_{j=1}^{J} \sum_{i \in I_j} c_j^i \overset{\circ}{x}{}_j^i. \tag{1.28}$$

We show that inequality (1.28) holds. We use the Theorem on the marginal value (1.11) from Chapter 1 taking into account the nonuniqueness of dual estimates $\{\overset{\circ}{\eta}{}_j^k, \overset{\circ}{\delta}{}^l\}$. By analogy with the deduction of relation (1.25), we obtain

$$(\overset{\circ}{g}(0))' = \min_{\{\overset{\circ}{\eta}{}_j^k,\overset{\circ}{\delta}{}^l\}\in\Omega_{\eta,\delta}} \left(\sum_{j=1}^{J}\sum_{i\in I_j} c_j^i \hat{x}_j^i - \sum_{(j,k)\in R_1}\sum_{i\in I_j} b_j^{ik}\bar{x}_j^i \overset{\circ}{\eta}{}_j^k - \sum_{l\in R_2}\sum_{j=1}^{J}\sum_{i\in I_j} \bar{b}_j^{il}\bar{x}_j^i \overset{\circ}{\delta}{}^l \right)$$

Let the minimum in the right-hand side of the latter equation be attained for some $\{\tilde{\eta}_j^k, \tilde{\delta}^l\} \in \Omega_{\eta,\delta}$. According to the first duality Theorem 1.3 of Chapter 1, the equality

$$\sum_{(j,k)\in R_1} p_j^k \tilde{\eta}_j^k + \sum_{l\in R_2} p^l \tilde{\delta}^l - \sum_{j=1}^{J}\sum_{i\in I_j} c_j^i \overset{\circ}{x}{}_j^i = 0.$$

holds for conjugate problems (1.7)–(1.10) and (1.12)–(1.14).

The last two equations imply

$$(\overset{\circ}{g}(0))' = \sum_{j=1}^{J}\sum_{i\in I_j} \left(c_j^i - \sum_{l\in R_2} \bar{b}_j^{il}\tilde{\delta}^l \right) \bar{x}_j^i + \sum_{l\in R_2} p^l \tilde{\delta}^l -$$

$$-\sum_{j=1}^{J}\sum_{i\in I_j} c_j^i \overset{\circ}{x}{}_j^i + \sum_{(j,k)\in R_1} \tilde{\eta}_j^k \left(p_j^k - \sum_{i\in I_j} b_j^{ik}\bar{x}_j^i \right). \qquad (1.29)$$

The first three terms of the right-hand side of (1.29) are greater or equal to $P(\bar{x}_j^i, \tilde{\delta}^l)$ (inequalities of Theorem 1.10 of Chapter 1). The last term of (1.29) is greater than or equal to zero according to (1.14), (1.16). Therefore, inequality (1.28) holds. Due to Theorem 1.3, the estimate $\overset{\circ}{g}(\rho) > \overset{\circ}{g}(0)$ applies in neighborhood of the point $\rho = 0$.

§2. Quadratic Programming

In this section, the decomposition method based on aggregated variables from different blocks is applied to block problems of quadratic programming. These problems are obtained from (1.1)–(1.14) by adding quadratic terms to the functional. These problems have the following form:

$$f = \sum_{j=1}^{J}\sum_{i\in I_j} c_j^i x_j^i + \sum_{j=1}^{J}\sum_{m\in I_j}\sum_{n\in I_j} d_j^{mn} x_j^m x_j^n \to \max, \qquad (2.1)$$

$$\sum_{i\in I_j} b_j^{ik} x_j^i \le p_j^k, \qquad k \in K_j, \tag{2.2}$$

$$x_j^i \ge 0 \text{ for } i \in I_j, \qquad x_j^i = 0 \text{ for } i \in \bar{I}\backslash I_j, \qquad j \in [1:J], \tag{2.3}$$

$$\sum_{j=1}^{J}\sum_{i\in I_j} \bar{b}_j^{il} x_j^i \le p^l, \qquad l \in [1:L]. \tag{2.4}$$

By assumption, matrices $\{d_j^{mn}\}$, $j \in [1:J]$ are symmetric and correspond to negative definite quadratic forms.

We consider the dual problem for the problem (2.1)–(2.4):

$$\varphi = -\sum_{j=1}^{J}\sum_{m\in I_j}\sum_{n\in I_j} d_j^{mn} x_j^m x_j^n + \sum_{j=1}^{J}\sum_{k\in K_j} p_j \xi_j^k + \sum_{l=1}^{L} p^l \nu^l \min, \tag{2.5}$$

$$-2\sum_{n\in I_j} d_j^{in} x_j^n + \sum_{k\in K_j} b_j^{ik}\xi_j^k + \sum_{l=1}^{L} \bar{b}_j^{il}\nu^l \ge c_j^i, \qquad i \in I_j, \tag{2.6}$$

$$x_j^i \ge 0 \text{ for } i \in I_j, \qquad x_j^i = 0 \text{ for } i \in \bar{I}\backslash I_j, \qquad j \in [1:J], \tag{2.7}$$

$$\xi_j^k \ge 0, \qquad j \in [1:J], \qquad k \in K_j, \qquad \nu^l \ge 0, \qquad l \in [1:L]. \tag{2.8}$$

We introduce the aggregated problem for (2.1)–(2.4). To this end, we substitute $x_j^i = \alpha_j^i X^i$ in (2.1)–(2.4) as before, taking the weights α_j^i to be fixed. Thus, we have:

$$g = \sum_{i=1}^{I} C^i X^i + \sum_{m=1}^{I}\sum_{n=1}^{I} D^{mn} X^m X^n \to \max, \tag{2.9}$$

$$\sum_{i=1}^{I} B_j^{ik} X^i \le p_j^k, \qquad j \in [1:J], \qquad k \in K_j, \tag{2.10}$$

$$\sum_{i=1}^{I} B^{il} X^i \le p^l, \qquad l \in [1:L], \tag{2.11}$$

$$X^i \ge 0, \qquad i \in [1:I]. \tag{2.12}$$

Here, we introduce notation additional to that from (1.1):

$$D^{mn} = \sum_{j=1}^{J} d_j^{mn} \alpha_j^m \alpha_j^n.$$

It is easily seen that the matrix $\{D^{mn}\}$ is symmetric and corresponds to a negative definite quadratic form.

The problem dual to (2.9)–(2.12) has the following form:

$$\psi = -\sum_{m=1}^{I}\sum_{n=1}^{I} D^{mn}X^m X^n + \sum_{j=1}^{J}\sum_{k\in K_j} p_j^k \eta_j^k + \sum_{l=1}^{L} p^l \delta^l \to \min, \tag{2.13}$$

$$-2\sum_{n=1}^{I} D^{in}X^n + \sum_{j=1}^{J}\sum_{k\in K_j} B_j^{ik}\eta_j^k + \sum_{l=1}^{L} B^{il}\delta^l \geq C^i, \qquad i \in [1:I], \tag{2.14}$$

$$\begin{gathered} X^i \geq 0, \qquad i\in[1:I], \qquad \eta_j^k \geq 0, \qquad i \in [1:J], \qquad k\in K_j, \\ \delta^l \geq 0, \qquad l \in [1:L]. \end{gathered} \tag{2.15}$$

Let $\overset{\circ}{\delta}{}^l$ be an optimal solution to the problem (2.13)–(2.15), which is supposed to be unique. We formulate problems for separate blocks, i.e., for fixed $j \in [1:J]$:

$$h_j = \sum_{i\in I_j}\left(c_j^i - \sum_{l=1}^{L} \bar{b}_j^{il}\overset{\circ}{\delta}{}^l\right)x_j^i + \sum_{m\subset I_j}\sum_{n\in I_j} d_j^{mn}x_j^m x_j^n \to \max, \tag{2.16}$$

$$\sum_{i\in I_j} b_j^{ik}x_j^i \leq p_j^k, \qquad k\in K_j, \tag{2.17}$$

$$x_j^i \geq 0 \text{ for } i \in I_j, \qquad x_j^i = 0 \text{ for } i\in \bar{I}\backslash I_j. \tag{2.18}$$

The problems dual to block problems (2.16)–(2.18) are written in the form

$$\chi_j = -\sum_{m\in I_j}\sum_{n\in I_j} d_j^{mn}x_j^m x_j^n + \sum_{k\in K_j} p_j^k\zeta_j^k \to \min, \tag{2.19}$$

$$-2\sum_{n\in I_j} d_j^{in}x_j^n + \sum_{k\in K_j} b_j^{ik}\zeta_j^k \geq c_j^i - \sum_{l=1}^{L}\bar{b}_j^{il}\overset{\circ}{\delta}{}^l, \qquad i\in I_j, \tag{2.20}$$

$$x_j^i \geq 0 \text{ for } i\in I_j, \qquad x_j^i = 0 \text{ for } i \in \bar{I}\backslash I_j, \qquad \zeta_j^k \geq 0, \qquad k\in K_j. \tag{2.21}$$

We establish an optimality criterion of an intermediate solution. Let, for some weights α_j^i, an optimality solution $\overset{\circ}{X}{}^i$ of the macroproblem (2.9)–(2.12) be obtained. Let the unique dual estimates $\overset{\circ}{\delta}{}^l$ correspond to the optimal solutions $\hat{x}_j^i$ of the problems for single blocks (2.16)–(2.18).

Theorem 2.1. *The satisfaction of the equality*

$$\sum_{j=1}^{J}\sum_{i\in I_j} c_j^i \hat{x}_j^i + \sum_{j=1}^{J}\sum_{m\in I_j}\sum_{n\in I_j} d_j^{mn}\hat{x}_j^m\hat{x}_j^n - \sum_{j=1}^{J}\sum_{i\in I_j}\sum_{l=1}^{L}\bar{b}_j^{il}\overset{\circ}{\delta}{}^l\hat{x}_j^i + \sum_{l=1}^{L} p^l\overset{\circ}{\delta}{}^l -$$
$$-\sum_{j=1}^{J}\sum_{i\in I_j} c_j^i \overset{\circ}{x}{}_j^i - \sum_{j=1}^{J}\sum_{m\in I_j}\sum_{n\in I_j} d_j^{mn}\overset{\circ}{x}{}_j^m\overset{\circ}{x}{}_j^n = 0 \qquad (2.22)$$

is a sufficient condition of optimality of the disaggregated solution $\overset{\circ}{x}{}_j^i = \alpha_j^i\overset{\circ}{X}{}^i$. *If problem (2.1)–(2.4) is solvable, and* $\overset{\circ}{x}{}_j^i$ *is not optimal for it, then the sign* $=$ *is replaced by* $>$ *in (2.22).*

Theorem 2.1 is proved analogously to Theorem 1.1.

We study the problem of monotonicity with respect to the functional of the iterative process. Let, for fixed α_j^i, the optimal solution $\overset{\circ}{X}{}^i$ of the aggregated problem (2.9)–(2.12) be obtained. Below, it is assumed that this solution is unique and all components $\overset{\circ}{X}{}^i$. $i \in [1:I]$ are strictly greater than zero. Let $\{\overset{\circ}{X}{}^i, \overset{\circ}{\eta}{}_j^k, \overset{\circ}{\delta}{}^l\}$ be a unique solution for the dual problem (2.13)–(2.15).

Theorem 2.2. *For the disaggregated solution* $\overset{\circ}{x}{}_j^i$ *nonoptimal for problem (2.1)–(2.4), there exists a solution that is feasible for (2.1)–(2.4) and has the value of the functional greater than* $\overset{\circ}{f} = f(\overset{\circ}{x}{}_j^i)$.

Proof. The formula for the marginal value $(\overset{\circ}{g}(0))'$ of the macroproblem (2.9)–(2.12) is established in the same way as in Theorem 1.2. This formula has the form

$$(\overset{\circ}{g}(0))' = \sigma_1 + \sigma_2 + \sigma_3, \qquad (2.23)$$

where $\sigma_1 > 0$, $\sigma_2, \sigma_3 \geq 0$.

Under the assumption of uniqueness of $(\overset{\circ}{g}(0))'$, according to the Theorem 1.16 on the marginal value from Chapter 1, the derivative $\{\overset{\circ}{X}{}^i, \overset{\circ}{\eta}{}_j^k, \overset{\circ}{\delta}{}^l\}$ is equal to the Lagrange function for the functional (2.9)–(2.12) under the given optimal values, if we replace B_j^{ik} by

$$b_j^{ik}\left(\hat{x}_j^i - \alpha_j^i\sum_{r=1}^{J}\hat{x}_r^i\right)(\overset{\circ}{X}{}^i)^{-1},$$

B^{il} by

$$\sum_{j=1}^{J} \bar{b}_j^{il} \left(\hat{x}_j^i - \alpha_j^i \sum_{r=1}^{J} \hat{x}_r^i \right) (\overset{\circ}{X}{}^i)^{-1},$$

C^l by

$$\sum_{j=1}^{J} c_j^i \left(\hat{x}_j^i - \alpha_j^i \sum_{r=1}^{J} \hat{x}_r^i \right) (\overset{\circ}{X}{}^i)^{-1},$$

and D^{mn} by the coefficient at the term linear in ρ in the expression

$$\sum_{j=1}^{J} d_j^{mn} \left[\alpha_j^m + \rho \left(\hat{x}_j^m - \alpha_j^m \sum_{r=1}^{J} \hat{x}_r^m \right) (\overset{\circ}{X}{}^m)^{-1} \right] \times$$
$$\times \left[\alpha_j^n + \rho \left(\hat{x}_j^n - \alpha_j^n \sum_{r=1}^{J} \hat{x}_r^n \right) (\overset{\circ}{X}{}^m)^{-1} \right].$$

As in Section 1, R_1 denotes the set of pairs of indices (j,k), for which $\overset{\circ}{\eta}{}_j^k \geq 0$, and R_2 denotes the set of indices l such that $\overset{\circ}{\delta}{}^l > 0$. Thus, we have

$$(\overset{\circ}{g}(0))' = \sum_{j=1}^{J} \sum_{i \in I_j} c_j^i \left(\hat{x}_j^i - \alpha_j^i \sum_{r=1}^{J} \hat{x}_r^i \right) (\overset{\circ}{X}{}^i)^{-1} (\overset{\circ}{X}{}^i) +$$
$$+ \sum_{m=1}^{I} \sum_{n=1}^{I} \sum_{j=1}^{J} d_j^{mn} \left[\left(\hat{x}_j^m - \alpha_j^m \sum_{r=1}^{J} \hat{x}_r^m \right) \alpha_j^n (\overset{\circ}{X}{}^m)^{-1} + \right.$$
$$\left. + \left(\hat{x}_j^n - \alpha_j^n \sum_{r=1}^{J} \hat{x}_r^n \right) \alpha_j^m (\overset{\circ}{X}{}^n)^{-1} \right] \overset{\circ}{X}{}^m \overset{\circ}{X}{}^n -$$
$$- \sum_{(j,k) \in R_1} \sum_{i \in I_j} b_j^{ik} \left(\hat{x}_j^i - \alpha_j^i \sum_{r=1}^{J} \hat{x}_r^i \right) (\overset{\circ}{X}{}^i)^{-1} \overset{\circ}{X}{}^i \overset{\circ}{\eta}{}_j^k -$$
$$- \sum_{l \in R_2} \sum_{j=1}^{J} \sum_{i \in I_j} d_j^{il} \left(\hat{x}_j^i - \alpha_j^i \sum_{r=1}^{J} \hat{x}_r^i \right) (\overset{\circ}{X}{}^i)^{-1} \overset{\circ}{X}{}^i \overset{\circ}{\delta}{}^l.$$

By partitioning the sums from the right-hand side of the latter equation into pairs and taking into account the definitions (1.11) together with notation D^{mn}, we obtain

$$(\overset{\circ}{g}(0))' = \sum_{j=1}^{J} \sum_{i \in I_j} c_j^i \hat{x}_j^i + \sum_{m \in I_j} \sum_{n \in I_j} d_j^{mn} \hat{x}_j^m \hat{x}_j^n +$$

$$+\sum_{m\in I_j}\sum_{n\in I_j}\sum_{j=1}^{J} d_j^{mn}\hat{x}_j^n \overset{\circ}{x}{}_j^m - \sum_{(j,k)\in R_1}\sum_{j=1}^{J} b_j^{ik}\hat{x}_j^i \overset{\circ}{\eta}{}_j^k -$$

$$-\sum_{l\in R_2}\sum_{j=1}^{J}\sum_{i\in I_j} \bar{b}_j^{il}\hat{x}_j^i \overset{\circ}{\delta}{}^l +$$

$$+\sum_{i=1}^{I}\sum_{r=1}^{I}\hat{x}_r^i\left[-C^i + \sum_{(j,k)\in R_1} B_j^{ik}\overset{\circ}{\eta}{}_j^k + \sum_{l\in R_2} B^{il}\overset{\circ}{\delta}{}^l\right] -$$

$$-\sum_{m=1}^{I}\sum_{n=1}^{I} D^{mn}\overset{\circ}{X}{}^m\sum_{r=1}^{J}\hat{x}_r^m - \sum_{m=1}^{I}\sum_{n=1}^{I} D^{mn}\overset{\circ}{X}{}^n\sum_{r=1}^{J}\hat{x}_r^n. \qquad (2.24)$$

By replacing summation indices in the latter two terms (2.24) and taking into account the symmetry of the matrix $\{D^{mn}\}$, we find that these terms are equal to

$$-2\sum_{i=1}^{I}\sum_{r=1}^{J}\hat{x}_r^i\sum_{n=1}^{I} D^{in}\overset{\circ}{X}{}^n. \qquad (2.25)$$

Since $\overset{\circ}{X}{}^i > 0$, the relations (2.14) hold as strict equalities. Then, taking into account (2.25), we find that the latter three sums of the right-hand side from (2.24) are equal to zero.

Let us write conditions of the equality of optimal values of functionals $\overset{\circ}{g}$ and $\overset{\circ}{\psi}$ of the pair of conjugated macroproblems (2.9)–(2.12) and (2.13)–(2.15):

$$\sum_{i=1}^{I} C^i\overset{\circ}{X}{}^i + \sum_{m=1}^{I}\sum_{n=1}^{I} D^{mn}\overset{\circ}{X}{}^m\overset{\circ}{X}{}^n =$$

$$= -\sum_{m=1}^{I}\sum_{n=1}^{I} D^{mn}\overset{\circ}{X}{}^m\overset{\circ}{X}{}^n + \sum_{(j,k)\in R_1} p_j^k\overset{\circ}{\eta}{}_j^k + \sum_{l\in R_2} p^l\overset{\circ}{\delta}{}^l.$$

The latter expression can be rewritten in the form

$$-\sum_{j=1}^{J}\sum_{i\in I_j} c_j^i\overset{\circ}{x}{}_j^i - \sum_{j=1}^{J}\sum_{m\in I_j}\sum_{n\in I_j} d_j^{mn}\overset{\circ}{x}{}_j^m\overset{\circ}{x}{}_j^n +$$

$$+\sum_{(j,k)\in R_1} p_j^k\overset{\circ}{\eta}{}_j^k + \sum_{l\in R_2} p^l\overset{\circ}{\delta}{}^l - \sum_{j=1}^{J}\sum_{m\in I_j}\sum_{n\in I_j} d_j^{mn}\overset{\circ}{x}{}_j^m\overset{\circ}{x}{}_j^n = 0. \qquad (2.26)$$

By adding to the right-hand side of (2.24) the left-hand side of (2.26) as well as the expression

$$\sum_{j=1}^{J}\sum_{m\in I_j}\sum_{n\in I_j} d_j^{mn}\hat{x}_j^m\hat{x}_j^n - \sum_{j=1}^{J}\sum_{m\in I_j}\sum_{n\in I_j} d_j^{mn}\hat{x}_j^m\hat{x}_j^n,$$

which is equal to zero, we obtain (2.23), where σ_1 is the left-hand side of (2.22),

$$\sigma_2 = \sum_{(j,k)\in R_1} \overset{\circ}{\eta}{}_j^k \left(p_j^k - \sum_{i\in I_j} b_j^{ik}\hat{x}_j^i \right),$$

and

$$\sigma_3 = -\sum_{j=1}^{J}\sum_{m\in I_j}\sum_{n\in I_j} d_j^{mn}(\hat{x}_j^m - \overset{\circ}{x}{}_j^m)(\hat{x}_j^n - \overset{\circ}{x}{}_j^n).$$

The negative definiteness of quadratic forms for matrices $\{d_j^{mn}\}$, $j \in [1 : J]$ implies that the written sum and, hence, σ_3 are greater or equal to zero. The relation (2.23) is established. It gives the estimate $\overset{\circ}{g}(\rho) > \overset{\circ}{g}(0)$ in a neighborhood of the point $\rho = 0$, which proves Theorem 2.2.

In the case of nonuniqueness of the dual estimates $\overset{\circ}{\delta}{}^l$, for the definition of the growth direction of the functional of macroproblem (2.9)–(2.12), we introduce the maximin problem: find

$$\max_{x_j^i\in M_x^j}\ \min_{\overset{\circ}{\delta}{}^l\in\Omega_\delta} P(x_j^i, \overset{\circ}{\delta}{}^l) \tag{2.27}$$

where

$$P(x_j^i, \overset{\circ}{\delta}{}^l) = \sum_{j=1}^{J}\sum_{i\in I_j} c_j^i x_j^i + \sum_{j=1}^{J}\sum_{m\in I_j}\sum_{n\in I_j} d_j^{mn}x_j^m x_j^n -$$
$$-\sum_{l\in R_2}\sum_{j=1}^{J}\sum_{i\in I_j} d_j^{il}x_j^i\overset{\circ}{\delta}{}^l + \sum_{l\in R_2} p^l\overset{\circ}{\delta}{}^l.$$

Here, the definitions of the sets M_x^j, Ω_δ coincide with those introduced in Section 1.

According to Theorem 1.14 of Chapter 1, the properties of convexity and concavity of function (2.27) and boundedness of the sets imply the existence of the saddle point $\bar{x}^i_j$, $\bar{\delta}^l$ of the problem (2.27). Under degeneracy, local monotonicity of the functional is the consequence of the estimate

$$(\overset{\circ}{g}(0))' \geq P(\bar{x}^i_j, \bar{\delta}^l) - \sum_{j=1}^{J} \sum_{i \in I_j} c^i_j \overset{\circ}{x}{}^i_j - \sum_{j=1}^{J} \sum_{m \in I_j} \sum_{n \in I_j} d^{mn}_j \overset{\circ}{x}{}^m_j \overset{\circ}{x}{}^n_j. \qquad (2.28)$$

The right-hand side of (2.28) for the disaggregated solution $\overset{\circ}{x}{}^i_j$, which is nonoptimal for problem (2.1)–(2.4), is strictly greater than zero. This fact is established analogously to inequality (1.28) in Section 1.

§3. Mathematical Programming

We apply the decomposition method using aggregation of variables for the block separable extremal problems of convex type in a finite-dimension space. According to the definition from Chapter 1, these problems have the following form:

$$f = \sum_{j=1}^{J} f_j(x_j) \rightarrow \max, \qquad (3.1)$$

$$g^k_j(x_j) \leq 0, \qquad k \in K_j, \qquad (3.2)$$

$$x_j \geq 0, \qquad j \in [1:J], \qquad (3.3)$$

$$\sum_{j=1}^{J} \tilde{g}^l_j(x_j) \leq 0, \qquad l \in [1:L]. \qquad (3.4)$$

Here, for each fixed $j \in [1:J]$, the vector x_j has components $\{x^1_j, \dots, x^I_j\}$. The functions $f_j(x_j)$, $g^k_j(x_j)$, $\tilde{g}^l_j(x_j)$ are supposed to be continuously differentiable, and their partial derivatives satisfy the Lipshitz condition. It is also assumed that the function $f_j(x_j)$ is concave, and the functions $g^k_j(x_j)$ and $\tilde{g}^l_j(x_j)$ are convex. The dual problem for (3.1)–(3.4) has the following form:

$$\varphi = \sum_{j=1}^{J} f_j(x_j) - \sum_{j=1}^{J} \sum_{k \in K_j} \xi^k_j g^k_j(x_j) - \sum_{l=1}^{L} \sum_{j=1}^{J} \nu^l \tilde{g}^l_j(x_j) \rightarrow \min, \qquad (3.5)$$

$$\sum_{j=1}^{J} \partial f_j(x_j)/\partial x^i_j - \sum_{j=1}^{J} \sum_{k \in K_j} \xi^k_j \partial g^k_j(x_j)/\partial x^i_j -$$

$$-\sum_{l=1}^{L}\sum_{j=1}^{J}\nu^l \partial \tilde{g}_j^l(x_j)/\partial x_j^i = 0 \text{ for } x_j^i > 0, \tag{3.6}$$

$$\sum_{j=1}^{J} \partial f_j(x_j)/\partial x_j^i - \sum_{j=1}^{J}\sum_{k\in K_j} \xi_j^k \partial g_j^k(x_j)/\partial x_j^i -$$
$$-\sum_{l=1}^{L}\sum_{j=1}^{J}\nu^l \partial \tilde{g}_j^l(x_j)/\partial x_j^i \le 0 \text{ for } x_j^i = 0, \quad j\in[1:J], \quad i\in[1:I], \tag{3.7}$$
$$\xi_j^k \ge 0, \quad j\in[1:J], \quad k\in K_j, \quad \nu^l\ge 0, \quad l\in[1:L].$$

We introduce the aggregated variables

$$X^i = \sum_{j=1}^{J} x_j^i, \qquad i\in[1:I]$$

and the aggregation weights $\alpha_j^i = x_j^i/X^i$. Then, we substitute $x_j^i = \alpha_j^i X^i$ in (3.1)–(3.4) for fixed α_j^i and obtain the aggregated problem:

$$g = F(X^i) \to \max, \tag{3.8}$$

$$G_j^k(X^i) \le 0, \qquad j\in[1:J], \qquad k\in K_j, \tag{3.9}$$

$$G^l(X^i)\le 0, \qquad l\in[1:L], \tag{3.10}$$

$$X^i \ge 0, \qquad i\in[1:I]. \tag{3.11}$$

Here, notation

$$F(X^i) = \sum_{j=1}^{J} f_j(\alpha_j^i X^i), \quad G_j^k(X^i) = g_j^k(\alpha_j^i X^i), \quad G^l(X^i) = \sum_{j=1}^{J}\tilde{g}_j^l(\alpha_j^i X^i), \tag{3.12}$$

is used, and the notation $f(\alpha_j^i X^i)$ is identified with $f(x_j)$. The same relation holds for the functions $g_j^k(\alpha_j^i X^l)$, $\tilde{g}_j^l(\alpha_j^i X^i)$. It is easily seen that the functions $-F(X^i)$, $G_j^k(X^i)$, $G^l(X^l)$ are convex with respect to variables X^i, $i\in[1:I]$. This fact is tested using the definition of the convex function with the notation (3.12).

We write the dual problem for the macroproblem (3.8)–(3.11):

$$\psi = F(X^i) - \sum_{j=1}^{J}\sum_{k\in K_j}\eta_j^k G_j^k(X^i) - \sum_{l=1}^{L}\delta^l G^l(X^i) \to \min, \tag{3.13}$$

$$\partial F(X^i)/\partial X^i - \sum_{j=1}^{J} \sum_{k \in K_j} \eta_j^k \partial G_j^k(X^i)/\partial X^i - \\ - \sum_{l=1}^{L} \delta^l \partial G^l(X^i)/\partial X^i = 0 \text{ for } X^i > 0$$

$$\partial F(X^i)/\partial X^i - \sum_{j=1}^{J} \sum_{k \in K_j} \eta_j^k \partial G_j^k(X^i)/\partial X^i - \\ - \sum_{l=1}^{L} \delta^l \partial G^l(X^i)/\partial X^i \leq 0 \text{ for } X^i = 0, \qquad i \in [1:I], \tag{3.14}$$

$$X^i \geq 0, \qquad i \in [1:I], \qquad \eta_j^k \geq 0, \qquad j \in [1:J], \qquad k \in K_j, \\ \delta^l \geq 0, \qquad l \in [1:L]. \tag{3.15}$$

Here, for some weights α_j^i, we obtain optimal Lagrange multipliers $\overset{\circ}{\delta}{}^l$, $l \in [1:L]$ for the problem (3.13)–(3.15). We consider local problems for each fixed $j \in [1:J]$:

$$h_j = f_j(x_j) - \sum_{l=1}^{L} \overset{\circ}{\delta}{}^l \tilde{g}_j^l(x_j) \to \max, \tag{3.16}$$

$$g_j^k(x_j) \leq 0, \qquad k \in K_j, \tag{3.17}$$

$$x_j \geq 0. \tag{3.18}$$

The dual problems for block ones are written in the following form:

$$\chi_j = f_j(x_j) - \sum_{l=1}^{L} \overset{\circ}{\delta}{}^l \tilde{g}_j^l(x_j) - \sum_{k \in K_j} \zeta_j^k g_j^k(x_j) \to \min, \tag{3.19}$$

$$\sum_{j=1}^{J} \partial f_j(x_j)/\partial x_j^i - \sum_{l=1}^{L} \overset{\circ}{\delta}{}^l \partial \tilde{g}_j^l(x_j)/\partial x_j^i - \\ - \sum_{k \in K_j} \zeta_j^k \partial g_j^k(x_j)/\partial x_j^i = 0 \text{ for } x_j^i > 0, \qquad i \in [1:I], \tag{3.20}$$

$$\sum_{j=1}^{J} \partial f_j(x_j)/\partial x_j^i - \sum_{l=1}^{L} \overset{\circ}{\delta}{}^l \partial \tilde{g}_j^l(x_j)/\partial x_j^i - \\ - \sum_{k \in K_j} \zeta_j^k \partial g_j^k(x_j)/\partial x_j^i \leq 0 \text{ for } x_j^i = 0, \qquad i \in [1:I],$$

$$x_j \geq 0, \qquad \zeta_j^k \geq 0, \qquad k \in K_j. \tag{3.21}$$

Below, we assume that problem (3.1)–(3.4) has a solution and the Slater condition holds for it with respect to constraints (3.2), (3.4). Then, this condition is satisfied for the local problems (3.16)–(3.18) with respect to constraints (3.17). The same is considered to hold for macroproblems (3.8)–(3.11) with respect to constraints (3.9), (3.10), where the weight coefficients satisfy the conditions

$$\alpha_j^i \geq 0, \qquad j \in [1:J], \qquad \sum_{j=1}^{J} \alpha_j^i = 1, \qquad i \in [1:I]. \tag{3.22}$$

We will consider the optimality criterion for the intermediate solution of the iterative process. Suppose, that, for some weights α_j^i, we obtained solution $\overset{\circ}{X}{}^i$ of the aggregated problem, and $\overset{\circ}{\delta}{}^l$ are unique optimal Lagrange multipliers that correspond to solutions $\hat{x}_j$, $j \in [1:J]$ of the local problems (3.16)–(3.18).

Theorem 3.1. *Satisfaction of the equality*

$$\sum_{j=1}^{J} \left[f_j(\hat{x}_j) - \sum_{l=1}^{L} \overset{\circ}{\delta}{}^l \tilde{g}_j^l(\hat{x}_j) - f_j(\overset{\circ}{x}_j) \right] = 0 \tag{3.23}$$

is a sufficient condition for optimality of the disaggregated solution $\overset{\circ}{x}{}_j^i$ for the problem (3.1)–(3.4). If solution $\overset{\circ}{x}{}_j^i$ is nonoptimal for the problem (3.1)–(3.4), then the sign = is replaced by > in relation (3.23).

Proof. The disaggregated solution is feasible for problem (3.1)–(3.4). This is established by the intermediate testing, using the notation of (3.12). The value of the functional (3.1) for this solution is $\overset{\circ}{f} = \sum_{j=1}^{J} f_j(\overset{\circ}{x}_j)$. Let $\hat{\zeta}_j^k$, $j \in [1:J]$, $k \in K_j$ be optimal solutions of the dual problems (3.19)–(3.21). Then, the set $\{\hat{x}_j, \hat{\zeta}_j^k, \overset{\circ}{\delta}{}^l\}$ satisfies conditions (3.6), (3.7) of the dual problem. The latter fact follows from (3.20), (3.21), (3.15). In other words, the vector $\hat{x}_j$, $j \in [1:J]$ is the solution that maximizes the functional (3.5) under fixed Lagrange multipliers ξ_j^k, ν^l equal to $\hat{\zeta}_j^k$, $\overset{\circ}{\delta}{}^l$.

The inequality of Theorem 1.5 of Chapter 1 holds. In this case, the inequality looks as follows:

$$\sum_{j=1}^{J} f_j(x_j) - \sum_{j=1}^{J} \sum_{k \in K_j} \hat{\zeta}_j^k g_j^k(\hat{x}_j) - \sum_{l=1}^{L} \sum_{j=1}^{J} \overset{\circ}{\delta}{}^l \tilde{g}_j^l(\hat{x}_j) \geq \sum_{j=1}^{J} f_j(\overset{\circ}{x}_j). \tag{3.24}$$

Here, the equality in (3.24) ensures optimality of the disaggregated solution $\overset{\circ}{x}_j$ for the problem (3.1)–(3.4). Under solvability conditions of this problem and nonoptimality of $\overset{\circ}{x}_j$ for it, relation (3.24) turns into the strict inequality.

The second term of the left-hand side of (3.24) is equal to zero due to Kuhn-Tucker Theorem 1.7 applied to local problems. The theorem is proved.

We establish that the functional of the iterative process is monotonic. Assume that, for some weights α_j^i, we obtained a unique optimal solution $\overset{\circ}{X}{}^i$ of the aggregated problem (3.8)–(3.11), and the optimal Lagrange multipliers $\overset{\circ}{\eta}{}_j^k$, $\overset{\circ}{\delta}{}^l$ are unique. In the sequel, as before, we assume that all components of the solution $\overset{\circ}{X}{}^i$ are strictly greater than zero. Let $\hat{x}_j$, $j \in [1:J]$ be solutions of the local problems with the values $\overset{\circ}{\delta}{}^l$ for their functionals.

Theorem 3.2. *For the disaggregated solution $\overset{\circ}{x}{}_j^i$ nonoptimal for problem (3.1)–(3.4), there exists a solution feasible for (3.1)–(3.4) with the functional value strictly greater than $\overset{\circ}{f} = f(\overset{\circ}{x}_j)$.*

Proof. We compute the marginal value of $(\overset{\circ}{g}(0))'$ when the macroproblem (3.8)–(3.11) depend on parameters ρ in the same way as in Chapter 1. We use Theorem 1.16 of Chapter 1. It is easily seen that all the assumptions of the theorem hold in the case considered. In particular, the statement on uniform convergence is deduced under the assumption that the derivatives of the input functions satisfy the Lipshitz condition. Thus, we have

$$(\overset{\circ}{g}(0))' = \partial \left[F(\overset{\circ}{X}) - \sum_{j=1}^{J} \sum_{k \in K_j} \overset{\circ}{\eta}{}_j^k G_j^k(\overset{\circ}{X}) - \sum_{l=1}^{L} \overset{\circ}{\delta}{}^l G^l(\overset{\circ}{X}) \right] / \partial\rho \bigg|_{\rho=0}.$$

Taking into account the formula for the derivatives of α_j^i

$$(\alpha_j^i(0))' = \left(\hat{x}_j^i - \alpha_j^i \sum_{r=1}^{J} \hat{x}_r^i \right) (\overset{\circ}{X})^{-1}$$

and notation (3.12), we transform the latter expression to the form

$$(\overset{\circ}{g}(0))' = \sum_{j=1}^{J} \sum_{i=1}^{I} \left(\hat{x}_j^i - \alpha_j^i \sum_{r=1}^{J} \hat{x}_r^i \right) (\partial f_j / \partial x_j^i) -$$

$$- \sum_{j=1}^{J} \sum_{k \in K_j} \sum_{i=1}^{I} \overset{\circ}{\eta}{}_j^k \left(\hat{x}_j^i - \alpha_j^i \sum_{r=1}^{J} \hat{x}_r^i \right) (\partial g_j^k / \partial x_j^i) -$$

$$-\sum_{l=1}^{L}\sum_{j=1}^{J}\sum_{i=1}^{I}\overset{\circ}{\delta}{}^{l}\left(\hat{x}_j^i - \alpha_j^i \sum_{r=1}^{J}\right)(\partial\tilde{g}_j^l/\partial x_j^i). \qquad (3.25)$$

Hereafter, the partial derivatives are taken for the disaggregated solution $\overset{\circ}{x}_j$.

Under assumption $\overset{\circ}{X}{}^i > 0$, conditions (3.14) of the dual macroproblem hold as equalities and, with the use of notation (3.12), are transformed as follows:

$$\sum_{j=1}^{J}\alpha_j^i(\partial f/\partial x_j^i) - \sum_{j=1}^{J}\sum_{k\in K_j}\overset{\circ}{\eta}{}_j^k\alpha_j^i(\partial g_j^k/\partial x_j^i) -$$

$$-\sum_{l=1}^{L}\sum_{j=1}^{J}\overset{\circ}{\delta}{}^l\alpha_j^i(\partial\tilde{g}_j^l/\partial x_j^i) = 0, \qquad i \in [1:I]. \qquad (3.26)$$

Grouping the terms that contain multiplier $\sum_{r=1}^{J}\hat{x}_r^i$ in the right-hand side of (3.25), we establish that they are equal to zero due to (3.26). Thus, we have

$$(\overset{\circ}{g}(0))' = \sum_{j=1}^{J}\sum_{i=1}^{I}\left[\hat{x}_j^i(\partial f_j/\partial x_j^i) -\right.$$

$$\left. - \sum_{k\in K_j}\overset{\circ}{\eta}{}_j^k\hat{x}_j^i(\partial g_j^k/\partial x_j^i) - \sum_{l=1}^{L}\overset{\circ}{\delta}{}^l\hat{x}_j^i(\partial\tilde{g}_j^l/\partial x_j^i)\right]. \qquad (3.27)$$

Applying Kuhn-Tucker Theorem 1.7 from Chapter 1 to the optimal solution of macroproblem (3.8)–(3.11), we obtain

$$\sum_{j=1}^{J}\sum_{k\in K_j}\overset{\circ}{\eta}{}_j^k G_j^k(\overset{\circ}{X}) + \sum_{l=1}^{L}\overset{\circ}{\delta}{}^l G^l(\overset{\circ}{X}) = 0.$$

The latter equality is transformed to the following one:

$$-\sum_{j=1}^{J}\left[\sum_{k\in K_j}\overset{\circ}{\eta}{}_j^k g_j^k(\overset{\circ}{x}_j) + \sum_{l=1}^{L}\overset{\circ}{\delta}{}^l\tilde{g}_j^l(\overset{\circ}{x}_j)\right] = 0. \qquad (3.28)$$

We add the left-hand side of (3.28) together with

$$\sum_{j=1}^{J}f_j(\hat{x}_j) - \sum_{j=1}^{J}f_j(\hat{x}_j) = 0, \qquad \sum_{j=1}^{J}f_j(\overset{\circ}{x}_j) - \sum_{j=1}^{J}f_j(\overset{\circ}{x}_j) = 0,$$

$$\sum_{j=1}^{J}\sum_{k\in K_j}\mathring{\eta}_j^k g_j^k(\hat{x}_j) - \sum_{j=1}^{J}\sum_{k\in K_j}\mathring{\eta}_j^k g_j^k(\hat{x}_j) = 0,$$

$$\sum_{l=1}^{L}\sum_{j=1}^{J}\mathring{\delta}^l \tilde{g}_j^l(\hat{x}_j) - \sum_{l=1}^{L}\sum_{j=1}^{J}\mathring{\delta}^l \tilde{g}_j^l(\hat{x}) = 0,$$

whose values are equal to zero, to the right-hand side of (3.27).

We subtract the expression

$$\sum_{j=1}^{J}\sum_{i=1}^{I}\left[\mathring{x}_j^i(\partial f_j/\partial x_j^i) - \sum_{k\in K_j}\mathring{\eta}_j^k\,\mathring{x}_j^i(\partial g_j^k/\partial x_j^i) - \sum_{l=1}^{L}\mathring{\delta}^l\,\mathring{x}_j^i(\partial \tilde{g}_j^l/\partial x_j^i)\right] = 0$$

from the right-hand side of (3.27). The latter expression is obtained from equalities (3.26), afterwards each of them for fixed j is multiplied by $\mathring{X}^i$ and then summed over $i \in I$. Finally, we obtain the following expression:

$$\begin{aligned}(\mathring{g}(0))' = {} & \sum_{j=1}^{J}\left[f_j(\hat{x}_j) - \sum_{l=1}^{L}\mathring{\delta}^l\tilde{g}_j^l(\hat{x}_j) - f_j(\mathring{x}_j)\right] - \sum_{j=1}^{J}\sum_{k\in K_j}\mathring{\eta}_j^k g_j^k(\hat{x}_j) + \\ & + \sum_{j=1}^{J}\left\{\left[f_j(\mathring{x}_j^i) + \sum_{i=1}^{I}(\hat{x}_j^i - \mathring{x}_j^i)\partial f_j/\partial x_j^i - f_j(\hat{x}_j)\right] + \right. \\ & + \sum_{k\in K_j}\mathring{\eta}_j^k\left[g_j^k(\hat{x}_j) - \sum_{i=1}^{I}(\hat{x}_j^i - \mathring{x}_j^i)\partial g_j^k/\partial x_j^i - g_j^k(\mathring{x}_j)\right] + \\ & \left. + \sum_{l=1}^{L}\mathring{\delta}^l\left[\tilde{g}_j^l(\hat{x}_j) - \sum_{i=1}^{I}(\hat{x}_j^i - \mathring{x}_j^i)\partial \tilde{g}_j^l/\partial x_j^i - \tilde{g}_j^l(\mathring{x}_j)\right]\right\}. \end{aligned} \tag{3.29}$$

The first sum in (3.29) is strictly greater than zero due to Theorem 3.1, since it is the left-hand side of (3.23). The second sum (3.29) is greater or equal to zero according to (3.15), (3.17). Finally, the third sum is also greater or equal to zero according to (3.15) and assumed properties of convexity or concavity of the input functions. Therefore, the estimate $\mathring{g}(\rho) > \mathring{g}(0)$ holds in a neighborhood of the point $\rho = 0$, which proves Theorem 3.1.

We study the nonuniqueness of optimal Lagrange multipliers $\mathring{\eta}_j^k$, $\mathring{\delta}^l$ of the macroproblem (3.8)–(3.11). Introducing the maximin problem, we find

$$\max_{x_j\in M_x^j}\ \min_{\mathring{\delta}^l\in\Omega_\delta} P(x_j, \mathring{\delta}^l), \tag{3.30}$$

where

$$P(x_j, \overset{\circ}{\delta}^l) = \sum_{j=1}^{J} \left[f_j(x_j) - \sum_{l=1}^{L} \overset{\circ}{\delta}^l \tilde{g}_j^l(x_j) \right].$$

Here, the meaning of the sets M_x^j, Ω_δ is the same as in Section 1. If these sets are restricted, then, according to Theorem 1.15 of Chapter 1, by convexity and concavity of the input functions, there exists a saddle point $\bar{x}_j, \bar{\delta}^l$ of problem (3.30). Components $\bar{x}_j$ are substituted in the formula for $\alpha_j^i(\rho_j)$ (2.33) of Chapter 1 for variables $\hat{x}_j$ and give the growth direction for the functional of the macroproblem.

The following statement is analogous to Theorem 3.1:

Theorem 3.3. *The satisfaction of the equality*

$$P(\bar{x}_j, \bar{\delta}^l) - \sum_{j=1}^{J} f_j(\overset{\circ}{x}{}_j^i) = 0 \tag{3.31}$$

is a sufficient optimality condition for the disaggregated solution $\overset{\circ}{x}_j$ of problem (3.1)–(3.4) under the nonuniqueness of optimal Lagrange multipliers $\overset{\circ}{\eta}{}_j^k, \overset{\circ}{\delta}^l$. If this solution is nonoptimal for (3.1)–(3.4), then in relation (3.31) the sign $=$ is replaced by $>$.

We study monotonicity of the functional of the iterative process under the nonuniqueness of $\overset{\circ}{\eta}{}_j^k, \overset{\circ}{\delta}^l$. According to Theorem 1.16, in this case, the marginal value $(\overset{\circ}{g}(0))'$ is computed in the form

$$(\overset{\circ}{g}(0))' = \min_{\{\overset{\circ}{\eta}{}_j^k, \overset{\circ}{\delta}^l\} \in \Omega_{\eta,\delta}} \sum_{j=1}^{J} \sum_{i=1}^{I} \Bigg[\bar{x}_j^i(\partial f_j / \partial x_j^i) - \\ - \sum_{k \in K_j} \overset{\circ}{\eta}{}_j^k \bar{x}_j^i (\partial g_j^k / \partial x_j^i) - \sum_{l=1}^{L} \overset{\circ}{\delta}^l \bar{x}_j^i (\partial \tilde{g}_j^l / \partial x_j^i) \Bigg]. \tag{3.32}$$

Assume that the minimum in the right-hand side of (3.32) is attained for some $\{\tilde{\eta}_j^k, \tilde{\delta}^l\} \in \Omega_{\eta,\delta}$. Then, the right-hand side of (3.32) is transformed, in the same way as in the proof of Theorem 3.2, to the right-hand side of (3.29), where $\hat{x}_j$ is replaced by $\bar{x}_j$, and $\{\overset{\circ}{\eta}{}_j^k, \overset{\circ}{\delta}^l\}$ is replaced by $\{\tilde{\eta}_j^k, \tilde{\delta}^l\}$. Due to the properties of the saddle point (inequality of Theorem 1.10 from Chapter 1), the first sum over j in (3.29) is greater than or equal to the value

$$P(\bar{x}_j, \bar{\delta}^l) - \sum_{j=1}^{J} f_j(\overset{\circ}{x}{}_j^i) > 0.$$

This implies the estimate $\overset{\circ}{g}(\rho) > \overset{\circ}{g}(0)$ in a neighborhood of the point $\rho = 0$.

§4. Classical Calculus of Variations

In this section, we consider the problem of classical calculus of variations, or finding a continuous differentiable function $x(t)$ with given boundary values $[t_1, t_2]$ that provides weak, i.e. in the norm C_1, extremum (maximum) to the integral functional

$$\int_{t_1}^{t_2} f(t, x(t), x'(t))dt, \qquad x(t_1) = a^1, \qquad x(t_2) = a^2.$$

Here, the norm in C_1 is defined as usual:

$$||x(t)||_{C_1} = \max_{t_1 \le t \le t_2} |x(t)| + \max_{t_1 \le t \le t_2} |x'(t)|.$$

It is assumed that $x(t)$ is a vector function, $x(t) = \{x_1(t), \dots, x_j(t), \dots x_J(t)\}$. In turn, each component $x_j(t)$ is also a vector function, $x_j(t) = \{x_j^1(t), \dots, x_j^i(t), \dots, x_j^I(t)\}$.

We consider the integral functional in the form

$$f = \int_{t_1}^{t_2} \left[\sum_{j=1}^{J} f_j(t, x_j(t), x'(t)) \right] dt. \tag{4.1}$$

Moreover, there are two types of integral constraints:

$$\int_{t_1}^{t_2} g_j^k(t, x_j(t), x'(t)dt \le p_j^k, \qquad j \in [1:J], \qquad k \in K_j, \tag{4.2}$$

and

$$\int_{t_1}^{t_2} \left[\sum_{j=1}^{J} \bar{g}_j^l(t, x_j(t), x_j'(t)) \right] dt \le p^l, \qquad l \in [1:L]. \tag{4.3}$$

For each fixed j, relations (4.2) give block integral constraints. Relations (4.3) are binding constraints and as well as being functional (4.1) are block separable. Thus, relations (4.1)–(5.3), together with

$$x_j(t_1) = a_j^1, \qquad x_j(t_2) = a_j^2 \tag{4.4}$$

comprise a block separable problem of the calculus of variations.

Further development of the decomposition method reduce the variation problem (4.1)–(4.4) to problems with variables of lesser dimensions. It is assumed that the functions $f_j(t, x_j, x'_j)$, $g_j^k(t, x, x'_j)$, $\bar{g}_j^l(t, x_j, x'_j)$ have continuous partial derivatives with respect to all variables up to the second order inclusive. This assumption is typical in the classical calculus of variations. Under the nonzero Hessian that corresponds to variables x'_j, this condition guarantees the existence of continuous second derivatives of the solution. We consider also that the functions $-f_j(t, x_j, x'_j)$, $g_j^k(t, x, x'_j)$, $\bar{g}_j^l(t, x, x'_j)$ are convex with respect to variables x_j and x'_j for each $t \in [t_1, t_2]$. The convexity conditions will be used in the following form:

$$f_j(t, \hat{x}_j, \hat{x}'_j) - f_j(t, \mathring{x}_j \mathring{x}'_j) \leq \sum_{i=1}^{I} \left[(\hat{x}_j^i - \mathring{x}_j^i) \frac{\partial f_j(t, \mathring{x}_j, \mathring{x}'_j)}{\partial x_j^i} + \right.$$

$$\left. + ((\hat{x}_j^i)' - (\mathring{x}_j^i)') \frac{\partial f_j(t, \mathring{x}_j, \mathring{x}'_j)}{\partial (x_j^i)'} \right]. \tag{4.5}$$

Analogous inequalities, but with the opposite sign, hold for the functions $g_j^k(t, x, x'_j)$ and $\bar{g}_j^l(t, x, x'_j)$. We write inequalities (4.5) in equivalent operator form

$$Z(t, \mathring{x}_j, \mathring{x}'_j, \hat{x}_j, \hat{x}'_j)[f_j] \geq 0,$$

where

$$Z(t, \mathring{x}_j, \mathring{x}'_j, \hat{x}_j, \hat{x}'_j)[f_j] = f_j(t, \mathring{x}_j, \mathring{x}'_j) + \sum_{i=1}^{I} \left[(\hat{x}_j^i - \mathring{x}_j^i) \frac{\partial f_j(t, \mathring{x}_j, \mathring{x}'_j)}{\partial x_j^i} + \right.$$

$$\left. + ((\hat{x}_j^i)' - (\mathring{x}_j^i)') \frac{\partial f_j(t, \mathring{x}_j, \mathring{x}'_j)}{\partial x_j^i} \right] - f_j(t, \hat{x}_j, \hat{x}'_j).$$

We consider the Wolfe dual of the problem (5.1)–(5.4). Let ξ_j^k and μ^l be Lagrange multipliers that correspond to (4.2) and (4.3). Then, we have

$$\varphi = \int_{t_1}^{t_2} \left[\sum_{j=1}^{J} f_j(t, x_j, x'_j \right] dt + \sum_{j=1}^{J} \sum_{k \in K_j} \xi_j^k \left(p_j^k - \int_{t_1}^{t_2} g_j^k(t, x_j, x'_j) dt \right) +$$

$$+\sum_{l=1}^{L}\sum_{j=1}^{J}\mu^l\left(p^l-\int_{t_1}^{t_2}\bar{g}_j^l(t,x_j,x_j')dt\right)\to\min, \tag{4.6}$$

$$\sum_{j=1}^{J}\nabla_j^i\left[f_j(t,x_j,x_j')-\sum_{k\in K_j}\xi_j^k g_j^k(t,x_j,x_j')-\sum_{l=1}^{L}\mu^l\bar{g}_j^l(t,x_j,x_j')\right]=0, i\in[1:I],$$

$$x_j(t_1)=a_j^1,\qquad x_j(t_2)=a_j^2,\qquad j\in[1:J],$$

$$\xi_j^k\geq 0,\qquad k\in K_j,\qquad j\in[1:J],\qquad \mu^l\geq 0,\qquad l\in[1:L].$$

Here, we introduce the differential operator

$$\nabla_j^i=\partial/\partial x_j^i-d(\partial/\partial(x_j^i)')/dt,$$

and the corresponding equalities in (4.6) are Euler equations.

We introduce the aggregated variables

$$X^i(t)=\sum_{j=1}^{J}x_j^i(t),\qquad (X^i)'(t)=\sum_{j=1}^{J}(x_j^i)'(t),\qquad i\in[1:I],$$

and the weight aggregation functions $\alpha_j^i(t)=x_j^i(t)/X^i(t)$ that satisfy the condition

$$\sum_{j=1}^{J}\alpha_j^i(t)=1,\qquad i\in[1:I].$$

We fix the twice continuously differentiable functions $\alpha_j^i(t)$ and, substituting $x_j^i(t)=\alpha_j^i(t)X^i(t)$ in (4.1)–(4.4), obtain the variational aggregated problem:

$$g=\int_{t_1}^{t_2}F(t,X^i,(X^i)')dt\to\max,$$

$$\int_{t_1}^{t_2}G_j^k(t,X^i,(X^i)')dt\leq p_j^k,\qquad j\in[1:J],\qquad k\in K_j \tag{4.7}$$

$$\int_{t_1}^{t_2}G^l(t,X^i,(X^i)')dt\leq p^l,\qquad l\in[1:L],$$

$$X^i(t_1)=\sum_{j=1}^{J}a_j^{i1}=A^{i1},\qquad X^i(t_2)=\sum_{j=1}^{J}a_j^{i2}=A^{i2},$$

with the following notation:

$$F(t, X^i, (X^i)') = \sum_{j=1}^{J} f_j(t, \alpha_j^i X^i, (\alpha_j^i X^i)') = \sum_{j=1}^{J} f_j(t, \alpha_j^i X^i, (\alpha_j^i)' X^i + \alpha_j^i (X^i)'),$$

$$G_j^k(t, X^i, (X^i)') = g_j^k(t, \alpha_j^i X^i, (\alpha_j^i X^i)') = g_j^k(t, \alpha_j^i X^i, (\alpha_j^i)' X^i + \alpha_j^i (X^i)'),$$

$$G^l(t, X^i, (X^i)') = \sum_{j=1}^{J} \bar{g}_j^l(t, \alpha_j^i X^i, (\alpha_j^i X^i)') = \sum_{j=1}^{J} \bar{g}_j^l(t, \alpha_j^i X^i, (\alpha_j^i)' X^i + \alpha_j^i (X^i)').$$

The functions that occur in (4.7) are convex with respect to variables X^i, $(X^i)'$. For example, the following chain of relations holds for the function F:

$$\begin{aligned}
&F(t, X_2^i, (X_2^i)') - F(t, X_1^i, (X_1^i)') = \\
&\quad = \sum_{j=1}^{J} [f_j(t, \alpha_j^i X_2^i, (\alpha_j^i X_2^i)' - f_j(t, \alpha_j^i X_1, (\alpha_j^i X_1^i)'] \le \\
&\quad \le \sum_{j=1}^{J} \sum_{i=1}^{I} \Bigg[(\alpha_j^i X_2^i - \alpha_j^i X_1^i) \frac{\partial f_j(t, \alpha_j^i X_1^i, (\alpha_j^i X_1^i)')}{\partial x_j^i} + \\
&\qquad + ((\alpha_j^i X_2^i)' - (\alpha_j^i X_1^i)') \frac{\partial f_j(t, \alpha_j^i X_1^i, (\alpha_j^i X_1^i)')}{\partial (x_j^i)'} \Bigg] = \\
&\quad = \sum_{j=1}^{J} \sum_{i=1}^{I} \Bigg\{ (X_2^i - X_1^i) \Bigg[\alpha_j^i \frac{\partial f_j(t, \alpha_j^i X_1^i, (\alpha_j^i X_1^i)')}{\partial x_j^i} + \\
&\qquad\qquad + (\alpha_j^i)' \frac{\partial f_j(t, \alpha_j^i X_1^i, (\alpha_j^i X_1^i)')}{\partial (x_j^i)'} \Bigg] + \\
&\qquad + ((X_2^i)' - (X_1^i)') \alpha_j^i \frac{\partial f_j(t, \alpha_j^i X_1^i, (\alpha_j^i X_1^i)')}{\partial (x_j^i)'} \Bigg\} = \\
&\quad = \sum_{i-1}^{I} \left[(X_2^i - X_1^i) \frac{\partial F(t, X_1, X_1')}{\partial X^i} + ((X_2^i)' - (X_1^i)') \frac{\partial F(t, X_1, X_1')}{\partial (X^i)'} \right]
\end{aligned}$$

Let η_j^k and δ^l be Lagrange multipliers that correspond to the second and third constraints (4.7). Then, the problem that is Wolfe dual to macroproblem (4.7) has the form

$$\int_{t_1}^{t_2} F(t, X^i, (X^i)') dt + \sum_{j=1}^{J} \sum_{k \in K_j} \eta_j^k \left(p_j^k - \int_{t_1}^{t_2} G_j^k(t, X^i, (X^i)') dt \right) +$$

$$+\sum_{l=1}^{L}\delta^{l}\left(p^{l}-\int_{t_1}^{t_2}G^{l}(t,X^{i},(X^{i})')dt\right)\to\min,\qquad(4.8)$$

$$\nabla^{i}\left[F(t,X^{i},(X^{i})')-\sum_{j=1}^{J}\sum_{k\in K_j}\eta_j^k G_j^k(t,X^{i},(X^{i})')-\sum_{l=1}^{L}\delta^{l}G^{l}(t,X^{i},(X^{i})')\right]=0,$$

$$X^{i}(t_1)=A^{i1},\qquad X^{i}(t_2)=A^{i2},\qquad i\in[1:I],$$

$$\eta_j^k\geq 0,\qquad j\in[1:J],\qquad k\in K_j,\qquad \delta^{l}\geq 0,\qquad l\in[1:L].$$

Here, we introduce the differential operator

$$\nabla^{i}=\partial/\partial X^{i}-d(\partial/\partial(X^{i})')/dt.$$

Let $\overset{\circ}{\delta}{}^{l}$ be optimal Lagrange multipliers that correspond to the extremal solution of the variational aggregated problem (5.7). Consider the local variational problems for each fixed j, $j\in[1:J]$:

$$h_j=\int_{t_1}^{t_2}\left[f_j(t,x_j,x_j')-\sum_{l=1}^{L}\overset{\circ}{\delta}{}^{l}\bar{g}_j^{l}(t,x_j,x_j')\right]dt\to\max,$$

$$\int_{t_1}^{t_2}\bar{g}_j^{l}(t,x_j,x_j')dt\leq p_j^k,\qquad k\in K_j,\qquad(4.9)$$

$$x_j(t_1)=a_j^1,\qquad x_j(t_2)=a_j^2.$$

The functionals in (4.9) are obviously convex.

By introducing Lagrange multipliers ζ_j^k, $j\in[1:J]$, $k\in K_j$ that correspond to integral constraints (4.9), we can consider Wolfe dual problems for the local variational problems. For these problems, the Euler equation takes the form

$$\nabla_j^{i}\left[f_j(t,x_j,x_j')-\sum_{l=1}^{L}\overset{\circ}{\delta}{}^{l}\bar{g}_j^{l}(t,x_j,x_j')-\sum_{k\in K_j}\zeta_j^k g_j^k(t,x_j,x_j')\right]=0,\qquad(4.10)$$

$$j\in[1:J],\qquad i\in[1:I].$$

The iterative process is constructed according to the scheme from Section 2 of Chapter 1. We fix weight functions $\alpha_j^i(t)$ and solve the variation aggregated problem (4.7). Let $\overset{\circ}{X}{}^{i}(t)$ be the twice continuously differentiable extremal solution of this problem and $\overset{\circ}{x}{}^i_j(t) = \alpha^i_j(t)\,\overset{\circ}{X}{}^i(t)$ be the corresponding twice differentiable disaggregated solution $\hat{x}^i_j(t)$ of the local problems, which also have continuous derivatives up to the second order inclusive. Then, the weight functions are taken in the form

$$\alpha^i_j(t,\rho_j) = \left[\overset{\circ}{x}{}^i_j(t) + \rho_j(\hat{x}^i_j(t) - \overset{\circ}{x}{}^i_j(t))\right] \Big/ \sum_{r=1}^{J} \left[\overset{\circ}{x}{}^t_r(t) + \rho_r(\hat{x}^i_r(t) - \overset{\circ}{x}{}^i_r(t))\right], \tag{4.11}$$

where $\rho_j \in [0,1]$, $j \in [1:J]$. The weight functions have continuous second derivatives with respect to t if the denominator in (4.11) does not vanish in the interval $[t_1, t_2]$. The latter condition is assumed for all steps of the iterative process. For example, this condition holds if all $\overset{\circ}{x}{}^i_j(t)$, $\hat{x}^i_j(t)$ are strictly greater or less than zero.

Below, we assume that the original problem (4.1)–(4.4) and all macroproblems (4.7) have solutions, and the Slater condition holds for all of them. Then, it holds for all local problems. The extremality condition of the disaggregated solution $\overset{\circ}{x}{}^i_j(t)$ to the initial problem is the satisfaction of the following equality:

$$\int_{t_1}^{t_2} \left[\sum_{j=1}^{J} f_j(t, \hat{x}_j, \hat{x}'_j)\right] dt + \sum_{l=1}^{L} p^l\,\overset{\circ}{\delta}{}^l - $$
$$- \int_{t_1}^{t_2} \left[\sum_{l=1}^{L}\sum_{j=1}^{J} \overset{\circ}{\delta}{}^l \bar{g}^l_j(t, \hat{x}_j, \hat{x}'_j)\right] dt - \int_{t_1}^{t_2} \left[\sum_{j=1}^{J} f_j(t, \overset{\circ}{x}_j, \overset{\circ}{x}{}'_j)\right] dt = 0. \tag{4.12}$$

For the solution nonoptimal to the main problem (4.1)–(4.4), the sign = is replaced by > in (4.12).

The proof is analogous to that of Theorem 3.1 from Chapter 1. It is based on the duality principles and on the Kuhn-Tucker theorem for the local problems, since the Slater condition holds, and the functionals and constraints of these problems are convex and concave, respectively. We study monotonicity of the functional in the iterative process. We restrict ourselves to the uniqueness of optimal Lagrange multipliers $\overset{\circ}{\eta}{}^k_j$, $\overset{\circ}{\delta}{}^l$ of the problem (4.7) on a step of the iterative process.

Theorem 4.1. *For the disaggregated solution $\overset{\circ}{x}{}^i_j t = \alpha^i_j(t)\,\overset{\circ}{X}{}^i(t)$ that is not extremal for the initial problem (4.1)–(4.4), there exists a solution that* *satisfies (4.2)–(4.4) with a value of the functional (4.1) strictly greater than* $\overset{\circ}{f} = f(\overset{\circ}{x}_j)$.

Proof. We introduce the notation

$$\Phi_j(t, \alpha_j^i X^i, (\alpha_j^i X^i)') = f_j(t, \alpha_j^i X^i, (\alpha_j^i X^i)') -$$
$$- \sum_{k \in K_j} \eta_j^k g_j^k(t, \alpha_j^i X^i, (\alpha_j^i X^i)') - \sum_{l=1}^{L} \delta^l \bar{g}_j^l(t, \alpha_j^i X^i, (\alpha_j^i X^i)'). \qquad (4.13)$$

We have the following lemma:

Lemma 4.1. *For the extremal solution* $\overset{\circ}{X}{}^i(t)$*, the Euler equations in (5.8) are reduced to the relations*

$$\sum_{j=1}^{J} \alpha_j^i(t) \nabla_j^i \Phi_j(t, \overset{\circ}{x}_j, \overset{\circ}{x}{}'_j) = 0, \qquad i \in [1 : I]. \qquad (4.14)$$

Proof of Lemma 4.1. We consider a term from (4.13), for example $f_j(t, \alpha_j^i \overset{\circ}{X}{}^i, (\alpha_j^i \overset{\circ}{X}{}^i)')$, and apply operator ∇^i to it. We have

$$\nabla^i f_j(t, \alpha_j^i \overset{\circ}{X}{}^i, (\alpha_j^i \overset{\circ}{X}{}^i)') = \alpha_j^i \frac{\partial f_j(t, \alpha_j^i \overset{\circ}{X}{}^i, (\alpha_j^i \overset{\circ}{X}{}^i)')}{\partial x_j^i} +$$
$$+ (\alpha_j^i)' \frac{\partial f_j(t, \alpha_j^i \overset{\circ}{X}{}^i, (\alpha_j^i \overset{\circ}{X}{}^i)')}{\partial x_j^i} - \frac{d}{dt} \left[\alpha_j^i \frac{\partial f_j(t, \alpha_j^i \overset{\circ}{X}{}^i, (\alpha_j^i \overset{\circ}{X}{}^i)')}{\partial (x_j^i)'} \right].$$

By differentiating the last term of this relation, we obtain

$$(\alpha_j^i)' \frac{\partial f_j(t, \overset{\circ}{x}_j, \overset{\circ}{x}{}'_j)}{\partial (x_j^i)'} - \alpha_j^i \frac{d}{dt} \left[\frac{\partial f_j(f, \overset{\circ}{x}_j, \overset{\circ}{x}{}'_j}{\partial (x_j^i)'} \right].$$

Similar consideration of other terms (4.13) finishes the proof of Lemma 4.1.

We continue the proof of Theorem 4.1. Consider the extremal value of the functional $\overset{\circ}{g}(\rho)$ of the aggregated problems (4.7) as a function of ρ when the weight functions are taken in the form (4.11), where $\rho_j = \rho$. The following lemma establishes the formula for the marginal value $(\overset{\circ}{g}(0))'$ of the variational aggregated problems:

Lemma 5.2. *The derivative* $(\overset{\circ}{g}(0))'$ *is computed in the form*

$$(\overset{\circ}{g}(0))' = \sigma_1 + \sigma_2 + \sigma_3, \qquad (4.15)$$

where the value σ_1 is equal to the left-hand side of (4.12) and is strictly greater than zero, the value σ_2 is

$$\sigma_2 = \sum_{j=1}^{J} \sum_{k \in K_j} \overset{\circ}{\eta}{}_j^k \left(p_j^k - \int_{t_1}^{t_2} g_j^k(t, \hat{x}_j, \hat{x}'_j) dt \right) \geq 0,$$

and the value σ_3 is equal to

$$\sigma_3 = \int_{t_1}^{t_2} \left\{ \sum_{j=1}^{J} Z(t, \overset{\circ}{x}_j, \overset{\circ}{x}{}'_j, \hat{x}_j, \hat{x}'_j) \times \left[f_j - \sum_{k \in K_j} \overset{\circ}{\eta}{}_j^k g_j^k - \sum_{l=1}^{L} \overset{\circ}{\delta}{}^l \bar{g}_j^l \right] \right\} dt \geq 0.$$

Proof of Lemma 4.2. For each ρ, the optimal values $\overset{\circ}{X}{}^i(t, \rho)$, $\overset{\circ}{\eta}{}_j^k(\rho)$, $\overset{\circ}{\delta}{}^l(\rho)$ are the saddle points of the Lagrange function for the macroproblem (4.7). Since solutions $\overset{\circ}{X}{}^i(t, \rho)$ are twice continuously differentiable, they form a compact set in C_1. Hence, Theorem 1.6 from Chapter 1 on marginal value is generalized. Thus, we have

$$\begin{aligned}
(\overset{\circ}{g}(0))' = \int_{t_1}^{t_2} \sum_{i=1}^{I} \Bigg\{ & \sum_{j=1}^{J} \left[\frac{\partial f_j(t, \overset{\circ}{x}_j, \overset{\circ}{x}{}'_j)}{\partial x_j^i} \frac{\partial \alpha_j^i(t, 0)}{\partial \rho} + \right. \\
& \left. + \frac{\partial f_j(t, \overset{\circ}{x}_j, \overset{\circ}{x}{}'_j)}{\partial (x_j^i)'} \frac{\partial (\alpha_j^i(t, 0) \overset{\circ}{X}{}^i)'}{\partial \rho} \right] - \\
& - \sum_{j=1}^{J} \sum_{k \in K_j} \overset{\circ}{\eta}{}_j^k \left[\frac{\partial g_j^k(t, \overset{\circ}{x}_j, \overset{\circ}{x}{}'_j)}{\partial x_j^i} \frac{\partial \alpha_j^i(t, 0)}{\partial \rho} \overset{\circ}{X}{}^i + \right. \\
& \left. + \frac{\partial g_j^k(t, \overset{\circ}{x}_j, \overset{\circ}{x}{}'_j)}{\partial (x_j^i)'} \frac{\partial (\alpha_j^i(t, 0) \overset{\circ}{X}{}^i)'}{\partial \rho} \right] - \\
& - \sum_{l=1}^{L} \sum_{j=1}^{J} \overset{\circ}{\delta}{}^l \left[\frac{\partial \bar{g}_j^l(t, \overset{\circ}{x}_j, \overset{\circ}{x}{}'_j)}{\partial x_j^i} \frac{\partial \alpha_j^i(t, 0)}{\partial \rho} \overset{\circ}{X}{}^i + \right. \\
& \left. + \frac{\partial \bar{g}_j^l(t, \overset{\circ}{x}_j, \overset{\circ}{x}{}'_j)}{\partial (x_j^i)'} \frac{\partial (\alpha_j^i(t, 0) \overset{\circ}{X}{}^i)'}{\partial \rho} \right] \Bigg\} dt.
\end{aligned}$$

Further transformations will be made for the first two sums of the right-hand side of the written expression. Relation (4.11) concerns the derivatives with respect to ρ of the weight functions at the point $\rho = 0$. Taking this into account, we obtain

$$\int_{t_1}^{t_2} \sum_{j=1}^{J} \sum_{i=1}^{I} \left\{ \frac{\partial f_j(t, \overset{\circ}{x}_j, \overset{\circ}{x}'_j)}{\partial (x_j^i)'} \left[\hat{x}_j^i - \alpha_j^i \sum_{r=1}^{J} \hat{x}_r^i \right] + \right.$$

$$\left. + \frac{\partial f_j(t, \overset{\circ}{x}_j, \overset{\circ}{x}'_j)}{\partial (x_j^i)'} \left[\hat{x}_j^i - \alpha_j^i \sum_{r=1}^{J} \hat{x}_r^i \right]' \right\} dt =$$

$$= \int_{t_1}^{t_2} \sum_{j=1}^{J} \sum_{i=1}^{I} \left\{ \left[\hat{x}_j^i \frac{\partial f_j(t, \overset{\circ}{x}_j, \overset{\circ}{x}'_j)}{\partial x_j^i} + (\hat{x}_j^i)' \frac{\partial f_j(t, \overset{\circ}{x}_j, \overset{\circ}{x}'_j)}{\partial (x_j^i)'} \right] - \right.$$

$$\left. - \frac{\partial f_j(t, \overset{\circ}{x}_j, \overset{\circ}{x}'_j)}{\partial x_j^i} \alpha_j^i \sum_{r=1}^{J} \hat{x}_r^i - \frac{\partial f_j(t, \overset{\circ}{x}_j, \overset{\circ}{x}'_j)}{\partial (x_j^i)'} \left(\alpha_j^i \sum_{r=1}^{J} \hat{x}_r^i \right)' \right\} dt. \tag{4.16}$$

Integrating by parts the expression that containing the last terms of the right-hand side of (4.16), we obtain

$$\int_{t_1}^{t_2} \left\{ \sum_{j=1}^{J} \sum_{i=1}^{I} \frac{d}{dt} \left[\frac{\partial f_j(t, \overset{\circ}{x}_j, \overset{\circ}{x}'_j)}{\partial (x_j^i)'} \right] \alpha_j^i \sum_{r=1}^{J} \hat{x}_r^i \right\} dt -$$

$$- \sum_{j=1}^{J} \sum_{i=1}^{I} \frac{\partial f_j(t, \overset{\circ}{x}_j, \overset{\circ}{x}'_j)}{\partial (x_j^i)'} \alpha_j^i \sum_{r=1}^{J} \hat{x}_r^i \Bigg|_{t=t_1}^{t=t_2}. \tag{4.17}$$

The first sum (4.17) together with the third sum in the right-hand side of (4.16) and the corresponding sums for g_j^k and $\bar{g}_j^l$ are equal to zero according to (4.14). Furthermore, we multiply each relation (4.14) for fixed i by $\overset{\circ}{X}{}^i$, sum them over i and take the integral from t_1 to t_2. For the terms that correspond to $f_j(t, \overset{\circ}{x}_j, \overset{\circ}{x}'_j)$, we obtain the following expression:

$$\int_{t_1}^{t_2} \left\{ \sum_{j=1}^{J} \sum_{i=1}^{I} \overset{\circ}{x}{}_j^i \frac{\partial f_j(t, \overset{\circ}{x}_j, \overset{\circ}{x}'_j)}{\partial x_j^i} - \overset{\circ}{x}{}_j^i \frac{d}{dt} \left[\frac{\partial f_j(t, \overset{\circ}{x}_j, \overset{\circ}{x}'_j)}{\partial (x_j^i)'} \right] \right\} dt =$$

$$= \int_{t_1}^{t_2} \left\{ \sum_{j=1}^{J} \sum_{i=1}^{I} \overset{\circ}{x}{}_j^i \frac{\partial f_j(t, \overset{\circ}{x}_j, \overset{\circ}{x}'_j)}{\partial x_j^i} + (\overset{\circ}{x}{}_j^i)' \frac{\partial f_j(t, \overset{\circ}{x}_j, \overset{\circ}{x}'_j)}{\partial (x_j^i)'} \right\} dt -$$

$$- \sum_{j=1}^{J} \sum_{i=1}^{I} \alpha_j^i \overset{\circ}{X}{}^i \frac{\partial f_j(t, \overset{\circ}{x}_j, \overset{\circ}{x}'_j)}{\partial x_j^i} \Bigg|_{t=t_1}^{t=t_2} = 0.$$

With due regard to (4.17) and

$$\overset{\circ}{X}{}^i(t_1) - \sum_{r=1}^{J} \hat{x}^i_r(t_1) = A^{i1}, \qquad \overset{\circ}{X}{}^i(t_2) - \sum_{r=1}^{J} \hat{x}^i_r(t_2) = A^{i2}, \qquad i \in [1:I],$$

we subtract the left-hand side of the latter expression from (4.16). We obtain

$$(\mathring{g}(0))' = \int_{t_1}^{t_2} \sum_{j=1}^{J} \sum_{i=1}^{I} \Bigg\{ \Bigg[(\hat{x}^i_j - \mathring{x}^i_j) \frac{\partial f_j(t, \mathring{x}_j, \mathring{x}'_j)}{\partial x^i_j} + ((\hat{x}^i_j)' - (\mathring{x}^i_j)') \frac{\partial f_j(t, \mathring{x}_j, \mathring{x}'_j)}{\partial (x^i_j)'} \Bigg] -$$
$$- \sum_{k \in K_j} \mathring{\eta}^k_j \Bigg[(\hat{x}^i_j - \mathring{x}^i_j) \frac{\partial g^k_j(t, \mathring{x}_j, \mathring{x}'_j)}{\partial x^i_j} + ((\hat{x}^i_j)' - (\mathring{x}^i_j)') \frac{\partial g^k_j(t, \mathring{x}_j, \mathring{x}'_j)}{\partial (x^i_j)'} \Bigg] -$$
$$- \sum_{l=1}^{L} \mathring{\delta}^l \Bigg[(\hat{x}^i_j - \mathring{x}^i_j) \frac{\partial \bar{g}^l_j(t, \mathring{x}_j, \mathring{x}'_j)}{\partial x^i_j} + +((\hat{x}^i_j)' - (\mathring{x}^i_j)') \frac{\partial \bar{g}^l_j(t, \mathring{x}_j, \mathring{x}'_j)}{\partial (x^i_j)'} \Bigg] \Bigg\} dt. \tag{4.18}$$

By the Kuhn-Tucker theorem, the value

$$\sum_{j=1}^{J} \sum_{k \in K_j} \mathring{\eta}^k_j \left(p^k_j - \int_{t_1}^{t_2} g^k_j(t, \mathring{x}_j, \mathring{x}'_j) dt \right) + \sum_{l=1}^{L} \sum_{j=1}^{J} \mathring{\delta}^l \left(p^l - \int_{t_1}^{t_2} \bar{g}^l_j(t, \mathring{x}_j, \mathring{x}'_j) dt \right)$$

is equal to zero in macroproblem (4.7). We add it to the right-hand side of (4.18). Calculation gives (4.15), which proves Lemma 4.2.

We continue the proof of Theorem 4.1. According to (4.15), we have $(\mathring{g}(0))' > 0$. Therefore, the estimate $\mathring{g}(\rho) > \mathring{g}(0)$ holds in a neighborhood of the point $\rho = 0$.

Theorem 4.1 is proved.

The constructions above related to the variational problem with one independent variable and fixed boundary values. Consider a problem with free boundary values. Then, additional boundary conditions should hold. For the variational aggregated problem, these conditions take the following form:

$$\partial \Bigg[F(t, X^i, (X^i)') - \sum_{j=1}^{J} \sum_{k \in K_j} \eta^k_j G^k_j(t, X^i, (X^i)') -$$
$$- \sum_{l=1}^{L} \delta^l G^l(t, X^i, (X^i)') \Bigg] \Big/ \partial X^i = 0, \qquad i \in [1:I] \tag{4.19}$$

for $t = t_1$ and $t = t_2$. Conditions (4.19) for the function $F(t, X^i, (X^i)')$ and the extremal solution $\overset{\circ}{X}{}^i$ are written in the following form:

$$\sum_{j=1}^{J} \alpha_j^i \frac{\partial f_j(t, \overset{\circ}{x}_j, \overset{\circ}{x}{}'_j)}{\partial x_j^i} = 0 \qquad i \in [1 : I] \tag{4.20}$$

for $t = t_1$ and $t = t_2$. Then, due to (4.20), the last sums in (4.17) are equal to zero. Other constructions also hold.

Consider the variational problem, with the functional and constraints depending on several functions of several independent variables. For simplicity, let us have two independent variables t and τ, and study the variational problem with given values on the boundary Γ of domain M. Our goal is to prove Lemma 4.2. All other constructions are made analogously.

By applying Lemma 4.1, the Euler equations for aggregated problem (4.7) are reduced to the following:

$$\begin{aligned} &\sum_{j=1}^{J} \alpha_j^i \nabla_j^i(t, \tau) \times \\ &\times [\phi_j(t, \tau, \overset{\circ}{x}_j(t, \tau), \partial \overset{\circ}{x}_j(t, \tau)/\partial t, \partial \overset{\circ}{x}_j(t, \tau)/\partial \tau)] = 0, \\ &i \in [1 : I]. \end{aligned} \tag{4.21}$$

Here, we introduce the operator

$$\nabla_j^i(t, \tau) = \partial/\partial x_j^i - \partial\left[\partial/\partial x_{jt}^i\right]/\partial t - \partial\left[\partial/\partial x_{j\tau}^i\right]/\partial \tau.$$

We consider one term $f_j(t, \tau, \overset{\circ}{x}_j, \overset{\circ}{x}_{jt}, \overset{\circ}{x}_{j\tau})$ that occurs in the function Φ_j. Multiplying (4.21) for each fixed i by $\overset{\circ}{X}{}^i(t, \tau)$ and summing over i, we integrate over the domain M. Thus, we obtain

$$\iint_M \sum_{j=1}^{J} \sum_{i=1}^{I} \left[\overset{\circ}{x}{}_j^i \left(\frac{\partial f_j}{\partial x_j^i} \right) - \overset{\circ}{x}{}_j^i \frac{\partial}{\partial t} \left(\frac{\partial f_j}{\partial x_{jt}^i} \right) - \overset{\circ}{x}{}_j^i \frac{\partial}{\partial \tau} \left(\frac{\partial f_j}{\partial x_{j\tau}^i} \right) \right] dt d\tau = 0. \tag{4.22}$$

Integrals that correspond to the second and third terms of (4.22) are transformed by the following:

$$\begin{aligned} &\iint_M \sum_{j=1}^{J} \sum_{i=1}^{I} \left[\frac{\partial}{\partial t} \left(\overset{\circ}{x}{}_j^i \frac{\partial f_j}{\partial x_{jt}^i} \right) + \frac{\partial}{\partial \tau} \left(\overset{\circ}{x}{}_j^i \frac{\partial f_j}{\partial x_{j\tau}^i} \right) \right] dt d\tau - \\ &\qquad - \iint_M \sum_{j=1}^{J} \sum_{i=1}^{I} \left(\overset{\circ}{x}{}_{jt}^i \frac{\partial f_j}{\partial x_{jt}^i} + \overset{\circ}{x}{}_{j\tau}^i \frac{\partial f_j}{\partial x_{j\tau}^i} \right) dt d\tau. \end{aligned} \tag{4.23}$$

Applying the Green formula to the first integral of (4.23), we obtain

$$\int_\Gamma \sum_{j=1}^{J}\sum_{i=1}^{I}\left[\overset{\circ}{x}{}_j^i\left(\frac{\partial f_j}{\partial x_{j\tau}^i}\right)dt-\overset{\circ}{x}{}_j^i\left(\frac{\partial f}{\partial x_{jt}^i}\right)d\tau\right]. \tag{4.24}$$

Further, by analogy with (4.16), we consider the terms that correspond to the functions $f_j(t,\tau,\overset{\circ}{x}_j,\overset{\circ}{x}_{jt},\overset{\circ}{x}_{j\tau})$ in the expression for the marginal value:

$$\int\int_M \sum_{j=1}^{J}\sum_{i=1}^{I}\left\{\frac{\partial f_j}{\partial x_j^i}\left[\hat{x}_j^i-\alpha_j^i\sum_{r=1}^{J}\hat{x}_r^i\right]+\frac{\partial f_j}{\partial x_{jt}^i}\frac{\partial}{\partial t}\left[\hat{x}_j^i-\alpha_j^i\sum_{r=1}^{J}\hat{x}_r^i\right]+\right.$$
$$\left.+\frac{\partial f_j}{\partial x_{j\tau}^i}\frac{\partial}{\partial \tau}\left[\hat{x}_j^i-\alpha_j^i\sum_{r=1}^{J}\hat{x}_r^i\right]\right\}dtd\tau.$$

The last integral is represented in the form

$$\int\int_M \sum_{j=1}^{J}\sum_{i=1}^{I}\left[\hat{x}_j^i\frac{\partial f_j}{\partial x_j^i}+\hat{x}_{jt}^i\frac{\partial f_j}{\partial x_{jt}^i}+\hat{x}_{j\tau}^i\frac{\partial f_j}{\partial x_{j\tau}^i}\right]dtd\tau-$$
$$-\int\int_M \sum_{j=1}^{J}\sum_{i=1}^{I}\left(\frac{\partial f_j}{\partial x_j^i}\alpha_j^i\sum_{r=1}^{J}\hat{x}_j^i\right)dtd\tau-$$
$$-\int\int_M \sum_{j=1}^{J}\sum_{i=1}^{I}\left\{\frac{\partial f_j}{\partial x_{jt}^i}\frac{\partial}{\partial t}\left[\alpha_j^i\sum_{r=1}^{J}\hat{x}_r^i\right]+\right.$$
$$\left.+\frac{\partial f_j}{\partial x_{j\tau}^i}\frac{\partial}{\partial \tau}\left[\alpha_j^i\sum_{r=1}^{J}\hat{x}_r^i\right]\right\}dtd\tau. \tag{4.25}$$

We integrate by parts the third integral in (4.25) and obtain

$$-\int\int_M \sum_{j=1}^{J}\sum_{i=1}^{I}\alpha_j^i\sum_{r=1}^{J}\hat{x}_r^i\left[\frac{\partial}{\partial t}\left(\frac{\partial f_j}{\partial x_{jt}^i}\right)+\frac{\partial}{\partial \tau}\left(\frac{\partial f_j}{\partial x_{j\tau}^i}\right)\right]dtd\tau+$$
$$+\int_\Gamma\left[\left(\alpha_j^i\sum_{r=1}^{J}\hat{x}_r^i\right)\left(\frac{\partial f_j}{\partial x_{j\tau}^i}\right)dt-\left(\alpha_j^i\sum_{r=1}^{J}\hat{x}_r^i\right)\left(\frac{\partial f_j}{\partial x_{jt}^i}\right)d\tau\right]. \tag{4.26}$$

The second sum (4.25) together with the first sum as well as with the corresponding explicitly nonwritten terms for g_j^k and $\bar{g}_j^l$ are cancelled according to (4.21). The integrals along the contour (4.24) and the last one in (4.26) are cancelled, since $\overset{\circ}{X}{}^i(t,\tau)$ and $\sum\limits_{r=1}^{J} \hat{x}_r^i(t,\tau)$ coincide on the boundary of Γ of the domain M. All other considerations are analogous to those applied in the case of an independent variable.

Finally, we consider the dependence of the functional and constraints on the higher order derivatives, up to the nth order inclusive, in the variational problem with fixed boundary values and one independent variable. We can show that the Euler equation for the aggregated problem (4.7) is reduced to the following equations (Lemma 4.1):

$$\sum_{j=1}^{J} \alpha_j^i \left[\frac{\partial \Phi_j}{\partial x_j^i} - \frac{d}{dt}\left(\frac{\partial \Phi_j}{\partial (x_j^i)'}\right) + \frac{d^2}{dt^2}\left(\frac{\partial \Phi_j}{\partial (x_j^i)''}\right) + \ldots + \right.$$
$$\left. +(-1)^n \frac{d^n}{dt^n}\left(\frac{\partial \Phi_j}{\partial (x_j^i)^{(n)}}\right)\right] = 0, \qquad i \in [1:I]. \qquad (4.27)$$

The terms that correspond to $f_j(t, \overset{\circ}{x}_j, \overset{\circ}{x}{}'_j, \ldots, \overset{\circ}{x}{}_j^{(n)})$ and occur in the expression for the marginal value $(\overset{\circ}{g}(0))'$, by analogy to (4.16), have the following form:

$$\int_{t_1}^{t_2} \sum_{j=1}^{J} \sum_{i=1}^{I} \left\{ \frac{\partial f_j}{\partial x_j^i} \left[\hat{x}_j^i - \alpha_j^i \sum_{r=1}^{J} \hat{x}_r^i \right] + \frac{\partial f_j}{\partial (x_j^i)'} \left[\hat{x}_j^i - \alpha_j^i \sum_{r=1}^{J} \hat{x}_r^i \right]' + \right.$$
$$+ \frac{\partial f_j}{\partial (x_j^i)''} \left[\hat{x}_j^i - \alpha_j^i \sum_{r=1}^{J} \hat{x}_r^i \right]'' + \ldots$$
$$\left. \ldots + \frac{\partial f_j}{\partial (x_j^i)^{(n)}} \left[\hat{x}_j^i - \alpha_j^i \sum_{r=1}^{J} \hat{x}_r^i \right]^{(n)} \right\} dt.$$

The latter integral as well as (4.16) is partitioned into two integrals. Finally, we obtain the formula analogous to (4.18):

$$(\overset{\circ}{g}(0))' = \int_{t_1}^{t_2} \sum_{j=1}^{J} \sum_{i=1}^{I} \left[(\hat{x}_j^i - \overset{\circ}{x}{}_j^i) \frac{\partial \Phi_j}{\partial x_j^i} + ((\hat{x}_j^i)' - (\overset{\circ}{x}{}_j^i)') \frac{\partial \Phi_j}{\partial (x_j^i)'} + \right.$$
$$\left. + ((\hat{x}_j^i)'' - (\overset{\circ}{x}{}_j^i)'') \frac{\partial \Phi_j}{\partial (x_j^i)''} + \ldots + ((\hat{x}_j^i)^{(n)} - (\overset{\circ}{x}{}_j^i)^{(n)}) \frac{\partial \Phi_j}{\partial (x_j^i)^{(n)}} \right] dt,$$

where all variables are taken at the point $\overset{\circ}{x}_j, \overset{\circ}{x}{}'_j, \overset{\circ}{x}{}''_j, \ldots, \overset{\circ}{x}{}_j^{(n)})$. The other construction are carried out in the same way as for the case of first order derivatives.

Comments and References to Chapter 2

The decomposition scheme of the iterative aggregation is generalized to the convex block separable problem of mathematical programming, without using the cumbersome linearization and approximation procedures as it was in Dantzig-Wolfe method (described in detail in the book of L.S. Lasdon [5]). Wolfe dual problems in mathematical programming are covered in the book of E.G. Gol'shtein [3]. The same problem for the calculus of variations is considered in the works by L. Bittner [1] and A.D. Ioffe and V.M. Tikhomirov [4]. The book of I.M. Gel'fand and S.V. Fomin [2] is a useful introduction to the classical calculus of variations. This chapter is based on the articles of V.I. Tsurkov [6–9].

References to Chapter 2

[1] Bittner L. , Abschätzungen bei Variationsmethoden mit Hilfe Dualitätssätzen I, *J. Numer. Math.*, 1968, vol.11, no. 2, pp. 129–143.

[2] Gel'fand I.M. and Fomin S.V., *Variatsionnoe ischislenie* (Calculus of Variations), Moscow, Fizmatgiz, 1961.

[3] Gol'shtein E.G., Teoriya dvoistvennosti v matematicheskom programmirovanii i eyo prilozheniya (Duality Theory in Mathematical Programming and Its Applications), Moscow: Nauka, 1971.

[4] Ioffe A.D. and Tikhomirov V.M., *Dvoistvennost' v zadachakh variatsionnogo ischisleniya* (Duality in Problems of Calculus of Variations), *Dokl. Akad. Nauk SSSR*, 1968, vol. 180, no. 4, pp. 789–792.

[5] Lasdon L.S., *Optimizatsiya bol'shikh sistem* (Optimization of Large Systems), Moscow, Nauka: 1975 [Russian translation].

[6] Tsurkov V.I., *Dekompozitsiya na osnove aggregirovaniya dlya funktsional'nykh sistem* (Aggregation-Based Decomposition for Functional Systems), *Avtom. Telemekh.*, 1980, no. 3, pp. 145-155.

[7] Tsurkov V.I., *Dekompozitsiya v klassicheskom variatsionnom ischislenii* (Decomposition in Classical Calculus of Variations), *Zh. Vych. Mat. Mat. Fiz.*, 1979, vol. 19, no. 6, pp. 1396–1413.

[8] Tsurkov V.I., *Dekompozitsiya v vypuklom matematicheskom programmirovanii* (Decomposition in Convex Mathematical Programming), in *Computational and Applied Mathematics*, vol. 28, Kiev, 1979.

[9] Tsurkov V.I., *Printsip decompozitsii dlya blocno-separabel'nykh sistem* (Decomposition Principle for Block Separable Systems), *Dokl. Akad. Nauk SSSR*, 1979, vol. 246, no. 1, pp. 27–31.

Chapter 3

Hierarchical Systems of Mathematical Physics

In this chapter, we consider the application of the iterative decomposition method based on aggregation of variables to block separable optimal control problems. We introduce a special class of hierarchical optimal control problems. These are two-level systems, whose subsystems are models of mathematical physics. First, we give a statement of dynamic separable block problems with ordinary differential equations and present the main constructions of the decomposition method. The decomposition uses the two-level scheme described above; controls are aggregated. After this, we present a series of simple analytical examples that show all the stages of the operation of the algorithm and outline its application for the extremal problems, for which duality principles hold. Then, the method of iterative aggregation is extended to block statements, where subsystems are described by partial differential equations of different types. To justify the method, we use the concept of weak solutions in functional spaces. Lastly, we study the block linear quadratic optimal control problems. We describe an analytical method to decrease dimensions, based on the reducing the systems of linear algebraic equations that are obtained after the application of the maximum principle to the initial problems.

§1. Construction of the Method for Block Separable Problems of Optimal Control

The decomposition method based on aggregation of variables from various blocks was given at length in Chapters 1 and 2. Below, we consider its particular application in block separable optimal control problems.

Let $\mathcal{L}_1[0,T]$ be a Banach space of functions summed in the interval $[0,T]$ and the norm be given in the form $||f||_{\mathcal{L}_1[0,T]} = \int\limits_0^T |f(t)|dt$. We also consider the corresponding space $\mathcal{L}_1^n[0,T]$ of n-dimensional vector functions with the norm

$$||f||_{\mathcal{L}_1^n[0,T]} = \sum_{r=1}^{n} ||f_r||_{\mathcal{L}_1[0,T]}.$$

Let $\mathcal{L}_\infty[0,T]$ be a Banach space of functions summed in the interval with any degree. It coincides with the space of functions measurable and bounded almost everywhere in the interval $[0,T]$. The norm has the following form:

$$||f||_{\mathcal{L}_\infty[0,T]} = \text{ vrai } \max_{0\le t\le T} |f(t)|,$$

where vrai is the so-called true maximum. The space $\mathcal{L}_\infty^n[0,T]$ of vector-functions is introduced similarly.

We consider the problem of optimal control with local and binding constraints and block separable functional of the form

$$f(u) = \sum_{j=1}^{J} w_j(x_j(T)) + \int_0^T \left\{ \sum_{j=1}^{J} c_j(x_j(t), u_j(t), t) \right\} dt \to \max, \tag{1.1}$$

$$dx_j(t)/dt = A_j(t)x_j(t) + b_j(u_j(t), t), u_j(t) \ge 0, \tag{1.2}$$

$$x_j(0) = \kappa_j, \quad v_j(x_j(T)) \le 0, \quad p_j(x_j(t), u_j(t), t) \le 0, \quad j \in [1:J], \tag{1.3}$$

$$\sum_{j=1}^{J} q_j(x_j(T)) \le 0, \qquad \sum_{j=1}^{J} d_j(x_j(t), u_j(t), t) \le 0. \tag{1.4}$$

Here, for each fixed number $j \in [1:J]$, the entering vector functions are of the following dimensions: x_j, b_j, κ_j – N_j, u_j – I, v_j – S_j, q_j – R, p_j – K_j, d_j – L. The matrices $A_j(t)$, $j \in [1:J]$ have dimensions $N_j \times N_j$.

Relations and inequalities in (1.2), (1.3) are block conditions (local constraints for subsystems). Sums in (1.4) give binding constraints of the dynamic system. They as well as functional (1.1) are separable with respect to blocks .

Problem (1.1)–(1.4) is related to optimal control problems with mixed constraints of Boltz type. This problem consists of finding controls $u_j(t) \in \mathcal{L}_\infty^T$, $j \in [1:J]$ and the corresponding (with respect to (1.2)) phase variables $x_j(t)$ that maximize functional (1.1) and satisfy constraints (1.2)–(1.4).

The initial problem has dimension $J \times I$ with respect to controls. We consider the main aspects of applying the decomposition method, using control aggregation following the scheme given at length in Chapters 1 and 2. The decomposition reduces dimensions of control variables, which are iteratively aggregated and disaggregated.

Assume that the following Ter-Krikorov conditions for the input functions hold. The functions $-p_j(x_j, u_j, t)$, $-d_j(x_j, u_j, t)$, $b_j(u_j, t)$,

$c_j(x_j, u_j, t)$, $j \in [1 : J]$ are continuously differentiable in the whole space of variables, convex with respect to x_j, u_j, and increase monotonically with respect to x_j. The functions $-v_j(\beta_j)$, $-q_j(\beta_j)$, $w_j(\beta_j)$, $j \in [1 : J]$ are also continuously differentiable, concave and monotonically increasing with respect to β_j. It is also considered that the bounded measurable components of matrices $A_j(t)$, $j \in [1 : J]$ are equal to or greater than zero (almost everywhere) in the interval $[0, T]$.

Moreover, it is assumed that, for the main problem (1.1)–(1.4) and for all intermediate optimal control problems that are introduced in constructing the decomposition method, the Ter-Krikorov conditions hold, reducing the problems to problems of convex programming in Banach spaces, subject to duality principles, namely the Kuhn-Tucker theorem, and the Lagrange saddle point theorem. For problem (1.1)–(1.4), the assumptions are reduced to satisfying of the Slater condition subject to the constraints of this problem and the inequalities

$$x_j(t) \geq \varepsilon > 0, \qquad p_j(0,0,t) \leq 0, \qquad j \in [1 : J], \qquad \sum_{j=1}^{J} d_j(0,0,t) \leq 0.$$

Note that these conditions entail the Pontryagin maximum principle for problems of optimal control with mixed constraints.

The method of decomposition introduces aggregated controls $U^i(t) = \sum_{j=1}^{J} u_j^i(t)$, $i \in [1 : I]$ and the aggregation weight functions $\alpha_j^i(t) = u_j^i(t)/U^i(t)$, $j \in [1 : J]$, $i \in [1 : I]$. By fixing the weights $\alpha_j^i(t)$ and substituting $u_j^i(t) = \alpha_j^i(t)U^i(t)$ into (1.1)–(1.4), we obtain the problem with unknown aggregated controls

$$g(U) = \sum_{j=1}^{J} w_j(x_j(T)) + \int_0^T C(x_j(t), U^i(t), t)dt \to \max,$$

$$dx_j(t)/dt = A_j(t)x_j(t) + B_j(U^i(t), t), x_j(0) = \kappa_j, v_j(x_j(T)) \leq 0, \qquad (1.5)$$

$$P_j(x_j(t), U^i(t), t) \leq 0, \qquad j \in [1 : J]; U^i(t) \geq 0, \qquad i \in [1 : I],$$

$$Q(x_j(T)) \leq 0, \qquad D(x_j(t), U^i(t), t) \leq 0,$$

where the obvious notation

$$B_j(U^i,t) = b_j(\alpha_j^i U^i,t), \qquad P_j(x_j,U^i,t) = p_j(x_j,\alpha_j U^i,t),$$
$$C(x_j,U^i,t) = \sum_{j=1}^{J} c_j(x_j,\alpha_j^i U^i,t), \qquad Q(x_j(T)) = \sum_{j=1}^{J} q_j(x_j(T)),$$
$$D(x_j,U^i,t) = \sum_{j=1}^{J} d_j(x_j,U^i,t)$$

is introduced.

A solution to the aggregated problem (1.5) is a control vector $\overset{\circ}{U}(t) \in \mathcal{L}_\infty^I[0,T]$. In what follows, we assume that all its components are strictly greater than zero almost everywhere on the interval $[0,T]$.

For problem (1.5), we consider the Wolfe dual problem:

$$\psi = \sum_{j=1}^{J} w_j(x_j(T)) + \int_0^T \Bigg\{ C(x_j,U^i,t) + \sum_{j=1}^{J} \Bigg[B_j(U^i,t) +$$
$$+ A_j x_j - \frac{dx_j(t)}{dt} \Bigg] \chi_j(t) - \sum_{j=1}^{J} P_j(x_j,U^i,t)\eta_j(t) - D(x_j,U^i t)\delta(t) \Bigg\} dt -$$
$$- \sum_{j=1}^{J} [v_j(x_j(T))\omega_j] - Q(x_j(T))\mu \to \min, \qquad (1.6)$$

$$-\frac{d\chi_j(t)}{dt} = A_j^T(\chi_j(t) - \frac{\partial P_j(x_j,U^i,t)}{\partial x_j}\eta_j(t) -$$
$$- \frac{\partial D(x_j,U^i,t)}{\partial x_j}\delta(t) + \frac{\partial C(x_j,U^i,t)}{\partial x_j}, \qquad j \in [1:J], \qquad (1.7)$$

$$-\sum_{j=1}^{J} \frac{\partial B_j(U^i,t)}{\partial U^i}\chi_j(t) + \sum_{j=1}^{J} \frac{\partial P_j(x_j,U^i,t}{\partial U^i}\eta_j(t) +$$
$$+ \frac{\partial D(x_j,U^i,t)}{\partial U^i}\delta(t) = \frac{\partial C(x_j,U^i,t)}{\partial U^i}, \qquad i \in [1:I], \qquad (1.8)$$

$$\chi_j(T) = -\frac{\partial v_j(x_j(T))}{\partial x_j}\omega_j - \frac{\partial q_j(x_j(T))}{\partial x_j}\mu +$$
$$+ \frac{\partial w_j(x_j(T))}{\partial x_j}, \qquad j \in [1:J], \qquad (1.9)$$

$$\eta_j(t) \geq 0, \qquad \omega_j \geq 0, \qquad j \in [1:J], \qquad \delta(t) \geq 0, \qquad \mu \geq 0. \qquad (1.10)$$

Here, vectors and vector functions have dimensions:

$$\omega_j - [S_j], \quad \chi_j - [N_j], \quad \eta_j - [K_j]; \quad j \in [1:J]; \quad \mu - [R], \delta - [L].$$

The dual functions and vectors are considered in the following spaces:

$$\chi_j(t) \in \mathcal{L}_1^{N_j}[0,T], \qquad \eta_j \in \mathcal{L}_1^{K_j}[0,T], \qquad j \in [1:J],$$

where $\mathcal{R}_+^R$ denotes the nonnegative octant in the n-dimensional Euclidean space.

For some fixed weights $\alpha_j^i(t)$, let us find the unique optimal solutions of the pairs of conjugated problems (1.5) and (1.6)–(1.10), respectively, $\overset{\circ}{U}{}^i(t)$, $\overset{\circ}{x}_j(t)$, $\overset{\circ}{\chi}(t)$, $\overset{\circ}{\eta}(t)$, $\overset{\circ}{\delta}(t)$, $\overset{\circ}{\omega}$, $\overset{\circ}{\mu}$. We consider local problems for subsystems (blocks). They fall into J independent problems, where each $j \in [1:J]$ relates to constraints (1.2), (1.3). The terms $-q_j(x_j(T))\,\overset{\circ}{\mu} - \int_0^T d_j(x_j(t), u_j(t), t)\,\overset{\circ}{\delta}(t)dt$ are added to the terms from functional (1.1).

Let $\hat{u}_j(t) \in \mathcal{L}_\infty^I[0,T]$, $j \in [1:J]$ be optimal controls of the local problems and $\overset{\circ}{u}_j(t)$ be disaggregated controls given in the following way: $\overset{\circ}{u}{}_j^i(t) = \alpha_j^i(t)\,\overset{\circ}{U}{}^i(t)$, $j \in [1:J]$, $i \in [1:I]$. In line with the decomposition method, we introduce a function of the weights $\alpha_j^i(t,\rho_j)$ that depend on parameters ρ_j according to the formula

$$\alpha_j^i(t,\rho_j) = \left[\hat{u}_j^i(t) + \rho_j(\hat{u}_j^i(t) - \overset{\circ}{u}{}_j^i(t))\right] \left[\sum_{j=1}^J (\overset{\circ}{u}{}_j^i(t) + \rho_j(\hat{u}_j^i(t) - \overset{\circ}{u}{}_j^i(t)))\right]^{-1},$$

$$j \in [1:J], \qquad i \in [1:I], 0 \leq \rho_j \leq 1. \qquad (1.11)$$

We then consider problems with aggregated variables (1.5) having weights (1.11). In this case, the optimal value of the functional of problem (1.5) is the function of parameters ρ_j, $j \in [1:J]$. We denote this function by $\overset{\circ}{g}(\rho_j)$ and state the problem about maximizing $\overset{\circ}{g}(\rho_j)$ in the unit cube $0 \leq \rho_j \leq 1$, $j \in [1:J]$. Let the maximum be attained for some values $\overset{\circ}{\rho}_j$. Then, the weights for the following step of the iterative process are computed by the formula (1.11), where $\overset{\circ}{\rho}_j$ are substituted for ρ_j. In particular cases, the maximization $\overset{\circ}{g}(\rho_j)$ can be replaced by a one-dimensional maximization $\overset{\circ}{g}(\rho)$, where $\rho_j = \rho$, $j \in [1:J]$.

Thus, we formally reduce the problems' dimensions: the main problem has $J \times I$ unknown controls; the aggregated problem has I unknown controls, and the local problem has I unknown controls each. Finally, the problem of maximization of the function $\overset{\circ}{g}(\rho_j)$ in the unit cube includes J variables or one variable.

Optimizing the feasible solution $\overset{\circ}{u}_j(t)$, $\overset{\circ}{x}_j$ of problem (1.1)–(1.4) (the criterion of termination of the iterative process) requires satisfaction of the equality

$$\sum_{j=1}^{J}\left[w_j(\hat{x}_j(T)) - q_j(\hat{x}_j(T))\,\overset{\circ}{\mu}\right] + \int_0^T \left[\sum_{j=1}^{J} c_j(\hat{x}_j, \hat{u}_j, t) - d_j(\hat{x}_j, \hat{u}_j, t)\,\overset{\circ}{\delta}(t)\right] dt -$$

$$-\sum_{j=1}^{J} w_j(\overset{\circ}{x}_j(T)) - \int_0^T \left[\sum_{j=1}^{J} c_j(\overset{\circ}{x}_j, \overset{\circ}{u}_j, t)\right] dt = 0. \qquad (1.12)$$

If problem (1.1)–(1.4) has a solution and the solution $\overset{\circ}{u}_j(t)$, $\overset{\circ}{x}_j(t)$ is not optimal for it, then the left-hand side of (1.12) is strictly greater than zero.

Criterion (1.12) follows from proof for the problems of mathematical programming in Chapter 2. Let $\hat{\lambda}_j(t)$, $\hat{\zeta}_j(t)$, $\hat{\omega}_j$, $j \in [1 : J]$ be extremal solutions of the dual local problems. Then, the set $\hat{\lambda}_j(t)$, $\hat{\zeta}_j(t)$, $\hat{\omega}_j$, $\overset{\circ}{\mu}$ together with $\hat{u}_j(t)$, $\hat{x}_j(t)$ is a feasible solution to the dual of the source problem (1.1)–(1.4). The following inequality holds for the values of the functionals of feasible solutions to the pairs of conjugated problems

$$\sum_{j=1}^{J} w_j(\hat{x}_j(T)) + \int_0^T \Bigg\{ \sum_{j=1}^{J} c_j(\hat{x}_j, \hat{u}_j, t) + \Bigg[b_j(\hat{u}_j, t) -$$

$$- A_j(t)\hat{x}_j(t) - \frac{d\hat{x}_j(t)}{dt}\Bigg] \hat{\lambda}_j(t) - p_j(\hat{x}_j, \hat{u}_j, t)\hat{\zeta}_j(t) -$$

$$- d_j(\hat{x}_j, \hat{u}_j, t)\,\overset{\circ}{\delta}(t)\Bigg\} dt - \sum_{j=1}^{J}\left[v_j(\hat{x}_j(T))\hat{\omega}_j + q_j(\hat{x}_j(T))\,\overset{\circ}{\mu}\right] \geq$$

$$\geq \sum_{j=1}^{J} w_j(\overset{\circ}{x}_j(T)) + \int_0^T \left[\sum_{j=1}^{J} c_j(\overset{\circ}{x}_j, \overset{\circ}{x}_j, \overset{\circ}{u}_j, t)\right] dt. \qquad (1.13)$$

The term in square brackets under the integral sign in the left-hand side of (1.13) is equal to zero according to (1.2). The third term in this integral and the first term in the last sum of (1.13) are equal to zero because of the conditions of complementary slackness (the Kuhn-Tucker theorem as applied to block problems). Thus, relation (6.13) implies (6.12).

The proof of monotonicity with respect to the functional of the iterative process uses the derivative $(\overset{\circ}{g}(0))'$ with due regard to $\partial\alpha_j^i(t,0)/\partial\rho = (\hat{u}_j(t) - \alpha_j^i(t)\sum_{r=1}^{J}\hat{u}_r^i(t))$. Here, we use the theorem on the marginal value and take into account the dependence of optimal solutions on $\overset{\circ}{x}_j(t)$. Thus, we have

$$(\overset{\circ}{g}(0))' = \sum_{j=1}^{J}\frac{\partial w_j}{\partial x_j(T)}\frac{\partial \overset{\circ}{x}_j(T,))}{\partial\rho} + \int_0^T\left\{\sum_{j=1}^{J}\left[\frac{\partial c_j}{\partial x_j}\frac{\partial\overset{\circ}{x}_j(t,0)}{\partial\rho} + \frac{\partial c_j}{\partial u_j}(\hat{u}_j -\right.\right.$$
$$-\alpha_j\sum_{r=1}^{J}\hat{u}_r) + \frac{\partial b_j}{\partial u_j}(\hat{u}_j - \alpha_j\sum_{r=1}^{J}\hat{u}_r)\overset{\circ}{\chi}_j +$$
$$\left.\left. + \left(A_j(t)\frac{\partial\overset{\circ}{x}_j(t,0)}{\partial\rho}\overset{\circ}{\chi}_j\right)\right]\right\}dt - \frac{\partial}{\partial\rho}\left[\int_0^T\frac{d\overset{\circ}{x}_j(t,\rho)}{dt}\overset{\circ}{\chi}_j dt\right]_{\rho=0} -$$
$$-\int_0^T\sum_{j=1}^{J}\left[\frac{\partial p_j}{\partial x_j}\frac{\partial\overset{\circ}{x}_j(t,0)}{\partial\rho} + \frac{\partial p_j}{\partial u_j}(\hat{u}_j - \alpha_j\sum_{r=1}^{J}\hat{u}_r)\overset{\circ}{\eta}_j(t)\right] +$$
$$\left. + \left[\frac{\partial d_j}{\partial x_j}\frac{\partial\overset{\circ}{x}_j(t,0)}{\partial\rho} + \frac{\partial d_j}{\partial u_j}(\hat{u}_j - \alpha_j\sum_{r=1}^{J}\hat{u}_r)\overset{\circ}{\delta}(t)\right]\right\}dt -$$
$$-\sum_{j=1}^{J}\left[\frac{\partial v_j}{\partial x_j(T)}\frac{\partial\overset{\circ}{x}_j(T,0)}{\partial\rho}\overset{\circ}{\omega}_j + \frac{\partial q_j}{\partial x_j(T)}\frac{\partial\overset{\circ}{x}_j(T,0)}{\partial\rho}\overset{\circ}{\mu}\right]. \tag{1.14}$$

The third term in the right-hand side of (1.14) takes the following form:

$$-\frac{\partial}{\partial\rho}\left[\int_0^T\frac{d\overset{\circ}{x}_j(t,\rho)}{dt}\overset{\circ}{\chi}_j(t)dt\right]_{\rho=0} =$$
$$= \frac{\partial\overset{\circ}{x}_j(T,0)}{\partial\rho}\overset{\circ}{\chi}_j(T) + \int_0^T\frac{\partial\overset{\circ}{x}_j(T,0)}{\partial\rho}\frac{d\overset{\circ}{\chi}_j(t)}{dt}dt. \tag{1.15}$$

The terms in (1.14) that contain the sum $\sum_{r=1}^{J}\hat{u}_r$ are cancelled due to (1.8). The integrals in (1.14), (1.15), where derivatives $\partial\overset{\circ}{x}_j(t,0)/\partial\rho$ occur are cancelled due to (1.7). We replace the values $\chi_i(T)$ by their expressions in relation (1.15) according to (1.9). Therewith, the first and the last sums in (1.14) are cancelled. Thus, we obtain

$$(\overset{\circ}{g}(0))' = \int_0^T \left\{ \sum_{j=1}^J \left(\frac{\partial c_j}{\partial u_j} \hat{u}_j + \frac{\partial b_j}{\partial u_j} \hat{u}_j \overset{\circ}{\chi}_j - \frac{\partial p_j}{\partial u_j} \hat{u}_j \overset{\circ}{\eta}_j - \frac{\partial d_j}{\partial u_j} \hat{u}_j \overset{\circ}{\delta} \right) \right\} dt. \quad (1.16)$$

Here, the partial derivatives are taken for the solution $\overset{\circ}{u}_j(t)$, $\overset{\circ}{x}_j(t)$.

Then, all terms are taken to the right-hand side, multiplied by $\overset{\circ}{U}(t)$, summed over i from 0 to I, and integrated from 0 to T. We add the obtained expression, which is equal to zero, to the right-hand side of (1.16) and finally obtain

$$(\overset{\circ}{g}(0))' = \int_0^T \left\{ \sum_{j=1}^J \left[\frac{\partial c_j}{\partial u_j} (\hat{u}_j - \overset{\circ}{u}_j) + \frac{\partial b_j}{\partial u_j} (\hat{u}_j - \overset{\circ}{u}_j) \overset{\circ}{\chi}_j - \right.\right.$$
$$\left.\left. - \frac{\partial p_j}{\partial u_j} (\hat{u}_j - \overset{\circ}{u}_j) \overset{\circ}{\eta}_j - \frac{\partial d_j}{\partial u_j} (\hat{u}_j - \overset{\circ}{u}_j) \overset{\circ}{\delta} \right] \right\} dt. \quad (1.17)$$

Relations (1.7) are multiplied by $(\hat{x}_j(t) - \overset{\circ}{x}_j(t))$. We sum them over $n \in [1 : N_j]$ and $j \in [1 : J]$ and integrate with respect to the differential connections (1.2), (1.5). Taking into account the initial values in (1.3), (1.5), we obtain the equality

$$\int_0^T \left\{ \sum_{j=1}^J \left[\frac{\partial c_j}{\partial x_j} (\hat{x}_j - \overset{\circ}{x}_j) - \frac{\partial p_j}{\partial x_j} (\hat{x}_j - \overset{\circ}{x}_j) \overset{\circ}{\eta}_j - \right.\right.$$
$$\left.\left. - \frac{\partial d_j}{\partial x_j} (\hat{x}_j - \overset{\circ}{x}_j) \overset{\circ}{\delta} + (b_j(\overset{\circ}{u}_j, t) - b_j(\hat{u}_j, t)) \overset{\circ}{\chi}_j \right] \right\} dt +$$
$$\sum_{j=1}^J (\hat{x}_j(T) - \overset{\circ}{x}_j(T)) \overset{\circ}{\chi}(T) = 0. \quad (1.18)$$

The left-hand side of (1.18) is added to the right-hand side of (1.17). Here the values $\chi_j(T)$ are replaced by their values according to (1.9). Moreover, the values

$$\int_0^T \left[\sum_{j=1}^J c_j(\hat{x}_j, \hat{u}_j, t) \right] dt, \qquad \int_0^T \left[\sum_{j=1}^J c_j(\overset{\circ}{x}_j, \overset{\circ}{u}_j, t) \right] dt,$$

$$\int_0^T \left[\sum_{j=1}^J p_j(\overset{\circ}{x}_j, \overset{\circ}{u}_j, t) \right] dt,$$

$$\int_0^T \left[\sum_{j=1}^J d_j(\hat{x}_j, \hat{u}_j, t) \right] dt, \qquad \sum_{j=1}^J v_j v_j(\hat{x}_j(T)) \, \overset{\circ}{\omega}_j,$$

$$\sum_{j=1}^J q_j(\hat{x}_j(T)) \, \overset{\circ}{\mu}, \qquad \sum_{j=1}^J \left[w_j(\hat{x}_j(T) - w_j(\overset{\circ}{x}_j(T)) \right]$$

are added to and subtracted from (1.17). We add the following terms to (1.17), which are equal to zero due to the conditions of complementary slackness:

$$\int_0^T \left[\sum_{j=1}^J p_j(\overset{\circ}{x}_j, \overset{\circ}{u}_j, t) \, \overset{\circ}{\eta}_j \right] dt, \qquad \int_0^T \left[\sum_{j=1}^J d_j(\overset{\circ}{x}_j, \overset{\circ}{u}_j, t) \, \overset{\circ}{\delta} \right] dt,$$

$$\sum_{j=1}^J v_j(\overset{\circ}{x}_j(T)) \, \overset{\circ}{\omega}_j, \qquad \sum_{j=1}^J q_j(\overset{\circ}{x}_j(T)) \, \overset{\circ}{\mu}_j.$$

The final expression for the marginal value looks as follows:

$$(\overset{\circ}{g}(0))' = \pi_1 + \pi_2 + \pi_3 + \pi_4 + \pi_5, \tag{1.19}$$

where

$\pi_1 > 0$ – is a left-hand side of (1.12);

$$\pi_2 = - \int_0^T \left[\sum_{j=1}^J p_j(\hat{x}_j, \hat{u}_j, t) \, \overset{\circ}{\eta}_j \right] dt,$$

$$\pi_3 = - \sum_{j=1}^J v_j(\hat{x}_j(T)) \, \overset{\circ}{\omega}_j,$$

$$\pi_4 = \int_0^T \left\{ \sum_{j=1}^J Z(\overset{\circ}{x}_j, \overset{\circ}{u}_j, \hat{x}_j, \hat{u}_j, t) \left[c_j + b_j \, \overset{\circ}{\chi}_j(t) - p_j \, \overset{\circ}{\eta}_j(t) - d_j \, \overset{\circ}{\delta}(t) \right] \right\} dt,$$

$$\pi_5 = \sum_{j=1}^J Z(\overset{\circ}{x}_j(T), \hat{x}_j(T)) \left[w_j - v_j \, \overset{\circ}{\omega}_j - q_j \, \overset{\circ}{\mu} \right].$$

(1.20)

where the concavity operators

$$Z(\overset{\circ}{x}_j, \overset{\circ}{u}_j, \hat{x}_j, \hat{u}_j, t)\left[f_j\theta(t)\right] = [f_j(\overset{\circ}{x}_j, \overset{\circ}{u}_j, t) +$$

$$+\sum_{n=1}^{N_j} \frac{\partial f_j}{\partial x_j^n}(\overset{\circ}{x}_j, \overset{\circ}{u}_j, t)(\hat{x}_j^n - \overset{\circ}{x}{}_j^n) +$$

$$\left. +\sum_{i=1}^{I} \frac{\partial f_j}{\partial u_j^i}(\overset{\circ}{x}_j, \overset{\circ}{u}_j, t)(\hat{u}_j^i - \overset{\circ}{u}{}_j^i) - f_j(\hat{x}_j, \hat{u}_j, t)\right] \theta(t),$$

$$Z(\overset{\circ}{x}_j(T), \hat{x}_j(T))[f_j\theta] =$$

$$= \left[f_j(\overset{\circ}{x}_j(T)) + \sum_{n=1}^{N_j} \frac{\partial f_j(\overset{\circ}{x}_j(T))}{\partial x_j^n(T)}(\hat{x}_j^n(T) - \overset{\circ}{x}{}_j^n(T)) - f_j(\hat{x}_j(T))\right] \theta$$

are introduced.

We analyze (1.19) and (1.20). The values π_2 and π_3 are greater than or equal to zero due to inequalities in (1.3), (1.10). The values π_4 and π_5 are greater than or equal to zero due to the supposed convexity properties of the input functions and due to (1.10). It is important to point out that the variables $\overset{\circ}{\chi}(t)$ are greater or equal to zero, since they are solutions of equations (1.7) with right boundary conditions (1.9), where $\overset{\circ}{\chi}_j(T) \geq 0$ and $d\overset{\circ}{\chi}_j(t)/dt \leq 0$, $j \in [1 : J]$ hold according to the assumed monotonicity properties of the input functions as well as $A_j(t) \geq 0$.

Consider the optimal control problem with discrete time.

Let l_1 be a Banach space of real sequences summed over the absolute value with the norm

$$||x||_{l_1} = \sum_{i=1}^{\infty} |x_i|.$$

Let l_∞ be a Banach space of real sequences summed over the absolute value with arbitrary degree $p \geq 1$, where the norm is given in the form

$$||x||_{l_\infty} = \max_i |x_i|.$$

By l_1^n, we denote the Banach space of sequences of n-dimensional vectors such that all sequences of k coordinates belong to the space l_1. The norm in l_1^n is defined as follows:

$$||x||_{l_1^n} = \sum_{k=1}^{n} ||x^k||_{l_1}.$$

The space l_∞^n is introduced similarly.

Consider the block separable problem of optimal control with discrete time on the infinite interval:

$$f(u) = \sum_{t=0}^{\infty}\sum_{j=1}^{J} c_j(x_j(t), u_j(t), t) q_j(t) \to \max, \tag{1.21}$$

$$q_j(t) \geq 0, \qquad \sum_{t=0}^{\infty} q_j(t) < \infty, \qquad j \in [1:J],$$

$$x_j(t+1) = A_j(t)x_j(t) + b_j(u_j(t), t), \qquad x_j(0) = \kappa_j, \qquad u_j(t) \geq 0, \tag{1.22}$$

$$p_j(x_j(t), u_j(t), t) \leq 0, \qquad j \in [1:J], \tag{1.23}$$

$$\sum_{j=1}^{J} d_j(x_j(t), u_j(t), t) \leq 0, \qquad t = 0, 1, \ldots. \tag{1.24}$$

Here, the vector functions have the same dimensions as the corresponding values in (1.1)–(1.4). A solution of problem (1.21)–(1.24) fines the sequence of controls $u_j(t) \in l_\infty^l$, $j \in [1:J]$ that satisfy (1.22)–(1.24) and maximize functional (1.21).

Discrete optimal control problems of the type discussed were studied by A.M. Ter-Krikorov without assumption of block separableness. Our goal is to show the applicability of the decomposition method based on aggregation of variables for problems of this class.

We assume that the following Ter-Krikorov conditions hold for the input functions. The vector functions $c_j(\beta, \gamma, t)$, $b_j(\gamma, t)$, $p_j(\beta, \gamma, t)$, $d_j(\beta, \gamma, t)$ are continuously differentiable in the whole space of variables and for every $t = 0, 1, \ldots$; moreover, they are convex and monotonically increasing with respect to β. Partial derivatives of these functions are uniformly bounded and equicontinuous on every bounded set in $\mathbf{R}^{N_j} \times \mathbf{R}^I$, $j \in [1:J]$. The norms of matrices $A_j(t)$ satisfy the inequalities $||A_j(t)|| \leq \varepsilon < 1$, $t = 0, 1, \ldots$, $j \in [1:J]$, and their components are greater than or equal to zero. Moreover, it is assumed that the Slater condition holds for the initial problem and all intermediate problems of the iterative method. Thus, discrete optimal control problems are reduced to the problems of convex programming in Banach spaces, and the duality theorems hold.

The decomposition is accomplished using the scheme described above. We discuss the main aspects of the iterative process for the case considered.

The problem with aggregated controls has the following form:

$$g(U) = \sum_{t=0}^{\infty} C(x_j(t), U^i(t), t) \to \max, \tag{1.25}$$

$$x_j(t+1) = A_j(t)x_j(t) + B_j(U^i(t), t), \qquad x_j(0) = \kappa_j,$$

$$P_j(x_j(t), U^i(t), t) \le 0, \qquad j \in [1:J], \tag{1.26}$$

$$D(x_j(t), U^i(t), t) \le 0, \qquad t = 0, 1, \dots, \tag{1.27}$$

where the notation is as introduced earlier. Moreover, for the sequence of optimal controls $\overset{\circ}{U}{}^i \in l^l_\infty$ of problem (8.5)–(8.7), it is assumed that $\overset{\circ}{U}{}^l(t) > 0$, $i \in [1:I]$, $t = 0, 1, \dots$.

We consider the Wolfe dual problem for problem (8.5)–(8.7):

$$\psi(\chi_j, \eta_j, \delta) = \sum_{t=0}^{\infty} \Bigg\{ \sum_{j=1}^{J} c_j(x_j, U^i, t) q_j(t) -$$

$$- \left[B_j(x_j, U^i, t) + A_j(t)x_j(t) - x_j(t+1)\right] \chi_j(t+1) -$$

$$- \sum_{j=1}^{J} P_j(x_j, U^i, t)\eta_j(t) - D(x_j, U^i(t), \delta(t) \Bigg\} \min, \tag{1.28}$$

$$\chi_j(t) = A_j^T(t)\chi_j(t+1) + \frac{\partial P_j(x_j, U^i, t)}{\partial x_j}\eta_j(t) +$$

$$+ \frac{\partial D(x_j, U^i, t)}{\partial x_j}\delta(t) - \frac{\partial C(x_j, U^i, t)}{\partial x_j}, \qquad j \in [1:J], \tag{1.29}$$

$$\eta_j(t) \ge 0, \qquad j \in [1:J], \delta(t) \ge 0, \tag{1.30}$$

$$\lim_{t\to\infty} \chi_j(t) = 0, \qquad j \in [1:J], \qquad \eta_j(t) \in l_1^{k_j}, \qquad \delta(t) \in l_1^L,$$

$$\sum_{j=1}^{J} \frac{\partial B_j(U^i, t)}{\partial U^i}\chi_j(t+1) + \sum_{j=1}^{J} \frac{\partial P_j(x_j, U^i, t)}{\partial U^i}\eta_j(t) +$$

$$+ \frac{\partial D(x, U^i, t)}{\partial U^i}\delta(t) = \frac{\partial C(x_j, U^i, t)}{\partial U^i}, \qquad t = 0, 1, \dots. \tag{1.31}$$

Assume that, for fixed weights $\alpha_j^i(t)$, $t = 0, 1, \ldots$, we have unique extremal solutions $\{\overset{\circ}{U}{}^i(t), \overset{\circ}{x}_j(t)\}$ and $\{\overset{\circ}{\chi}_j(t), \overset{\circ}{\eta}_j(t), \overset{\circ}{\delta}(t)\}$ of the pair of conjugated problems (1.25)–(1.27) and (1.28)–(1.31). The local problems fall into J problems, where, for each $j \in [1 : J]$, we have conditions (1.22)–(1.23), and the terms

$$-\sum_{t=0}^{\infty} d_j(x_j(t), u_j(t), t)\, \overset{\circ}{\delta}(t)$$

are added to the functionals.

If $\hat{u}_j(t)$, $t = 0, 1, \ldots$, $j \in [1 : J]$ are optimal solutions of local problems, and $\overset{\circ}{u}_j(t)$, $t = 0, 1, \ldots$, $j \in [1 : J]$ are disaggregated controls, then the optimality criterion of the solution $\{\overset{\circ}{u}_j(t), \overset{\circ}{x}_j(t)\}$, $j \in [1 : J]$ for the initial problem (1.21)–(1.24) takes the following form:

$$\sum_{t=0}^{\infty}\sum_{j=1}^{J} \Big[c_j(\hat{x}_j, \hat{u}_j, t) q_j(t) - d_j(\hat{x}_j, \hat{u}_j, t)\, \overset{\circ}{\delta}(t) - \\ - c_j(\overset{\circ}{x}_j, \overset{\circ}{u}_j, t) q_j(t)\Big] = 0. \qquad (1.32)$$

Local monotonicity for the functional of the iterative process is derived from the computation of the marginal value for the functional of the macroproblem (1.25)–(1.27) with weight functions of the form (1.11). By analogy with (1.14), we have

$$(\overset{\circ}{g}(0))' = \\ = \sum_{t=0}^{\infty}\sum_{j=1}^{J} \Bigg\{ \left[\frac{\partial c_j}{\partial x_j}\frac{\partial \overset{\circ}{x}_j(t,0)}{\partial \rho} + \frac{\partial c_j}{\partial u_j}\left(\hat{u}_j - \alpha_j \sum_{r=1}^{J} \hat{u}_r\right)\right] q_j(t) - \\ - \frac{\partial b_j}{\partial u_j}\left(\hat{u}_j - \alpha_j \sum_{r=1}^{J} \hat{u}_r\right) \overset{\circ}{\chi}_j(t+1) - \\ - A_j(t)\frac{\partial \overset{\circ}{x}_j(t,0)}{\partial \rho}\overset{\circ}{\chi}_j(t+1) + \frac{\partial \overset{\circ}{x}_j(t+1,0)}{\partial \rho}\overset{\circ}{\chi}_j(t+1) - \\ - \left[\frac{\partial p_j}{\partial u_j}\left(\hat{u}_j - \alpha_j \sum_{r=1}^{J} \hat{u}_r\right) + \frac{\partial p_j}{\partial x_j}\frac{\partial \overset{\circ}{x}_j(t,0)}{\partial \rho}\right]\overset{\circ}{\eta}_j(t) - \\ - \left[\frac{\partial d_j}{\partial u_j}\left(\hat{u}_j - \alpha_j \sum_{r=1}^{J} \hat{u}_r\right) + \frac{\partial d_j}{\partial x_j}\frac{\partial \overset{\circ}{x}_j(t,0)}{\partial \rho}\right]\overset{\circ}{\delta}(t). \qquad (1.33)$$

According to (1.31), the terms containing $\sum_{r=1}^{J} \hat{u}_r$ are cancelled in the right-hand side of (1.33). Since $\partial x_j(0,0)/\partial\rho = 0$, due to the initial condition, we have the following identity:

$$\sum_{t=0}^{\infty}\sum_{j=1}^{J} \frac{\partial \overset{\circ}{x}_j(t+1,0)}{\partial\rho} \overset{\circ}{\chi}_j(t+1) = \sum_{t=0}^{\infty}\sum_{j=1}^{J} \frac{\partial \overset{\circ}{x}_j(t,0)}{\partial\rho} \overset{\circ}{\chi}_j(t).$$

Therefore, taking account of (1.29), the terms with $\partial \overset{\circ}{x}_j(y,0)/\partial\rho$ cancel in the right-hand side of (1.33). Thus, we obtain (1.33)

$$\overset{\circ}{g}(0))' = \sum_{t=0}^{\infty}\sum_{j=1}^{J}\left[\frac{\partial c_j}{\partial u_j}\hat{u}_j q_j(t) - \frac{\partial b_j}{\partial u_j}\hat{u}_j \overset{\circ}{\chi}_j(t+1) - \right.$$
$$\left. -\frac{\partial p_j}{\partial u_j}\hat{u}_j \overset{\circ}{\eta}_j(t) - \frac{\partial d_j}{\partial u_j}\hat{u}_j \overset{\circ}{\delta}(t)\right].$$

Furthermore, we deduce the equalities

$$\sum_{t=0}^{\infty}\sum_{j=1}^{J}\left[\frac{\partial c_j}{\partial u_j}\overset{\circ}{u}_j q_j(t) - \frac{\partial b_j}{\partial u_j}\overset{\circ}{u}_j \overset{\circ}{\chi}_j(t+1) - \frac{\partial p_j}{\partial u_j}\overset{\circ}{u}_j \overset{\circ}{\eta}_j(t) - \frac{\partial d_j}{\partial u_j}\overset{\circ}{u}_j \overset{\circ}{\delta}(t)\right] = 0,$$

$$\sum_{t=0}^{\infty}\sum_{j=1}^{J}\left[\frac{\partial c_j}{\partial u_j}(\hat{x}_j - \overset{\circ}{x}_j)q_j(t) - \frac{\partial p_j}{\partial x_j}(\hat{x}_j - \overset{\circ}{x}_j)\overset{\circ}{\eta}_j - \frac{\partial d_j}{\partial x_j}(\hat{x}_j - \overset{\circ}{x}_j)\overset{\circ}{\delta}(t) + \right.$$
$$\left. +(b_j(\hat{u}_j(t),t) - b_j(\overset{\circ}{u}_j(t),t)\overset{\circ}{\chi}_j(t+1)\right] = 0$$

in the usual way.

Due to the conditions of complementary slackness for problem (1.25)–(1.27), the following expressions vanish:

$$\sum_{t=0}^{\infty} p_j(\overset{\circ}{x}_j, \overset{\circ}{u}_j, t)\,\overset{\circ}{\eta}_j(t), \qquad \sum_{t=0}^{\infty} d_j(\overset{\circ}{x}_j, \overset{\circ}{u}_j, t)\,\overset{\circ}{\delta}(t).$$

Finally, we obtain the expression for the derivative

$$(\overset{\circ}{g}(0))' = \pi_1 + \pi_2 + \pi_3,$$

where the value π_1 is the left-hand side of (8.12), which is strictly greater than zero, if the solution $\{\overset{\circ}{u}_j(t), \overset{\circ}{x}_j(t)\}$, $t = 0, 1, \ldots$, $j \in [1 : J]$ is not optimal for the initial problem (8.1), (8.4); the value π_2 is equal to

$$-\sum_{t=0}^{\infty}\sum_{j=1}^{J} p_j(\hat{x}_j, \hat{u}_j, t)\, \mathring{\eta}_j(t)$$

or greater than or equal to zero according to (8.3), (8.10). Finally, for π_3, we have

$$\pi_3 = \sum_{t=0}^{\infty}\sum_{j=1}^{J} Z(\mathring{x}_j, \mathring{u}_j, \hat{x}_j, \hat{u}_j, t)\left[c_j q_j(t) - b_j \mathring{\chi}_j(t+1) - p_j \mathring{\eta}_j(t) - d_j \mathring{\delta}(t)\right].$$

The value π_3 is greater than or equal to zero due to the assumed properties of the convexity of the input functions and relation (1.30). It is important to note that the inequality $\mathring{\chi}_j \geq 0$, $t = 0, 1, \ldots$, $j \in [1 : J]$ holds, since the values $\mathring{\chi}_j(t)$ satisfy the relations (1.29) and the condition at infinity (1.30), the input functions are monotone, and the components of the matrix $A_j(t)$ are nonnegative.

§2. Analytical Examples

In this section, we present the simplest statements of hierarchical problems of optimal control, in which all the intermediary problems of decomposition using aggregation of variables have analytical solutions. Some considered statements do not belong to the class of block separable problems of optimal control stated in the previous section. However, the duality principles that underlie the decomposition method hold for them. Thus, these models show all the stages of the iterative algorithm as well as illustrating its use for the broad class of block separable optimal control problems.

In the analysis of the after effect of an explosion, the optimal control problem for harmonic oscillators is studied. This problem is formulated in the following form:

$$x(T) \to \max, \qquad m d^2 x(t)/dt^2 + kx(t) = u(t), \qquad 0 \leq u(t) \leq w,$$

where m is the mass of oscillator, k is its rigidity, $u(t)$ is a control function, T is finite time. For $t = 0$, the oscillator is in the origin and is motionless.

Let us have a system that consists, for simplicity, of two harmonic oscillators. We state the following problem of optimal control:

$$f = c_1 x_1(T) + c_2 x_2(T) \to \max,$$

$$dx_1(t)/dt = z_1(t), \qquad dx_2(t)/dt = z_2(t),$$
$$dz_1(t)/dt = -\omega_1^2 x_1(t) + u_1(t)/m_1, \quad dz_2(t)/dt = -\omega_2^2 x_2(t) + u_2(t)/m_2, \tag{2.1}$$
$$0 \le u_1(t) \le w_1, \qquad 0 \le u_2(t) \le w_2, \qquad u_1(t) + u_2(t) \le w,$$

where ω_1^2, ω_2^2, w_1, w_2, w, c_1, c_2 are positive constants. In problem (2.1), there is a pair of local conditions for subsystems. There is also a binding constraint on the controls $u_1(t)$, $u_2(t)$, that unites these two oscillators in one system.

First, we solve problem (2.1) by using directly the Pontryagin maximum principle, then we construct the solution by means of the iterative decomposition method. The Hamiltonian of problem (2.1) is

$$H = \varphi_1 z_1 + \psi_1(-\omega_1^2 x_1 + u_1/m_1) + \varphi_2 z_2 + \psi_2(-\omega_2^2 x_2 + u_2/m_2),$$

and the conjugated variables satisfy the following equations and boundary conditions:

$$d\varphi_j(t)/dt = \omega_j^2 \psi_j(t), \qquad d\psi_j(t)/dt = -\varphi_j(t),$$
$$\varphi_j(T) = c_j, \qquad \psi_j(T) = 0, \qquad j = 1, 2. \tag{2.2}$$

Thus, after integrating (2.2), we obtain

$$\varphi_j(t) = c_j \cos \omega_j (T - t),$$
$$\psi_j(t) = (c_j/\omega_j) \sin \omega_j (T - t), \qquad j = 1, 2.$$

According to the maximum principle, we come to the following problem:

$$[(c_1/m_1\omega_1) \sin \omega_1 (T - t)]\, u_1(t) +$$
$$+ [(c_2/m_2\omega_2) \sin \omega_2 (T - t)]\, u_2(t) \to \max, \tag{2.3}$$
$$0 \le u_1(t) \le w_1, \qquad 0 \le u_2(t) \le w_2, \qquad u_1(t) + u_2(t) \le w.$$

Assume that $w_1 > w$, $w_2 > w$, then, the second inequalities in the local constraints (2.3) are inessential. We will also consider

$$(c_1 \sin \omega_1 T)/(m_1 \omega_1) > (c_2 \sin \omega_2 T)/(m_2 \omega_2);$$

$$\omega_1 T, \omega_2 T \le \pi/2, \qquad c_1/m_1 \le c_2/m_2.$$

For small t, the first coefficient for control of the functional of problem (2.3) is greater than the second coefficient. For t close to T, we have $\sin\omega_j(T-t) \sim w_j(T-t)$, $j = 1, 2$, and the first coefficient is less than the second. Thus, there is a switching point t^*, which is the root of the equation

$$c_1 \sin\omega_1(T-t)/(m_1\omega_1) = c_2 \sin\omega_2(T-t)/(m_2\omega_2),$$

and the optimal control of problem (2.1) is represented in the form

$$u_1(t) = w, \qquad t \in [0, t^*), \qquad u_1(t) = 0, \qquad t \in (t^*, T],$$

$$u_2(t) = 0, \qquad t \in [0, t^*), \qquad u_2(t) = w, \qquad t \in (t^*, T].$$

By integrating ordinary differential equations in (2.1) for discontinuous optimal control, we obtain

$$\begin{aligned} \overset{*}{x}_1(t) &= w\left[\cos\omega_1(t-t^*) - \cos\omega_1 t\right]/(m_1\omega_1^2); \\ \overset{*}{x}_2(t) &= w\left[1 - \cos\omega_2(t-t^*)\right]/(m_2\omega_2^2); \end{aligned} \tag{2.4}$$

and the optimal value of the functional of problem (2.1) takes the form

$$\begin{aligned} f^* = c_1 w\left[\cos\omega_1(T-t^*) - \cos\omega_1 T\right]/(m_1\omega_1^2) + \\ + c_2 w\left[1 - \cos\omega_2(T-t^*)\right]/(m_2\omega_2^2). \end{aligned} \tag{2.5}$$

The decomposition method assumes consideration of the aggregated problem, which, in this case, has the following form:

$$\begin{aligned} &g(U) = c_1 x_1(T) + c_2 x_2(T) \to \max, \\ &dx_1(t)/dt = z_1(t), \qquad dx_2(t)/dt = z_2(t), \\ &dz_1(t)/dt = -\omega_1^2 x_1(t) + \alpha_1 U_1(t)/m_1, \\ &dz_2(t)/dt = -\omega_2^2 x_2(t) + \alpha_2(t) U(t)/m_2, \\ &U(t) \leq w, \end{aligned} \tag{2.6}$$

where nonnegative aggregation weights $\alpha_1(t)$, $\alpha_2(t)$ were given, and $\alpha_1(t) + \alpha_2(t) = 1$.

The dual variables φ_j, ψ_j, $j = 1, 2$ of problem (2.6) coincide with analogous variables for the initial problem (2.1). For all weights, the optimal control in (2.6) takes the form $U(t) = w$. The dual function $\overset{\circ}{\delta}(t)$ for the constraint $U(t) \leq w$ is computed from the condition $\overset{\circ}{\delta}(t) = \psi_1\alpha_1/m_1 + \psi_2\alpha_2/m_2$ and is equal to

$$\overset{\circ}{\delta}(t) = c_1\alpha_1(t)\sin\omega_1(T-t)/(m_1\omega_1) + c_2\alpha_2(t)\sin\omega_2(T-t)/(m_2\omega_2).$$

For the initial iteration, we assign the weights α_1, $\alpha_2 > 0$ and consider the local problems. The first of them is formulated in the form

$$h_1(u_1) = c_1x_1(T) - \int_0^T \overset{\circ}{\delta}(\tau)u_1(\tau)d\tau \to \max, \qquad \frac{dx_1(t)}{dt} = z_1(t),$$

$$\frac{dz_1(t)}{dt} = -\omega_1^2x_1(t) + \frac{u_1(t)}{m_1}, \qquad 0 \leq u_1(t) \leq w.$$

According to the maximum principle for this problem, we have

$$\{c_1\sin\omega_1(T-t)/(\omega_1m_1) - [\alpha_1c_1\sin\omega_1(T-t)/(\omega_1m_1)+ \\ + \alpha_2c_2\sin\omega_2(T-t)/(\omega_2m_2)]\}\, u_1(t) \to \max,$$

or, which is the same,

$$\alpha_2\left[c_1\sin\omega_1(T-t)/(\omega_1m_1) - c_2\sin\omega_2(T-t)/(\omega_2m_2)\right]u_2(t) \to \max.$$

Thus, we obtained a discontinuous optimal control $\hat{u}_1(t)$ with the switching point t^*: $\hat{u}_1(t) = w_1$, $t \in [0, t^*]$, $\hat{u}_1(t) = 0$, $t \in (t^*, T]$.
Analogous consideration of the second local problem leads to the optimal control

$$\hat{u}_1(t) = 0, \qquad t \in [0, t^*), \qquad \hat{u}_2(t) = w_2, \qquad t \in (t^*, T].$$

Moreover, according to the general scheme of the decomposition method, we construct the aggregation weights $\alpha_1(t, \rho_1, \rho_2)$, $\alpha_2(t, \rho_1, \rho_2)$ for $0 \leq \rho_j$, $j = 1, 2$:

$$\begin{aligned}
\alpha_1(t, \rho_1, \rho_2) &= \left[\overset{\circ}{u}_1(t) + \rho_1(\hat{u}_1(t) - \overset{\circ}{u}_1(t))\right]\Big[\overset{\circ}{U} + \rho_1(\hat{u}_1(t) - \\
&\quad - \overset{\circ}{u}_1(t)) + \rho_2(\hat{u}_2(t) - \overset{\circ}{u}_2(t))\Big]^{-1}, \\
\alpha_2(t, \rho_1, \rho_2) &= \left[\overset{\circ}{u}_2(t) + \rho_2(\hat{u}_2(t) - \overset{\circ}{u}_2(t))\right]\Big[\overset{\circ}{U} + \rho_1(\hat{u}_1(t) - \\
&\quad - \overset{\circ}{u}_1(t)) + \rho_2(\hat{u}_2(t) - \overset{\circ}{u}_2(t))\Big]^{-1},
\end{aligned} \tag{2.7}$$

where $\overset{\circ}{u}_j(t) = \alpha_j(t)\overset{\circ}{U}(t)$, $j = 1,2$. These weights have discontinuity at the point t^* for all $\rho_j \in [0,1]$, $j = 1,2$. If we introduce notation

$$\bar{\alpha}_j(t,\rho_1,\rho_2) = \alpha_j(t,\rho_1,\rho_2), \qquad t \in [0,t^*),$$

$$\tilde{\alpha}_j(t,\rho_1,\rho_2) = \alpha_j(t,\rho_1,\rho_2), \qquad t \in (t^*,T), \qquad j = 1,2,$$

then, by integrating ordinary differential equations in (2.6) with discontinuous right-hand sides $\alpha_j(t,\rho_1,\rho_2)\overset{\circ}{U}/m_j$, by analogy with (2.5), we obtain

$$\begin{aligned}\overset{\circ}{g}(\rho_1,\rho_2) = c_1 w(m_1\omega_1^2)^{-1}\,[\bar{\alpha}_1(\rho_1,\rho_2)(\cos\omega_1(T-t^*) - \cos\omega_1 T) + \\ +\tilde{\alpha}_1(\rho_1,\rho_2)(1-\cos\omega_1(T-t^*))] + \\ +c_2 w(m_2\omega_2^2)^{-1}\,[\bar{\alpha}_2(\rho_1,\rho_2)(\cos\omega_2(T-t^*) - \\ -\cos\omega_2 T) + \tilde{\alpha}_2(\rho_1,\rho_2)(1-\cos\omega_2(T-t^*))]\,. \qquad (2.8)\end{aligned}$$

We assume $\rho_1 = \rho_2 = \rho$, then the fractional linear functions $\bar{\alpha}_j(\rho)$, $\tilde{\alpha}_j(\rho)$, $j = 1,2$ increase monotonically in the interval $[0,1]$. Therefore, the maximum of $\overset{\circ}{g}(\rho)$ is attained for $\rho = 1$, where $\bar{\alpha}_1 = \tilde{\alpha}_2 = 1$, $\tilde{\alpha}_1 = \bar{\alpha}_2 = 0$, and function (2.8) gives the optimal value of functional (2.4) of the initial problem (2.1). Thus, the optimal solution is obtained in one iteration. It remains to test the optimality criterion (1.12), which looks as follows:

$$\begin{aligned}\pi_1 = c_1\hat{x}_1(T) + c_2\hat{x}_2(T) + \int_0^T \overset{\circ}{\delta}(t) w\,dt - \\ - \int_0^T \overset{\circ}{\delta}(t)(\hat{u}_1(t) + \hat{u}_2(t))dt - c_1\overset{\circ}{x}_1(T) - c_2\overset{\circ}{x}_2(T) = 0,\end{aligned}$$

where the values $\overset{\circ}{x}_j(T)$, $j = 1,2$ correspond to the disaggregated controls $\overset{\circ}{u}_j(t)$, $j = 1,2$, and, for the optimal weights, they are computed according to (2.4). The values $\hat{x}_j(T)$, $j = 1,2$ are found in the same way as relation (2.4) is obtained. For optimal weights, we have the following expression for the conjugated variable

$$\overset{\circ}{\delta}(t) = c_1(\omega_1 m_1)^{-1}\sin\omega_1(T-t), \qquad t \in [0,t^*),$$

$$\overset{\circ}{\delta}(t) = c_2(\omega_2 m_2)^{-1}\sin\omega_2(T-t), \qquad t \in [t^*,T),$$

from where we deduce $\pi_1 = 0$.

We give one more example. The problem has a minimax criterion and is a formulation of the model of optimal control under uncertainty. Consider a dynamic system with control, and let the uncertainty consist in the fact that the initial phase variables are not given precisely and can belong to a convex set M_x. In this case, if feasible controls are substituted in the right-hand sides of the differential equations, then the obtained ends of trajectories form set M_z in the space of phase coordinates. The problem is to find the optimal control that minimizes the maximal deviation of the points of the set M_z from the given one.

As in the previous example, we consider the system that consists of two subsystems and denote by z_1, z_2 and M_{z_1}, M_{z_2} the coordinates and the sets of the trajectories ends, respectively. We state the following problem:

$$\begin{aligned} &f(u_1,u_2) = \max_{z_1\in M_{z_1}} \varphi_1(z_1) + \max_{z_2\in M_{z_2}} \varphi_2(z_2) \to \min, \\ &dx_1(t)/dt = b_1u_1(t), \qquad x_1 \in M_{x_1}, \\ &dx_2(t)/dt = b_2u_2(t), \qquad x_2 \in M_{x_2}, \qquad 0 \le u_1 \le w_1, \\ &0 \le u_2 \le w_2, \qquad u_1(t)+u_2(t) \ge w. \end{aligned} \tag{2.9}$$

Here, b_1, b_2, w_1, w_2, w are positive constants, the sets $M_{x_j} = \{x_j | \, |x_j| \le 1\}$, $j=1,2$. The functions $\varphi_j(z_j)$, $j=1,2$ have the form $0,5(z_j)^2$ and characterize the distances from the ends of trajectories to the origin.

First, we give a direct solution of problem (2.9) applying the necessary Kurzhanskii conditions. To this end, we introduce the conjugated equations with parameters l_1, l_2:

$$d\psi_1(t)/dt = 0, \qquad \psi_1(T) = l_1, \qquad d\psi_2(t)/dt = 0, \qquad \psi_2(T) = l_2,$$

from where we have $\psi_1(t) = l_1$, $\psi_2(t) = l_2$, $t \in [0,T]$.

For the definiteness, let the coefficients of the stated problem (2.9) satisfy the following inequalities:

$$w_1 > w, \qquad w_2 > w, \qquad b_1 > b_2, \qquad b_2^2 wT + b_2 < b_1. \tag{2.10}$$

We introduce the Lagrange function of the form

$$\mathcal{L}(l_1,l_2,u_1,u_2) = \int_0^T (l_1b_1u_1 + l_2b_2u_2)dt + \tilde{\psi}_1(l_1) + \tilde{\psi}_2(l_2), \tag{2.11}$$

where the concave functions $\psi_j(l_j)$ are given as follows:

$$\psi_j(l_j) = 0,5 \qquad |l_j| \leq 1,$$

$$\psi_j(l_j) = |l_j| = 0,5(l_j)^2, \qquad |l_j| > 1, \ j = 1, 2.$$

First, following Kurzhanskii method, we establish the form of the functions

$$w(l_1, l_2) = \min \left\{ \int_0^T (l_1 b_1 u_1 + l_2 b_2 u_2) dt | u_1 + u_2 \geq w, \right.$$

$$\left. 0 \leq u_1 \leq w_1, \qquad 0 \leq u_2 \leq w_2 \right\}.$$

Thus, we have

$$\omega(l_1, l_2) = l_1 b_1 w T \ \text{ for } \ l_1 b_1 \geq b_2 b_2,$$

$$\omega(l_1, l_2) = l_2 b_2 w T \ \text{ for } \ l_1 b_1 < l_2 b_2.$$

Further, we should look for the maximum of the concave function $p(l_1, l_2) = \omega(l_1, l_2) + \psi_1(l_1) + \psi_2(l_2)$. Taking into account the third inequality in (2.10), we consider this function for $l_1 = 1$ and $l_2 \in [1, b_1/b_2]$. We have $0,5 + l_2 b_2 w T + l_2 - 0,5 l_2^2$. The maximum of this function is attained at the point $l_2^* = b_2 w T + 1$, where, due to the fourth inequality in (2.10), we obtain $l_2^* < b_1/b_2$. Obviously, this local extremum maximizes the function $p(l_1, l_2)$.

After that, the optimal controls of problem (2.9) are computed from the minimum condition

$$b_1 u_1 + b_2(b_2 w T + 1) u_2 \to \min$$

under constraints $u_1 + u_2 \geq w$, $0 \leq u_1 \leq w_1$, $0 \leq u_2 \leq w_2$. Due to the fourth inequality in (2.10), this implies $\overset{*}{u}_1 = 0$, $\overset{*}{u}_2 = w$.

The optimal value of the functional of problem (2.9) is equal to the Lagrangian (2.11) at the point $(1, b_2 w T + 1, 0, w)$. We have

$$f^* = 1 + 0,5(b_2 w T)^2 + b_2 w T.$$

Now, we present the construction of the decomposition method, applying variable aggregation to problem (2.9). We fix the weights $\alpha_1, \alpha_2 > 0$ and consider the problem with aggregated variables

$$g(U) = \max_{z_1 \in M_{z_1}} \varphi_1(z_1) + \max_{z_2 \in M_{z_2}} \varphi_2(z_2) \to \min,$$

$$dx_1(t)/dt = \alpha_1 b_1 U, \qquad x_1 \in M_{x_1},$$

$$dx_2(t)/dt = \alpha_2 b_2 U, \qquad x_2 \in M_{x_2},$$

$$w \leq U \leq w_1 + w_2.$$

The Lagrange function will take the form

$$\mathcal{L}(l_1, l_2, U) = \int_0^T (l_1 \alpha_1 b_1 + l_2 \alpha_2 b_2) U dt + \psi_1(l_1) + \psi_2(l_2).$$

We consider the function

$$\Omega(l_1, l_2, U) = \min \left\{ \int_0^T (l_1 \alpha_1 b_1 + l_2 \alpha_2 b_2) U dt | w \leq U \leq w_1 + w_2 \right\}.$$

Thus, we obtain

$$\Omega(l_1, l_2) = (\alpha_1 b_1 l_1 + \alpha_2 b_2 l_2) wT \text{ for } \alpha_1 b_1 l_1 + \alpha_2 v_2 l_2 \geq 0,$$

$$\Omega(l_1, l_2) = (\alpha_1 b_1 l_1 + \alpha_2 b_2 l_2)(w_1 + w_2) T \text{ for } \alpha_1 b_1 l_1 + \alpha_2 v_2 l_2 < 0.$$

The maximum of the function $\Omega(l_1, l_2) = \psi_1(l_1) + \psi_2(l_2)$ is attained at the stationary points of the function $\alpha_j b_j w T l_j + l_j - 0,5 l_j^2$, $j = 1, 2$. The coordinates of these points are $\overset{\circ}{l}_j = (\alpha_j b_j w T + 1)$, $j = 1, 2$.

The optimal solution of the problem

$$[\alpha_1 b_1 (\alpha_1 b_1 w T + 1) + \alpha_2 b_2 (\alpha_2 b_2 w T + 1)] U \to \min,$$

$$w \leq U \leq w_1 + w_2$$

is $\overset{\circ}{U} = w$. The optimal value of the functional $\overset{\circ}{g}(\alpha_1, \alpha_2)$ of the aggregated problem, as a function of weights α_1, α_2, has the form

$$\overset{\circ}{g}(\alpha_1, \alpha_2) = 0,5(\alpha_1 b_1 w T)^2 + \alpha_1 b_1 w T + 0,5+ \\ +0,5(\alpha_2, b_2 w T)^2 + \alpha_2 b_2 w T + 0,5. \qquad (2.12)$$

The dual variable $\overset{\circ}{\delta}$ for the condition $U \geq w$ in the aggregated problem is computed as follows:

$$\overset{\circ}{\delta} = \alpha_1 b_1 \overset{\circ}{l}_1 + \alpha_2 b_2 \overset{\circ}{l}_2 = \alpha_1 b_1 (\alpha_1 b_1 w T + 1) + \alpha_2 b_2 (\alpha_2 b_2 w T + 1).$$

Consider the first local problem

$$\left[\max_{z_1 \in M_{z_1}} \varphi_1(z_1) - \int_0^T \overset{\circ}{\delta}\, u_1 dt \right] \to \min,$$

$$dx_1(t)/dt = b_1 u_1, \qquad x_1 \in M_{x_1}, \qquad 0 \leq u_1 \leq w_1.$$

We write the Lagrange function for it

$$\mathcal{L}_1(l_1, u_1) = \int_0^T (b_1 l_1 - \overset{\circ}{\delta}) u_1 dt + \psi_1(l_1).$$

Consider the function

$$\omega_1(l_1) = \min \left\{ \int_0^T (b_1 l_1 - \overset{\circ}{\delta}) u_1 dt, \qquad 0 \leq u_1 \leq w_1 \right\}.$$

We have

$$\omega_1(l_1) = w_1 T(l_1 - \overset{\circ}{\delta}/b_1), \qquad l_1 \leq \overset{\circ}{\delta}/b_1,$$

$$\omega_1(l_1) = 0, \qquad l_1 > \overset{\circ}{\delta}/b_1.$$

The coordinate of the break $\bar{l}_1$ of the function $\omega_1(l_1)$ is equal to

$$\alpha_1(\alpha_1 b_1 w T + 1) + \alpha_2 (b_2/b_1)(\alpha_2 b_2 w T + 1).$$

We assume that the weight coefficients α_1, α_2 are such that $l_1 < 1$. This can always be attained, since, for $\alpha_1 = 0$, $\alpha_2 = 1$, we have $l_1 < 1$ due to the third inequality in (2.10). Under the given assumption, the maximum of the function $\omega_1(l_1) + \psi_1(l_1)$ is attained at any point $\bar{l}_1$ of the interval $[\overset{\circ}{\delta}/b_1, 1]$, and the optimal solution of the local problem is as follows: $\hat{u}_1(t) = 0$, $t \in [0, T]$. The Lagrange function is equal to $\hat{\mathcal{L}}_2 = 0,5$.

Consider the second local problem

$$\left[\max_{z_2 \in M_{z_2}} \varphi_2(z_2) - \int_0^T \overset{\circ}{\delta}\, u_2 dt \right] \to \min,$$

$$dx_2(t)/dt = b_2 u_2, \qquad x_2 \in M_{x_2}, \qquad 0 \le u_1 \le w_2.$$

As well as for the first local problem, we have

$$\omega_2(l_2) = w_2 T(l_2 - \overset{\circ}{\delta}/b_2), \qquad \text{for} \quad l_2 \le \overset{\circ}{\delta}/b_2,$$

$$\omega_1(l_1) = 0, \qquad \text{for} \quad l_2 > 0.$$

The coordinate of the break point of the function $\omega_2(l_2)$ is computed in the form

$$\bar{l}_2 = \alpha_1(b_1/b_2)(\alpha_1 b_1 wT + 1) + \alpha_2(\alpha_2 b_2 wT + 1).$$

According to the third inequality in (2.10), we have $\hat{l}_2 > 1$.

Consider the function $\omega_2(l_2) + \psi_2(l_2)$ in the interval $[1, \overset{\circ}{\delta}/b_2]$. Under sufficiently large values of w_2, the stationary point of this function $l_2 = (w_2 T + 1)$ satisfies the condition $\hat{l}_2 > \bar{l}_2$. Thus, the maximum of the function $\omega_2(l_2) + \psi_2(l_2)$ is attained for $l_2 = \overset{\circ}{\delta}/b_2$. The optimal solution of the second local problem is $\hat{u}_2 = \tilde{w}_2 t \in [0, T]$, where $\tilde{w}_2$ is an arbitrary value in the interval $[0, w_2]$. The Lagrange function is computed in the form

$$\mathcal{L}_2(\alpha_1, \alpha_2) = 0,5 - 0,5\left[\alpha_1(b_1/b_2)(\alpha_1 b_1 wT + 1)\right]^2 - \\ -0,5\left[\alpha_2(\alpha_2 b_2 wT + 1)\right]^2. \qquad (2.13)$$

For the iterative process, the new weights are taken in the form

$$\alpha_1(\rho) = \left[\alpha_1 w(1-\rho)\right]\left[w(1-\rho) + \rho\tilde{w}_2\right]^{-1},$$

$$\alpha_2(\rho) = \left[\alpha_2 w(1-\rho) + \rho\tilde{w}_2\right]\left[w(1-\rho) + \rho\tilde{w}_2\right]^{-1}.$$

By substituting these weights in the right-hand side of equality (2.11), we obtain the function $\overset{\circ}{g}(\rho)$, which is monotonically decreasing with respect to ρ. Its minimum is attained for $\rho = 1$. This corresponds to the optimal values $\overset{*}{\alpha}_1 = 0$, $\overset{*}{\alpha}_2 = 1$. Thus, problem (2.9) is solved by the decomposition method in one iteration. It remains to test the optimality condition (1.12). According to (2.12), (2.13), we finally obtain

$$\pi_1 = \hat{\mathcal{L}}_1 + \hat{\mathcal{L}}_2(0,1) - \overset{\circ}{g}(0,1) - \overset{\circ}{\delta}(w - w) = 0.$$

In previous examples, we considered systems that contain two subsystems. We now consider the model with an arbitrary number of subsystems. We formulate the block separable statement on the basis of optimizing noise in dynamic systems under random perturbation. We have

$$\begin{aligned}
&f = \sum_{j=1}^{J} c_j x_j(T) \to \min, \\
&dx_j(t)/dt = 1/u_j(t), \qquad x_j(0) = \kappa_j, \qquad u_j(t) \geq 0, \qquad j \in [1:J], \\
&\sum_{j=1}^{J} u_j(t) \leq w.
\end{aligned} \tag{2.14}$$

Here, c_j, $j \in [1:J]$, w are constants strictly greater than zero.

First, we solve problem (2.14) using the maximum principle. The conjugated functions have the form $\psi_j(t) = c_j$, $j \in [1:J]$. Assume that the binding constraint in (2.14) is effective. Then, we come to an unconstrained maximization of the function

$$-\sum_{j=1}^{J} c_j/u_j + v(w - \sum_{j=1}^{J} u_j) \to \max,$$

where v is the Lagrange multiplier of the binding constraint. By differentiating with respect to u_j, we come to the system of equations with respect to unknowns u_j, $j \in [1:J]$, and v:

$$c_j/(u_j)^2 - v = 0, \qquad j \in [1:J], \qquad \sum_{j=1}^{J} u_j = w. \tag{2.15}$$

Using the first J control identities (2.15), we express $u_j = \sqrt{c_j/v}$ as a function of v and substitute the result in the binding condition. After that, we find $\sqrt{v} = \left(\sum_{j=1}^{J} \sqrt{c_j}\right) \Big/ w$ and obtain the final formula for optimal controls

$$\overset{*}{u}_j = w\sqrt{c_j}\left(\sum_{s=1}^{J} \sqrt{c_s}\right)^{-1}, \qquad j \in [1:J]. \tag{2.16}$$

Furthermore, we apply the decomposition method to problem (2.14) on the basis of aggregated variables. We fix the weights $[0,T]$ constant on the interval $\alpha_j > 0$, $j \in [1:J]$, thus obtaining the following problem with aggregated controls:

$$\begin{aligned}
&g = \sum_{j=1}^{J} c_j x_j(T) \to \min, \\
&dx_j(t)/dt = 1/(\alpha_j U(t)), \qquad x_j(0) = \kappa_j, \qquad 0 \leq U(t) \leq w.
\end{aligned}$$

The Hamiltonian of this problem is $H = -\sum_{j=1}^{J} c_j/(\alpha_j U)$, therefore its solution is $\overset{\circ}{U} = w$. The dual variable $\overset{\circ}{\delta}$ for the condition $U(t) \le w$ of the aggregated problem is computed in the form

$$\overset{\circ}{\delta} = \sum_{j=1}^{J} c_j \Big/ (\alpha_j w^2).$$

Consider the local problem with the number j

$$c_j x_j(T) + \int_0^T \overset{\circ}{\delta}\, u_j dt \to \min,$$

$$dx_j(t)/dt = 1/u_j(t), \qquad x_j(0) = \kappa_j, \qquad u_j(t) > 0.$$

The Hamiltonian of the local problem $H_j = -c_j/u_j - \overset{\circ}{\delta}\, u_j$ and the optimal controls $\hat{u}_j$ are expressed by the formula

$$\hat{u}_j = w\sqrt{c_j \Big/ \left(\sum_{s=1}^{J} (c_s/\alpha_s)\right)}. \tag{2.17}$$

Consider the optimal value of the functional of the aggregated problem as a function of parameter α_j. We have

$$\overset{\circ}{g}(\alpha_j) = \sum_{j=1}^{J} [c_j \kappa_j - T c_j/(\alpha_j w)]$$

and substitute a formula for new weights of the form (1.11) for α_j. Let these weights depend on one parameter ρ. We then have

$$\overset{\circ}{g}(\rho) = \sum_{j=1}^{J} \left[c_j \kappa_j - \frac{T(w(1-\rho) + \rho \sum_{s=1}^{J} \hat{u}_s c_j)}{(\alpha_j w(1-\rho) + \rho \hat{u}_j) w} \right]. \tag{2.18}$$

Using the decomposition method, we try to find the maximum with respect to ρ in the interval $[0, 1]$ of function $\overset{\circ}{g}(\rho)$. The right-hand side in (2.18) depending on ρ is a sum of linear fractional functions. The maximum of this sum is attained at the point $\rho = 1$, therefore these weights are equal to

$$\alpha_j = \hat{u}_j \Big/ \left(\sum_{s=1}^{J} \hat{u}_s\right), \qquad j \in [1 : J].$$

We substitute expressions (2.17) in the right-hand sides of these equalities, thus obtaining

$$\alpha_j = \sqrt{c_j}\left(\sum_{s=1}^{J}\sqrt{c_s}\right)^{-1}, \qquad j \in [1:J]. \tag{2.19}$$

This corresponds to the weights computed by optimal controls (2.16). Thus, the point $\rho = 1$ corresponds to the optimum of the initial problem (2.14), and its solution by the decomposition method was obtained in one iteration. It remains to test the optimality criterion (1.12). To this end, it suffices to prove that the equality $\overset{\circ}{u}_j = \alpha_j w = \hat{u}_j$, $j \in [1:J]$ holds for optimal weights. The conclusion is tested after substitution of optimal weights (2.19) in the right-hand sides of (2.17).

§3. Block Problems of Optimal Control with Partial Differential Equations

In this section, we give a statement of block separable problems of optimal control in mathematical physics. The method of iterative aggregation is applied to problems, whose subsystems are described by various partial differential equations of classical types.

Let Ω be a bounded open set of points $x = (x_1, \ldots, x_n)$ of Euclidean space $\mathcal{R}^n$, which has a sufficiently smooth boundary Γ, where the set Ω is situated at one side from Γ. Now, let $\mathcal{L}_2(\Omega)$ be the space of functions measurable and square integrable on Ω, and let $\mathcal{L}_2^m(\Omega)$ be the space of m-dimensional vector functions with components from $\mathcal{L}_2(\Omega)$. We consider the first-order Sobolev space $W_2^{(1)}$, i.e., the space of functions that have square integrable first-order generalized derivatives. The norms of the elements have the form

$$||y||_{\mathcal{L}_2(\Omega)} = \left[\int_\Omega |y|^2 dx\right]^{1/2}, \qquad y \in \mathcal{L}_2(\Omega),$$

$$||w||_{\mathcal{L}_2^m(\Omega)} = \max_{r=1,2,\ldots,m} ||w^r||_{\mathcal{L}_2(\Omega)}, \qquad w \in \mathcal{L}_2^m(\Omega),$$

$$||z||_{m_2^{(1)}(\Omega)} = \left[\int_\Omega (|z|^2 + |\nabla z|^2) dx\right]^{1/2}, \qquad z \in W_2^{(1)}(\Omega),$$

where the vector gradient $\nabla z = (z_{x_1}, \dots, z_{x_n})$ is introduced in the ordinary sense. The introduced space $\overset{\circ}{W}{}_2^{(1)}(\Omega)$ is the closure, with respect to the norm $\|\cdot\|_{W_2^{(1)}(\Omega)}$, of the set of infinitely differentiable functions in Ω that have carriers compact in Ω. As is known, we have embedding $\overset{\circ}{W}{}_2^{(1)}(\Omega) \subset \mathcal{L}_2(\Omega)$, which is a direct consequence of the Poincare-Friedrichs inequality

$$\|y\|_{\mathcal{L}_2(\Omega)} \le A\|y\|_{W_2^{(1)}(\Omega)}, \qquad \forall y \in \overset{\circ}{W}{}_2^{(1)}(\Omega), \qquad A = \text{const}.$$

Consider the following block separable optimal control problem:

$$f = \sum_{j=1}^{J} \left[\int_{\Omega} c_j(z_j, u_j, x)dx \right] \to \max, \tag{3.1}$$

$$-\Delta z_j(x) = b_j(u_j, x) \text{ in } \Omega, \qquad z_j|_{\Gamma} = 0, \tag{3.2}$$

$$p_j(z_j, u_j, x) \le 0, \qquad u_j \ge 0, \qquad j \in [1:J]; \qquad \sum_{j=1}^{J} d_j(z_j, u_j, x) \le 0, \tag{3.3}$$

where, for each $j \in [1:J]$, the value u_j is I-dimensional vector function of x with components $(u_j^1, \dots, u_j^i, \dots, u_j^I)$; $\Delta = \sum_{s=1}^{n} (\partial^2/\partial x_s^2)$ is the Laplace operator, b_j and c_j are scalar, and p_j and d_j are vector functions of their arguments of dimensions K_j and L, respectively.

We will assume that the functions $c_j(z_j, u_j, x)$, $b_j(u_j, x)$, $-p_j(z_j, u_j, x)$, $-d_j(z_j, u_j, x)$, $j \in [1:J]$ are continuously differentiable in the whole space of its variables, concave with respect to u_j and z_j and increase monotonically with respect to z_j. The partial derivatives of these functions with respect to u_j and z_j are assumed to be bounded with respect to the whole space of variables and to satisfy the Lipshitz condition.

Problem (3.1)–(3.3) consists in finding the control vectors $u_j(x) \in \mathcal{L}_2^I(\Omega)$, $j \in [1:J]$ and the weak solutions $z_j(x) \in \overset{\circ}{W}{}_2^{(1)}(\Omega)$ of the Dirichlet problem (3.2) that correspond to them, so that constraints (3.3) are satisfied, and the maximum of functional (3.1) is attained. For given controls $u_j(x) \in \mathcal{L}_2^I(\Omega)$, the weak (generalized) solution of problems (3.2) for each $j \in [1:J]$ means, as usual, the functions $z_j(x) \in \overset{\circ}{W}{}_2^{(1)}(\Omega)$ if the equalities

$$\int_{\Omega} [(-\nabla z_j(z), \nabla y(x) + b_j(u_j, x)y(x)]\, dx = 0 \tag{3.4}$$

are attained for any element $y(x) \in \overset{\circ}{W}{}_2^{(1)}(\Omega)$. Note that the inequalities in (3.3) are understood as partial ordering in $\mathcal{L}_2(\Omega)$ by means of the cone of nonnegative functions.

It is known that, for given controls $u_j(x) \in \mathcal{L}_2^I(\Omega)$, there exist unique weak solutions of each Dirichlet problem (3.2). This fact is expressed in the form $z_j(x) = \mathcal{N}u_j(x) \in \mathcal{L}_2(\Omega)$, where $\mathcal{N}$ is a totally continuous operator from $\mathcal{L}_2^I(\Omega)$ into $\mathcal{L}_2(\Omega)$. We substitute $z_j(x)$, $j \in [1:J]$ in the constraints and in the functional of problem (3.1)–(3.3) as did Ter-Krikorov in the cited papers. We obtain the reduction of problem (3.1)–(3.3) to the concave programming in the Banach space.

We assume that the Slater condition holds, and the block conditions in (3.2)–(3.3) for each $j \in [1:J]$ give the sets in the space $\mathcal{L}_2^I(\Omega)$ bounded by control. We assume that the following Ter-Krikorov inequalities hold almost everywhere in Ω:

$$z_j(x) > 0, \qquad p_j(0,0,x) \le 0, \qquad d_j(0,0,x) \le 0, j \in [1:J]. \tag{3.5}$$

Then, the boundedness of the sequences of dual variables under necessary optimality conditions written in the limit form is established. Using the boundedness of these sequences and weak compactness of the set bounded in $\mathcal{L}_2(\Omega)$, we deduce conditions of complementary slackness for the considered extremal problem. The Kuhn-Tucker theorems and the Pontryagin maximum principle follow from these conditions. In what follows, we assume that the Ter-Krikorov conditions hold for the main and all intermediary problems in the decomposition method. Note that the first inequality in (3.5) hold if $b_j(u_j, x) > 0$, since it follows from the positivity of the Green function of the Dirichlet problem for the Laplace equation.

Decomposition based on the aggregation of variables from various blocks used the method described in Section 1. We introduce aggregated controls $U^i(x)$, $i \in [1:I]$ and, for fixed weights $\alpha_j(x)$, $j \in [1:J]$, we obtain the following problem with aggregated variables:

$$\begin{gathered} g = \int_\Omega C(z,U,x)\,dx \to \max, \\ -\Delta z_j(x) = B_j(U,x), \qquad x \in \Omega, \qquad z_j|_\Gamma = 0, \qquad P_j(z_j,U,x) \le 0, \\ D(z,U,x) \le 0, \end{gathered} \tag{3.6}$$

where notation is analogous to that from Section 1.

We formulate the conjugated problem for (3.3):

$$\mathcal{L}(z_j(x), U(x), y_i(x), \eta_j(x), \delta(x)) \to \min, \tag{3.7}$$

$$-\Delta y_j(x) = (\partial/\partial z_j)\left[c_j - p_j \eta_j(x) - d_j \delta(x)\right], \quad x \in \Omega, \quad y_i|_\Gamma = 0, \tag{3.8}$$

$$(\partial/\partial U^i)\left\{C + \sum_{j=1}^{J}\left[B_j y_i(x) - P_j \eta_j(x)\right] - D\delta(x)\right\} \leq 0, \qquad i \in [1:I], \tag{3.9}$$

$$U(x) \geq 0, \qquad \eta_j(x) \geq 0, \qquad j \in [1:J], \qquad \delta(x) \geq 0. \tag{3.10}$$

Here, we introduce the vector functions of dual variables $\eta_j(x) \in \mathcal{L}_2^{K_j}(\Omega)$, $j \in [1:J]$, $\delta(x) \in \mathcal{L}_2^L(\Omega)$. The functional (3.7), as usual in Wolfe dual problems, is the Lagrangian of the problem with aggregated controls.

Assume that, for given normed weights $\alpha_j(x)$, we found a unique solution $\overset{\circ}{U}(x) \in \mathcal{L}_2^I(\Omega)$, $\overset{\circ}{U}{}^i(x) > 0$, $i \in [1:I]$ of the problem with aggregated variables (3.6), corresponding to unique weak solutions $\overset{\circ}{z}_j(x) \in \overset{\circ}{W}{}_2^{(1)}(\Omega)$, $j \in [1:J]$ of the Dirichlet problems in (3.6). Assume also that we find the unique solution $\overset{\circ}{\eta}_j(x) \in \mathcal{L}_2^{K_j}$, $j \in [1:J]$, $\overset{\circ}{\delta}(x) \in \mathcal{L}_2^L(\Omega)$ of the conjugated problem (3.7)–(3.10), corresponding to the weak solutions $\overset{\circ}{y}_j(x) \in \overset{\circ}{W}{}_2^{(1)}(\Omega)$ of the Dirichlet problem (3.8). Here, we use the assumption introduced earlier about the boundedness of partial derivative functions c_j, p_j, d_j in the whole space, so that the right-hand sides in (3.8) belong to the space $\mathcal{L}_2(\Omega)$. We formulate local problems for subsystems, which are given by block conditions from (3.2), (3.3) for every $j \in [1:J]$, and the terms $-\int_\Omega d_j(z_j, u_j, x)\, \overset{\circ}{\delta}(x) dx$. are added to functionals from (3.1). If $\hat{u}_j(x) \in \mathcal{L}_2^I(\Omega)$, $j \in [1:J]$ are optimal solutions of local problems that correspond to weak solutions $\hat{z}_j(x) \in \overset{\circ}{W}{}_2^{(1)}(\Omega)$ of the Dirichlet problem (3.2), then, by method from Section 1, we establish an optimality criterion for the disaggregated control $\overset{\circ}{u}_j(x) = \alpha_j(x)\, \overset{\circ}{U}(x)$ and $\overset{\circ}{z}_j$ for the initial problem (3.1)–(3.3). Thus, we have

$$\int_\Omega \sum_{j=1}^{J}\left[c_j(\hat{z}_j, \hat{u}_j, x) - d_j(\hat{z}_j, \hat{u}_j, x)\, \overset{\circ}{\delta}(x) - c_j(\overset{\circ}{z}_j, \overset{\circ}{u}_j, x)\right] dx = 0. \tag{3.11}$$

Local monotonicity of the functional of the iterative process, as in Section 1, is deduced from the formula for the marginal value of parametric problems with aggregated controls. Upon differentiating the Lagrangian, we obtain

$$(\overset{\circ}{g}(0))' = \int_{\Omega} \sum_{j=1}^{J} \left[\frac{\partial c_j}{\partial u_j} (\hat{u}_j - \alpha_j \sum_{r=1}^{J} \hat{u}_r) + \frac{\partial b_j}{\partial u_j} (\hat{u}_j - \alpha_j \sum_{r=1}^{J} \hat{u}_r)\, \overset{\circ}{y}_j - \right.$$

$$- \left(\nabla y_j, \nabla \frac{\partial \overset{\circ}{z}_j(x,0)}{\partial \rho} \right) - \frac{\partial p_j}{\partial u_j} (\hat{u}_j - \alpha_j \sum_{r=1}^{J} \hat{u}_r)\, \overset{\circ}{\eta}_j -$$

$$\left. - \frac{\partial d_j}{\partial u_j} (\hat{u}_j - \alpha_j \sum_{r=1}^{J} \hat{u}_r)\, \overset{\circ}{\delta} + \frac{\partial}{\partial z_j} (c_j - p_j \overset{\circ}{\eta}_j - d_j \delta) \frac{\partial \overset{\circ}{z}_j}{\partial \rho}(x,0) \right] dx.$$

We can easily see that $(\partial/\partial\rho)\, \overset{\circ}{z}_j(x,0) \in \overset{\circ}{W}{}_2^{(1)}(\Omega)$. Also, in Section 1, due to $U(x) > 0$ and (3.9), all terms with the sum $\sum_{r=1}^{J} \hat{u}_r$ are cancelled. By definition of the weak solution of the Dirichlet problem (3.4) for $\overset{\circ}{y}_j$, taking into account (3.8), the third and the last terms are also cancelled. Then, as in Section 1, we obtain

$$(\overset{\circ}{g}(0))' = \int_{\Omega} \left\{ \sum_{j=1}^{J} \frac{\partial}{\partial u_j} (c_j + b_j \overset{\circ}{y}_j - p_j \overset{\circ}{\eta}_j - d_j \overset{\circ}{\delta})(\hat{u}_j - \overset{\circ}{u}_j) \right\} dx. \tag{3.12}$$

From the definition of the weak solution (3.4), as applied to $\overset{\circ}{y}_j(x)$, we have the following equality:

$$\int_{\infty} \sum_{j=1}^{J} \left\{ (-\nabla \overset{\circ}{y}_j, \nabla(\hat{z}_j - \overset{\circ}{z}_j)) + \frac{\partial}{\partial z_j} (c_j - p_j \overset{\circ}{\eta}_j - d_j \overset{\circ}{\delta})(\hat{z}_j - \overset{\circ}{z}_j) \right\} dx = 0. \tag{3.13}$$

The first term in the left-hand side of (3.13), using the same definition, but applied to $(\hat{z}_j - \overset{\circ}{z}_j)$, is equal to

$$\int_{\Omega} \sum_{j=1}^{J} \left[b_j(\overset{\circ}{u}_j, x) - b_j(\hat{u}_j, x) \right] \overset{\circ}{y}_j dx. \tag{3.14}$$

We add the left-hand side of (3.13) to (3.12) taking account of (3.14). We also add the values

$$-\int_{\Omega} \left[\sum_{j=1}^{J} p_j(\overset{\circ}{z}_j, \overset{\circ}{u}_j, x)\, \overset{\circ}{\eta}_j(x) \right] dx,$$

$$-\int_{\Omega}\left[\sum_{j=1}^{J} d_j(\overset{\circ}{z}_j,\overset{\circ}{u}_j,x)\,\overset{\circ}{\delta}(x)\right] dx,$$

which are equal to zero due to the condition of complementary slackness for the aggregated problem (3.6). Finally, we obtain the formula

$$(\overset{\circ}{g}(0))' = \pi_1 + \pi_2 + \pi_3,$$

where $\pi_1 > 0$ is the left-hand side of (3.11). For π_2 and π_3, we have

$$\pi_2 = -\int_{\Omega}\left[\sum_{j=1}^{J} p_j(\hat{z}_j,\hat{u}_j,x)\,\overset{\circ}{\eta}_j(x)\right] dx \geq 0,$$

$$\pi_3 = \int_{\Omega}\left\{\sum_{j=1}^{J} Z(\overset{\circ}{z}_j,\overset{\circ}{u}_j\hat{z}_j,\hat{u}_j,x)\left[c_j + b_j\,\overset{\circ}{y}_j - p_j\,\overset{\circ}{\eta}_j - d_j\,\overset{\circ}{\delta}\right]\right\} dx \geq 0.$$

Here, we should point out that $\overset{\circ}{y}_j(x) \geq 0$, $j \in [1:J]$ follows from the non-negativity of the right-hand sides in (3.8) due to the assumed monotonicity with respect to z_j from the output functions and the above positiveness of the Green function of the Dirichlet problem for the Laplace equation.

We mention some generalizations. The previous considerations were related to the Dirichlet problem. However, the proposed method is applied to the second and third boundary problems, i.e., when the conditions

$$\partial z_j/\partial N + \lambda_i z_j|_{\Gamma} = 0, \qquad j \in [1:J] \tag{3.15}$$

hold on the boundary, where $\lambda_j = \lambda_j(\sigma)$ are given functions on Γ, $\partial z_j/\partial N$ are so-called derivatives along the normal. In this case, the weak solution is sought in $W_2^{(1)}(\Omega)$ and is defined by the identity

$$\int_{\Omega} [\nabla z_j, \nabla y) + b_j(u_j,x)y]\,dx + \int_{\Gamma} \lambda_j z_j y\,d\sigma = 0, \qquad \forall y \in W_2^{(1)}. \tag{3.16}$$

The boundary conditions in the conjugated problems will be the same as in (3.15). In the integral of the right-hand side for $(\overset{\circ}{g}(0))'$, the sum

$$\int_{\Gamma}\left[\sum_{j=1}^{J} \lambda_j \frac{\partial z_j(\sigma,0)}{\partial \rho}\,\overset{\circ}{y}_j(\sigma)\right] d\sigma$$

will also appear. According to (3.16), this sum will be cancelled by the third and the last term. The additional expression

$$\int_{\Gamma} \left[\sum_{j=1}^{J} \lambda_j \overset{\circ}{y}_j (\hat{z}_j - \overset{\circ}{z}_j) \right] d\sigma$$

will appear in the left-hand side of (3.13), which finally gives (3.14) for $(\hat{z}_j - \overset{\circ}{z}_j)$ for the definition of the generalized solution (3.15). The other results remain unchanged.

Consider the parabolic case. We introduce some more notation. Let $t \in [0, T]$ be a time variable, and let $Q = \Omega \times [0, T]$ be a cylinder with the lateral area $\Sigma = \Gamma \times (0, T)$. We introduce the Hilbert space $\mathcal{L}_2(S)$ of functions square integrable in the area S (in particular, Q, Σ). Let $W_2^{(1,1)}$ be a Hilbert space consisting of elements $\mathcal{L}_2(Q)$ that have general first-order derivatives with the scalar product and the norm

$$(w, v)_{W_2^{(1,1)}(Q)} = \int_Q (w, v + w_t, v_t + (\nabla w, \nabla v)) dx dt,$$

$$\|w\|_{W_2^{(1,1)}(Q)} = \sqrt{(w, w)_{W_2^{(1,1)}(Q)}}.$$

Let $\bar{W}_2^{(1)}(Q)$ be the class of functions that are square integrable over the cylinder Q and have the same square integrable first generalized derivatives with respect to their space coordinates.

Let, for some $j \in [1 : J]$, the quadratic forms under conditions

$$\sum_{s,r=1}^{n} a_{sr}^j(x) \xi_s \xi_r \geq \beta \sum_{r=1}^{n} \xi_r^2, \qquad \beta > 0, \qquad \xi_s, \xi_r \in \mathcal{R},$$

$$a_{sr}^j(x) = a_{rs}^j, \qquad \forall r, x \in [1 : n]$$

be given, where $a_{sr}^j(x)$ are bounded functions for $x \in \Omega \cup \Gamma$. Then, we introduce the operators

$$A_j w = - \sum_{s,r=1}^{n} \frac{\partial}{\partial x_s} a_{sr}^j(x) \left(\frac{\partial w}{\partial x_r} \right) \cos(N, x_s), \qquad j \in [1 : J]$$

and the derivatives along normals that are related to these operators,

$$\frac{\partial w}{\partial N_{A_j}} = \sum_{s,r=1}^{n} a_{sr}^j(x) \left(\frac{\partial w}{\partial x_r} \right) \cos(N, x_s), \qquad j \in [1 : J],$$

where $\cos(N, x_s)$ is the s-th direction cosine of the outer normal N to the boundary Σ.

Consider the block separable optimal control problem:

$$f = \sum_{j=1}^{J} \left[\int_{\Omega} q_j(z_j(x,T), x)dx + \int_{\Gamma} \int_{0}^{T} c_j(z_j(\sigma,t), u_j(\sigma,t)\sigma,t)d\sigma dt \right] \to \max, \tag{3.17}$$

$$\partial z_j(x,t)/\partial t + A_j z_j = f_j(x,t) \in \mathcal{L}_2(Q), \tag{3.18}$$

$$z_j(x,0) = z_j^{(0)}(x) \in \mathcal{L}_2(\Omega), \tag{3.19}$$

$$\partial z_j/\partial N_{A_j} = a_j(b_j(u_j(\sigma,t),\sigma,t) - z_j(\sigma,t)), \quad \sigma \in \Gamma, \quad t \in [0,T], \quad a_j > 0, \tag{3.20}$$

$$u_j(\sigma,t) \geq 0, \qquad p_j(z_j(\sigma,t), u_j(\sigma,t),\sigma,t) \leq 0, \qquad j \in [1:J], \tag{3.21}$$

$$\sum_{j=1}^{J} d_j(z_j(\sigma,t), u_j(\sigma,t),\sigma,t) \leq 0. \tag{3.22}$$

Here, the entering functions q_j, c_j, b_j $-d_j$ satisfy the condition of concavity, monotonicity, and differentiability introduced above.

A solution to problem (3.17)–(3.22) is a vector function of controls $u_j(\sigma,t) \in \mathcal{L}_2^I(\Sigma)$ with the corresponding weak solutions of mixed problems (3.18)–(3.20) $z_j(x,t) \in \bar{W}_2^{(1)}(Q)$, $j \in [1:J]$, so that the identities

$$\int_{\Omega} z_j y \Big|_{t_2}^{t_1} dx - \int_{\Omega} \int_{t_1}^{t_2} z_j \frac{\partial y}{\partial t} dx dt + \int_{\Omega} \int_{t_1}^{t_2} \sum_{s,r=1}^{n} a_{sr}^j(x) \frac{\partial z_j}{\partial x_r} \frac{\partial y}{\partial x_s} -$$
$$- \int_{\Omega} \int_{t_1}^{t_2} f_j y dx dt - \int_{\Gamma} \int_{t_1}^{t_2} a_j(b_j - z_j) y d\sigma dt = 0, \qquad j \in [1:J] \tag{3.23}$$

hold for every functions $y(x,t) \in W_2^{(1,1)}(Q)$ and every $t_1, t_2 \in [0,T]$. The satisfaction of initial conditions (3.19) is understood in the weak sense, i.e.,

$$\lim_{t \to 0} \int_{\Omega} (z_j(x,t) - z_j^{(0)}(x)) y(x,t) dx = 0, \quad \forall y \in W_2^{(1.1)}(Q), \quad j \in [1:J]. \tag{3.24}$$

The method of decomposition is constructed according to the scheme described above. Here, it is assumed that the Ter-Krikorov conditions indicated above are satisfied. These conditions reduce the main and all intermediate problems to the concave programming in the Banach space and establish the validity of duality theorems.

The aggregated problem is obtained from the initial one upon substituting $u_j(\sigma,t) = \alpha_j(\sigma,t)U(\sigma,t)$ with the corresponding notation. The dual of the aggregated one looks as follows:

$$\begin{aligned}
&\mathcal{L}\left(z_j(x,t), U(\sigma,t), y_j(x,t), \eta_j(\sigma,t), \delta(\sigma,t)\right) \to \min,\\
&\partial y_j(x,t)/\partial t - A_j y_j(x,t) = 0,\\
&y_j(x(T)) = \partial q_j(z_j(x,T),x)/\partial z_j,\\
&\partial y_j/\partial N_{A_j} = -a_j y_j(\sigma,t) + (\partial/\partial z_j)\left[C - P_j\eta_j(\sigma,t) - D\delta(\sigma,t)\right], \qquad (3.25)\\
&\sigma \in \Gamma, \qquad t \in [0,T], \qquad j \in [1:J];\\
&(\partial/\partial U^i)\left[C + a_j B_j y_j(\sigma,t) - P_j\eta_j(\sigma,t) - D\delta(\sigma,t)\right] \le 0, \quad i \in [1:I],\\
&U^i(\sigma,t) \ge 0, \quad i \in [1:I], \quad \eta_j(\sigma,t) \ge 0, \quad j \in [1:J], \quad \delta(\sigma,t) \ge 0.
\end{aligned}$$

Here, $\eta_j(\sigma,t) \in \mathcal{L}_2^{K_j}(\Sigma)$, $j \in [1:J]$, $\delta(\sigma,t) \in \mathcal{L}_2^L(\Sigma)$, and the weak solution $y_j(x,t)$ is sought in the space $W_2^{(1.1)}(Q)$ and is determined by the identity

$$\int_Q \left(\frac{\partial y_j}{\partial t} z - \sum_{s,r=1}^{n} a_{sr}^j(x) \frac{\partial y_j}{\partial x_r}\frac{\partial z}{\partial x_s}\right) dt dx - \\
- \int_\Sigma \left[a_j y_j - \frac{\partial}{\partial z_j}(C - P_j\eta_j - D\delta)\right] z d\sigma dt = 0, \qquad (3.26)$$

which should hold for any function $z(x,t) \in \bar{W}_2^{(1)}(Q)$. The initial condition in (3.25) should be satisfied in the weak sense, i.e.,

$$\lim_{t\to T-0} \int_\Omega \left(y_j(x,t) - \partial q_j(z_j(x,T),x)/\partial z_j\right) z(x,t) dx = 0, \\
\forall z(x,t) \in \bar{W}_2^{(1)}. \qquad (3.27)$$

By analogy with (1.12), (3.11), we deduce the optimality criterion of the intermediate disaggregated solution, which, in this case, has the form

$$\sum_{j=1}^{J} \left\{ \int_\Omega q_j(\hat{z}_j(x,T),x) dx + \int_\Sigma \left[c_j(\hat{z}_j(\sigma,t), \hat{u}_j(\sigma,t), \sigma, t) - \right.\right.$$

$$-d_j(\hat{z}_j(\sigma,t),\hat{u}_j(\sigma,t),\sigma,t)\,\mathring{\delta}(\sigma,t) - c_j(\mathring{z}_j(\sigma,t),\mathring{u}_j(\sigma,t),\sigma,t)\Big]\,d\sigma dt -$$

$$-\int\limits_{\Omega} q_j(\mathring{z}_j(x,T),x)dx \Bigg\} = 0.$$

Local monotonicity with respect to the functional of the iterative process is established using the formula for the marginal value of parametrically aggregated problems. We consider the central aspects in the deduction of this formula. The Lagrangian has the form

$$\mathcal{L} = \sum_{j=1}^{J}\Bigg\{\int\limits_{\Omega} q_j(z_j(x,T),x)dx + \int\limits_{\Sigma} c_j(z_j,\alpha_j U,\sigma,t)d\sigma dt -$$

$$-\int\limits_{\Omega} z_j y_j\Big|_{+0}^{T-0} dx + \int\limits_{\Omega} z_j y_j z_j \frac{\partial y_j}{\partial t} dx dt - \int\limits_{Q}\sum_{s,r=0}^{n} a_{sr}^{j}(x)\frac{\partial z_j}{\partial x_r}\frac{\partial y_j}{\partial x_s} dx dt +$$

$$+\int\limits_{Q} f_j y_j dx dt + \int\limits_{\Sigma} a_j(b_j(\alpha_j U,\sigma,t) - z_j)y_j d\sigma dt -$$

$$-\int\limits_{\Sigma} p_j(z_j,\alpha_j U,\sigma,t)\eta_j d\sigma dt - \int\limits_{\Sigma} d_j(z_j,\alpha_j U,\sigma,t)\delta d\sigma dt\Bigg\}.$$

By the theorem of marginal value from Chapter 1, we have

$$(g(0))' = \sum_{j=1}^{J}\Bigg(\int\limits_{\Omega}\frac{\partial q_j}{\partial z_j}\frac{\partial \mathring{z}_j(x,T,0)}{\partial\rho}dx + \int\limits_{\Sigma}\Bigg(\frac{\partial c_j}{\partial z_j}\frac{\partial \mathring{z}_j(\sigma,t,0)}{\partial\rho} +$$

$$+\frac{\partial c_j}{\partial u_j}\frac{\partial \alpha_k(\sigma,t,0)}{\partial\rho}\mathring{U}\Bigg)\Bigg) d\sigma dt - \int\limits_{\Omega}\frac{\partial \mathring{z}_j(x,t,0)}{\partial\rho}\mathring{y}_j(x,t)\Big|_{+0}^{T-0} dx +$$

$$+\int\limits_{\Omega}\frac{\partial \mathring{z}_j(x,t,0)}{\partial\rho}\frac{\partial \mathring{y}_j}{\partial t}dx dt - \int\limits_{\Omega}\sum_{s,r=0}^{n} a_{sr}^{j}(x)\frac{\partial}{\partial x_r}\frac{\partial \mathring{z}_j(x,t,0)}{\partial\rho}\frac{\partial \mathring{y}_j}{\partial x_s}dx dt +$$

$$+\int\limits_{\Sigma} a_j\left(\frac{\partial b_j}{\partial u_j}\frac{\partial \alpha_j(\sigma,t,0)}{\partial\rho}\mathring{U} - \frac{\partial \mathring{z}_j(\sigma,t,0)}{\partial\rho}\right)\mathring{y}_j d\sigma dt -$$

$$-\int\limits_{\Sigma}\left(\frac{\partial p_j}{\partial z_j}\frac{\partial \mathring{z}_j(\sigma,t,0)}{\partial\rho} + \frac{\partial p_j}{\partial u_j}\frac{\partial \alpha_j(\sigma,t,0)}{\partial\rho}\mathring{U}\right)\mathring{\eta}_j -$$

$$-\left(\frac{\partial d_j}{\partial z_j}\frac{\partial \overset{\circ}{z}_j(\sigma,t,0)}{\partial \rho}+\frac{\partial d_j}{\partial u_j}\frac{\partial \alpha_j(\sigma,t,0)}{\partial \rho}\overset{\circ}{U}\right)d\sigma dt. \qquad (3.28)$$

The first integral and the upper limit of the third integral in the right-hand side of (3.28) are cancelled due to (3.27). The lower limit of the third integral is equal to zero according to (3.24). All integrals that contain the factor $\partial \overset{\circ}{z}_j/\partial p$ are cancelled due to the definition (3.26). If we write down the expression for $\partial\alpha_j(\sigma,t,0)\partial p$ in detail, then the terms with the sum $\sum_{r=1}^{J}\hat{u}_r(\sigma,t)$ are cancelled as before. Finally, we obtain the expression analogous to (3.12):

$$(\overset{\circ}{g}(0))' = \sum_{j=1}^{J}\left\{\int_{\Sigma}\frac{\partial}{\partial u_j}\left[c_j+b_j\overset{\circ}{y}_j-p_j\overset{\circ}{\eta}_j-d_j\overset{\circ}{\delta}\right](\hat{u}_j-\overset{\circ}{u}_j)d\sigma dt\right\}. \qquad (3.29)$$

Consider the definition of the weak solution (3.26) for the functions $\overset{\circ}{y}_j(x,t)$, $j\in[1:J]$. As test functions, we take $(\hat{z}_j(x,t)-\overset{\circ}{z}_j(x,t))\in \bar{W}_2^{(1)}(Q)$:

$$\int_Q\left(\frac{\partial \overset{\circ}{y}_j}{\partial r}(\hat{z}_j-\overset{\circ}{z}_j)-\sum_{s,r=1}^{n}a^i_{sr}(x)\frac{\partial \overset{\circ}{y}_j}{\partial x_r}\frac{\partial}{\partial x_r}\frac{\partial}{\partial x_s}(\hat{z}_j-\overset{\circ}{z}_j)\right)dxdt-$$

$$-\int_{\Sigma}\left[a_j\overset{\circ}{y}_j-\frac{\partial}{\partial z_j}(c_j-p_j\overset{\circ}{\eta}_j-d_j\overset{\circ}{\delta})\right](\hat{z}_j-\overset{\circ}{z}_j)d\sigma dt=0, \qquad j\in[1:J]. \qquad (3.30$$

We add the following terms, equal to zero due to (3.24) and (3.27), to the left-hand side of (3.30):

$$\int_{\Omega}(\partial q_j(\overset{\circ}{z}_j(x,T),x)/\partial z_j(\hat{z}_j-\overset{\circ}{z}_j)dx-$$

$$-\int_{\Omega}\overset{\circ}{y}_j(x,T-0)(\hat{z}_j(x,T-0)-\overset{\circ}{z}_j(x,T-0))dx=$$

$$-\int_{\Omega}\overset{\circ}{y}_j(x,T-0)(\hat{z}_j(x,t+0)-\overset{\circ}{z}_j(x,t+0))dx.$$

If we apply the definition of the weak solution (3.23) for the functions $(\hat{z}_j-\overset{\circ}{z}_j)$ to the obtained expression and consider $\overset{\circ}{y}_j$ as test functions, we obtain

$$\sum_{j=1}^{J}\left\{\int_{\Sigma} a_j(b_j(\hat{u}_j,\sigma,t)-b_j(\overset{\circ}{u}_j,\sigma,t))\,\overset{\circ}{y}_j d\sigma dt+\right.$$

$$\left.+\int_{\Omega}\frac{\partial q_j}{\partial z_j}(\overset{\circ}{z}_j(x,T),x)(\hat{z}_j-\overset{\circ}{z}_j)dx+\int_{\Sigma}\left[c_j-p_j\overset{\circ}{\eta}_j-d_j\overset{\circ}{\delta}\right](\hat{z}_j-\overset{\circ}{z}_j)d\sigma dt\right\}=0.$$

We add the left-hand side of this relation, together with the terms

$$\sum_{j=1}^{J}\left[\int_{\Sigma} p_j(\overset{\circ}{z}_j(\sigma,t),\overset{\circ}{u}_j(\sigma,t),\sigma,t)\,\overset{\circ}{\eta}_j d\sigma dt\right],$$

$$\sum_{j=1}^{J}\left[\int_{\Sigma} d_j(\overset{\circ}{z}_j(\sigma,t),\overset{\circ}{u}_j(\sigma,t),\sigma,t)\,\overset{\circ}{\delta}\, d\sigma dt\right]$$

equal to zero due to the conditions of complementary slackness, to the right-hand side of (3.29) and finally obtain the formula for the marginal value analogous to (1.19). Here, it is important to note that $\overset{\circ}{y}_j(x,t)\geq 0$, $j\in[1:J]$. This follows from the nonnegativity of functions in finite and boundary conditions of problem (3.25) due to the assumed monotonicity of the entering functions as well as from the positiveness of Green function of the corresponding mixed problems.

Consider the hyperbolic case. Let $W_2^{(1,1)}(Q)$ be a Hilbert space obtained by complementing the set of infinitely differentiable functions equal to zero in the neighborhood of the lateral area Σ of the cylinder Q with respect to the norm $||\cdot||_{W_2^{(1,1)}(Q)}$

We introduce the block separable optimal control problem with partial differential equations of the hyperbolic type under distributed control:

$$f=\sum_{j=1}^{J}\left[\int\limits_{Q} c_j(z_j, z_{jx}, z_{jt}, u_j, x, t)dxdt+\right.$$
$$\left.+\int\limits_{\Omega} q_j(z_j(x,T), z_{jx}(x,T), z_{jt}(x,T), x)dx\right]\to \max,$$
$$\partial^2 z_j(x,t)/\partial t^2 + A_j z_j(x,t) = b_j(u_j(x,t), x, t),$$
$$z_j(x,0)=z_j^{(0)}(x)\in \overset{\circ}{W}{}_2^{(1)}(\Omega), \qquad z_{jt}(x,0)=z_j^{(1)}(x)\in \mathcal{L}_2(\Omega), \tag{3.31}$$
$$z_j(\sigma,t)=0, \qquad \sigma,t\in\Sigma,$$
$$u_j(x,t)\geq 0, \qquad p_j(z_j(x,t), u_j(x,t), x, t)\leq 0, \qquad j\in[1:J],$$
$$\sum_{j=1}^{J} d_j(z_j(x,t), u_j(x,t), x, t)\leq 0.$$

Here, in addition to the previous assumptions, we consider that the functions c_j, q_j are concave and monotonic with respect to variables z_{jx}, z_{jt}.

Optimal controls $u_j(x,t)$, $j\in[1:J]$ of problem (3.31) are sought in the space $\mathcal{L}_2^1(\Omega)$. They correspond to the weak solutions $z_j(x,t)\in \overset{\circ}{W}{}_2^{(1,1)}(Q)$, $j\in[1:J]$ of mixed problems in (3.31) that are defined by the identities

$$\int\limits_{\Omega}\frac{\partial z_j}{\partial t}\bigg|_{t_1}^{t_2}dx-\int\limits_{\Omega}\int\limits_{t_1}^{t_2}\left(\frac{\partial z_j}{\partial t}\frac{\partial y}{\partial t}-\sum_{r,s=1}^{n}a_{r,s}^{j}(x)\frac{\partial z_j}{\partial x_r}\frac{\partial y}{\partial x_s}+\right.$$
$$\left.+\,b_j y\right)dxdt=0, \qquad j\in[1:J] \tag{3.32}$$

for all test functions $y(x,t)\in \overset{\circ}{W}{}_2^{(1,1)}(Q)$ and $0\leq t_1\leq t_2\leq T$. The satisfaction of the second initial condition is understood in the weak sense

$$\lim_{\tau\to+0}\int\limits_{\Omega}(z_{jt}(x,t)-z_j^{(1)}(x))y(x,t)dx=0, \quad \forall y\in \overset{\circ}{W}{}_2^{(1,1)}(Q), \quad j\in[1:J]. \tag{3.33}$$

The method of decomposition is constructed as described above. The aggregated problem is formed after the substitution $u_j(x,t)=\alpha_j(x,t)U(x,t)$ and fixing the weights $\alpha_j(x,t)$. The dual of the aggregated problem one has the form

$$
\begin{aligned}
&\mathcal{L}(z_j(x,t), U(x,t), y_i(x,t), \eta_j(x,t), \delta(x,t)) \to \min, \\
&\partial^2 y_j(x,t)/\partial t^2 + A_j y_j(x,t) = (\partial/\partial z_j)\,[c_j - p_j\eta_j - d_j\delta] - \\
&-\frac{\partial}{\partial t}\left(\frac{\partial c_j}{\partial z_{jt}}\right) - \sum_{r=1}^{n} \frac{\partial}{\partial x_r}\left(\frac{\partial c_j}{\partial z_{jx_r}}\right), \\
&y_i(x,T) = \frac{\partial q_j}{\partial z_{jt}}, \qquad y_{it}(x,T) = -\frac{\partial q_j}{\partial z_j} + \sum_{r=1}^{n} \frac{\partial}{\partial x_r}\left(\frac{\partial q_j}{\partial z_{jx_r}}\right) - \\
&\left.\frac{\partial x_j}{\partial z_{jt}}\right|_{T-0}, \quad x \in \Omega, \quad y_j(\tau, t) = 0, \quad \sigma, t \in \Sigma, \quad j \in [1:J]; \\
&\frac{\partial}{\partial U^i}\left[C + \sum_{j=1}^{J} B_j y_j - P_j \eta_j - D\delta\right] \leq 0, \qquad i \in [1:I], \\
&U^i(x,t) \geq 0, \quad i \in [1:I], \quad \eta_j(x,t) \geq 0, \quad j \in [1:J], \quad \delta(x,t) \geq 0.
\end{aligned}
\tag{3.34}
$$

Here, $\eta_j(x,t) \in \mathcal{L}_2^{K_j}(Q)$, $\delta(x,t) \in \mathcal{L}_2^{L}(Q)$, $U(x,t) \in \mathcal{L}_2^{I}(Q)$, and the functions $y_i(x,t) \in W_2^{(1,1)}(Q)$ are weak solutions of mixed problems from (3.34) and satisfy the identities

$$
\int_{\Omega} \left.\frac{\partial y_j}{\partial t}\right|_{t_1}^{t_2} dx - \int_{\Omega}\int_{t_1}^{t_2} \left(\frac{\partial y_j}{\partial t}\frac{\partial z}{\partial t} - \sum_{r,s=1}^{n} a_{r,s}^{j}(x) \frac{\partial y_j}{\partial x_r}\frac{\partial z}{\partial x_s} + \right. \tag{3.35}
$$

$$
\left. + \sum_{r=1}^{n} \frac{\partial c_j}{\partial z_{jx_r}} \frac{\partial z}{\partial x_r}\right) dx dt + \int_{\Omega} \left.\frac{\partial c_j}{\partial z_{jt}} z\right|_{t_1}^{t_2} dx = 0
$$

for any test functions $z(x,t) \in \overset{\circ}{W}{}_2^{(1,1)}(Q)$ and any $t_1, t_2 \in [0,T]$. Satisfaction of the second final condition in (3.34) is understood in the weak sense

$$
\lim_{t \to T-0} \int_{\Omega} \left(y_{jt}(x,t) + \frac{\partial c_j}{\partial z_{jt}} + \frac{\partial q_j(x,T)}{\partial z_j} - \sum_{r=1}^{n} \frac{\partial}{\partial x_r}\left(\frac{\partial q_j(x,T)}{\partial z_j x_r}\right)\right) z(x,t) dx = 0
$$

$$
\forall z(x,t) \in \overset{\circ}{W}{}_2^{(1,1)}(Q), \qquad j \in [1:J]. \tag{3.36}
$$

The optimality criterion for the disaggregated solution is deduced as for the elliptic and parabolic cases. We will discuss the main aspects of the deduction of the formula for the marginal value of parametric aggregated problems. The Lagrangian in (3.34) looks as follows:

$$\mathcal{L} = \sum_{j=1}^{J} \left\{ \int_{\Omega} c_j(z_j, z_{jx}, z_{jt}, u_j, x, t) dx dt + \int_{\Omega} q_j(z_j(x,T), \right.$$

$$\left. z_{jx}(x,T), z_{jt}(x,T)x) dx \right\} + \int_{Q} \left(\frac{\partial z_j}{\partial t} \frac{\partial y_j}{\partial t} - \right.$$

$$\left. - \sum_{r,s=1}^{n} a_{rs}^{j}(x) \frac{\partial z_j}{\partial x_r} \frac{\partial y_j}{\partial x_s} + b_j y_j \right) dx dt -$$

$$- \int_{\Omega} \frac{\partial z_j}{\partial t} y_i \Big|_{+0}^{T-0} dx - \int_{Q} (p_j(z_j, u_j, x, t)\eta_j(x,t) + d_j(z_j, u_j, x, t)\delta(x,t)) dx dt.$$

The marginal value is computed in the following way:

$$(g(0))' = \sum_{j=1}^{J} \left\{ \int_{\Omega} \left(\frac{\partial c_j}{\partial z_j} \frac{\partial \overset{\circ}{z}_j(x,t,0)}{\partial \rho} + \sum_{r=1}^{n} \frac{\partial c_j}{\partial z_{jx_r}} \frac{\partial \overset{\circ}{z}_{jx_r}(x,t,0)}{\partial \rho} + \right. \right.$$

$$\left. + \frac{\partial c_j}{\partial z_{jt}} \frac{\partial \overset{\circ}{z}_{jt}(x,t,0)}{\partial \rho} + \frac{\partial c_j}{\partial u_j} \frac{\partial \alpha_j(x,t,0)}{\partial \rho} \overset{\circ}{U} \right) dx dt +$$

$$\int_{\Omega} \left[\frac{\partial q_j}{\partial z_j} \frac{\partial \overset{\circ}{z}_j(x,T,0)}{\partial \rho} + \sum_{r=1}^{n} \frac{\partial q_j}{\partial z_{jx_r}} \frac{\partial \overset{\circ}{z}_{jx_r}(x,T,0)}{\partial \rho} + \frac{\partial q_j}{\partial z_{jt}} \frac{\partial \overset{\circ}{z}_{jt}(x,T,0)}{\partial \rho} \right] dx +$$

$$+ \int_{Q} \left(\frac{\partial}{\partial t} \left(\frac{\partial z_j(x,t,0)}{\partial \rho} \right) \frac{\partial \overset{\circ}{y}_j}{\partial t} - \sum_{r,s=1}^{n} a_{rs}^{j}(x) \frac{\partial}{\partial x_r} \left(\frac{\partial \overset{\circ}{z}_j(x,t,0)}{\partial \rho} \right) \frac{\partial \overset{\circ}{y}}{\partial x_s} + \right.$$

$$\left. + \frac{\partial b_j}{\partial u_j} \frac{\partial \alpha_j(x,t,0)}{\partial \rho} \overset{\circ}{U} \overset{\circ}{y}_j \right) dx dt - \int_{\Omega} \frac{\partial}{\partial t} \left(\frac{\partial \overset{\circ}{z}_j(x,t,0)}{\partial \rho} \right) \overset{\circ}{y}_j \Big|_{+0}^{T-0} dx -$$

$$- \int_{Q} \left[\left(\frac{\partial p_j}{\partial z_j} \frac{\partial \overset{\circ}{z}_j(x,t,0)}{\partial \rho} + \frac{\partial p_j}{\partial u_j} \frac{\partial \alpha_j(x,t,0)}{\partial \rho} \overset{\circ}{U} \right) \overset{\circ}{\eta}_j + \right.$$

$$\left. \left. + \left(\frac{\partial d_j}{\partial z_j} \frac{\partial \overset{\circ}{z}_j(x,t,0)}{\partial \rho} + \frac{\partial d_j}{\partial u_j} \frac{\partial \alpha_j(x,t,0)}{\partial \rho} \overset{\circ}{U} \right) \overset{\circ}{\delta} \right] dx dt \right\}. \quad (3.37)$$

The lower limit of the fourth integral in the right-hand side of (3.37) is equal to zero due to the second initial condition in (3.31) and the definition (3.33). The upper limit of this integral is cancelled with the third term of the second integral according to the first initial condition in (3.34). We group the first three terms of the first integral in the right-hand side of (3.37), the two first terms of the third integral, and the first and the third terms of the last integral. According to definition (3.35) with the test function $\partial \overset{\circ}{z}_j(x,t,0)/\partial\rho$, these grouped terms give

$$\int_\Omega \left(\frac{\partial \overset{\circ}{y}_j}{\partial t} + \frac{\partial c_j}{\partial z_{jt}}\right) \frac{\partial \overset{\circ}{z}_j(x,t,0)}{\partial \rho}\Big|_{+0}^{T-0} dx. \tag{3.38}$$

We transform the second term in the second integral of the right-hand side of (3.37) as follows:

$$\int_\Omega \left[\sum_{r=1}^n \frac{\partial q_j}{\partial z_{jx_r}} \frac{\partial z_{jx_r}(x,T,0)}{\partial \rho}\right] dx =$$
$$= \int_\Omega \left[\sum_{r=1}^n \frac{\partial}{\partial x_r}\left(\frac{\partial q_j}{\partial z_{x_r}} \frac{\partial \overset{\circ}{z}_j(x,T,0)}{\partial \rho}\right)\right] dx -$$
$$- \int_\Omega \left[\sum_{r=1}^n \frac{\partial}{\partial x_r}\left(\frac{\partial q_j}{\partial z_{x_r}}\right) \frac{\partial \overset{\circ}{z}_j(x,T,0)}{\partial \rho}\right] dx. \tag{3.39}$$

The first integral in the right-hand side of (3.39) is equal to zero due to the Gauss-Ostrogradskii formula and the boundary condition in (3.31). The lower limit in (3.38) is equal to zero due to the initial condition for $z_j(x,t)$. The upper limit in (3.38) is cancelled with the second integral in the right-hand side of (3.39) as well as with the first term in the second integral of the right-hand side of (3.37). This follows from definition (3.36), where $\partial \overset{\circ}{z}(x,t,0)/\partial\rho$ plays the role of the test function. Finally, we obtain the expression analogous to (3.12), (3.29)

$$(\overset{\circ}{g}(0))' = \sum_{j=1}^J \left\{ \int_Q \frac{\partial}{\partial u_j}[c_j + b_j \overset{\circ}{y}_j - p_j \overset{\circ}{\eta}_j - d_j \overset{\circ}{\delta}](\hat{u}_j - \overset{\circ}{u}_j) dx dt \right\}. \tag{3.40}$$

Consider the definitions of weak solutions for the functions $\overset{\circ}{y}_j(x,t)$, $j \in [1:J]$, where the solutions $(\hat{z}_j(x,t) - \overset{\circ}{z}_j(x,t))$ are taken as test functions. We have

$$-\int_\Omega \frac{\partial \overset{\circ}{y}_j}{\partial t}(\hat{z}_j - \overset{\circ}{z}_j)\Big|_{+0}^{T-0} dx + \int_Q \left[\frac{\partial y_j}{\partial t}\frac{\partial}{\partial t}(\hat{z}_j - \overset{\circ}{z}_j)\right.$$

$$-\sum_{r,s=1}^{n} a_{r,s}^{j}(x)\frac{\partial y_j}{\partial x_r}\frac{\partial}{\partial x_s}(\hat{z}_j - \overset{\circ}{z}_j)+$$

$$+\frac{\partial}{\partial z_j}(c_j - p_j\overset{\circ}{\eta}_j - d_j\overset{\circ}{\delta})(\hat{z}_j - \overset{\circ}{z}_j) + \frac{\partial c_j}{\partial z_{jt}}\frac{\partial}{\partial t}(\hat{z}_j - \overset{\circ}{z}_j)+$$

$$+\sum_{r=1}^{n}\frac{\partial c_j}{\partial z_{jx_r}}\frac{\partial}{\partial x_r}(\hat{z}_j - \overset{\circ}{z}_j)\Bigg]\,dxdt - \int_{\Omega}\frac{\partial c_j}{\partial z_{jt}}(\hat{z}_j - \overset{\circ}{z}_j)\Bigg|_{+0}^{T-0}dx = 0. \tag{3.41}$$

The lower limits of the first and last integrals in the left-hand sides of (3.41) are equal to zero due to the boundary conditions for the functions $\hat{z}_j(x,t)$, $\overset{\circ}{z}_j(x,t)$. According to definition (3.36), the upper limits of these integrals are equal to

$$\int_{\Omega}\left[\frac{\partial q_j(x,T)}{\partial z_j} - \sum_{r=1}^{n}\frac{\partial}{\partial x_r}\left(\frac{\partial q_j(x,T)}{\partial z_{jx_r}}\right)\right](\hat{z}_j(x,T) - \overset{\circ}{z}_j(x,T))dx. \tag{3.42}$$

Because of the Gauss-Ostrogradskii formula and boundary conditions for $\overset{\circ}{z}_j(x,T)$, $\hat{z}_j(x,T)$, the second integral in (3.42) is transformed to the form

$$\int_{\Omega}\sum_{r=1}^{n}\left[\frac{\partial q_j(x,T)}{\partial z_{jx_r}}\frac{\partial}{\partial x_r}(\hat{z}_j(x,T) - \overset{\circ}{z}_j(x,T))\right]dx. \tag{3.43}$$

We add the expression

$$-\int_{\Omega}\frac{\partial}{\partial t}(\hat{z}_j - \overset{\circ}{z}_j)\,\overset{\circ}{y}_j\Bigg|_{+0}^{T-0}dx + \int_{\Omega}\frac{\partial}{\partial t}(\hat{z}_j - \overset{\circ}{z}_j)\,\overset{\circ}{y}_j\Bigg|_{+0}^{T-0}dx. \tag{3.44}$$

which is identical to zero, to the left-hand side of (3.41). According to the definition of the weak solution (3.32) as applied to $(\hat{z}_j(x,t) - \overset{\circ}{(}_jx,t))$ with test functions $\overset{\circ}{y}_j(x,t)$, the second integral in (3.44), together with the first two terms of the second integral in the left-hand side of (3.41), give

$$\int_{Q}(b_j(\overset{\circ}{u}_j,x,t) - b_j(\hat{u}_j,x,t))\,\overset{\circ}{y}_j(x,t)dxdt. \tag{3.45}$$

The lower limit in the second integral (3.44) is equal to zero due to the initial conditions for the functions $\partial\hat{z}_j(x,0)/\partial t$, $\partial\,\overset{\circ}{z}_j(x,0)/\partial t$. Due to the first initial condition in (3.34), the upper limit of this integral is equal to

$$\int_{\Omega}\frac{\partial q_j(x,T)}{\partial z_{jt}}\frac{\partial}{\partial t}(\hat{z}_j(x,T) - \overset{\circ}{z}_j(x,T))dx. \tag{3.46}$$

Thus, we obtain

$$(\overset{\circ}{g}(0))' = \sum_{j=1}^{J} \left\{ \int_{Q} \left[\frac{\partial}{\partial u_j}(c_j + b_j \overset{\circ}{y}_j - p_j \overset{\circ}{\eta}_j - d_j \overset{\circ}{\delta})(\hat{u}_j - \overset{\circ}{u}_j) + \right.\right.$$

$$+ \frac{\partial}{\partial z_j}(c_j - p_j \overset{\circ}{\eta}_j - d_j \overset{\circ}{\delta})(\hat{z}_j - \overset{\circ}{z}_j) + \frac{\partial c_j}{\partial z_{jt}}(\hat{z}_{jt} - \overset{\circ}{z}_{jt}) +$$

$$+ \sum_{r=1}^{n} \frac{\partial c_j}{\partial z_{jx_r}}(\hat{z}_{jx_r} - \overset{\circ}{z}_{jx_r}) +$$

$$\left. + (b_j(\overset{\circ}{u}_j, x, t) - b_j(\hat{u}, x, t))\, \overset{\circ}{y}_j(x, t) \right] dx dt +$$

$$+ \int_{\Omega} \left[\frac{\partial q_j(x, T)}{\partial z_j}(\hat{z}_j(x, T) - \overset{\circ}{z}_j(x, T)) + \sum_{r=1}^{n} \frac{\partial q_j(x, T)}{\partial z_{jx_r}}(\hat{z}_{jx_r}(x, T) - \right.$$

$$\left.\left. - \overset{\circ}{z}_{jx_r}(x, T)) + \frac{\partial q_j(x, T)}{\partial z_{jt}}(\hat{z}_{jt}(x, T) - \overset{\circ}{z}_{jt}(x, T)) \right] dx \right\}.$$

Further transformations according to scheme of Section 1 lead to formulas for marginal value analogous to (1.19). Here, we need to additionally require the nonnegativity of solutions $\overset{\circ}{y}_j(x, t)$, $j \in [1 : J]$. This assumption becomes unnecessary when the functions $b_j(u_j, x, t)$ are linear with respect to control.

Consider the block separable problem of optimal control, where phase variables satisfy the sets of first-order hyperbolic partial differential equations:

$$f = \sum_{j=1}^{J} \left\{ \int_0^l \int_0^T c_j(z_j(x,t), u_j(x,t), x, t) dx dt + \right.$$
$$\left. \int_0^T w_j(z_j(l,t), t) dt + \int_0^l q_j(z_j(x,T), x) dx \right\} \to \max,$$

$$\partial z_j(x,t)/\partial t + \xi_j(x,t) \partial z_j(x,t)/\partial x =$$
$$= A_j(x,t) z_j(x,t) + b_j(u_j(x,t), x, t), \tag{3.47}$$

$$z_j(x,0) = z_j^{(0)}(x) \qquad x \in [0,l], \qquad z_j(t,0) = z_j^{(0)}(t), \qquad t \in [0,T],$$
$$u_j(x,t) \geq 0, \qquad p_j(z_j(x,t), u_j(x,t), x, t) \leq 0, \qquad j \in [1:J],$$

$$\sum_{j=1}^{J} d_j(z_j(x,t), u_j(x,t), x, t) \leq 0.$$

Here, x is a one-dimensional space coordinate that varies in the range $[0,l]$. For every $j \in [1:J]$, the vector functions z_j, b_j, ξ_j have dimensions N_j; A_j are $N_j \times N_j$-matrices with nonnegative components. Characteristic directions $\xi_j(x,t)$ are considered positive and are monotonic in x. The vectors of control $u_j(x,t)$, $j \in [1:J]$ with dimension I are sought in the class of functions that are measurable and almost everywhere limited along any characteristic curve that corresponds to equations (3.47) and that belong to the rectangular $[0,l] \times [0,T]$. Assumptions on the concavity, monotony and differentiability of entering functions c_j, w_j, q_j, b_j, p_j, d_j remain the same. Note that linear first-order hyperbolic equations in (3.47) are written in the form of Riemann invariants.

The conjugate problem of the aggregated problem in the iterative process has the following form:

$$
\begin{aligned}
&\mathcal{L}(z_j(x,t), U(x,t), y_j(x,t), \eta_j(x,t), \delta(x,t)) \to \min,\\
&\partial y_j(x,t)/\partial t + \xi_j(x,t)\partial y_j(x,t)/\partial x = -(\partial \xi_j(x,t)/\partial x)y_j -\\
&-A_j^T(x,t)y_j - (\partial/\partial z_j)(c_j - p_j\eta_j - d_j\delta),\\
&y_j(x,T) = \frac{\partial q_j}{\partial z_j}(x,T), \qquad x \in [0,l],\\
&y_j(l,t) = [\partial w_j/\partial z_j(l,t)]\,[\xi_j(x,t)]^{-1}, \qquad t \in [0,T], \qquad j \in [1:J],\\
&\frac{\partial}{\partial U^i}\left[C + \sum_{j=1}^{J}(B_j y_j - P_j\eta_j) + D\delta\right] \le 0, \qquad i \in [1:I],\\
&U^i(x,t) \ge 0, \quad i \in [1:I], \quad \eta_j(x,t) \ge 0, \quad j \in [1:J], \quad \delta(x,t) \ge 0.
\end{aligned}
\tag{3.48}
$$

We summarize deduction of the formula for the marginal value of parametrically aggregated problems. The Lagrangian in (3.48) has the following form:

$$
\begin{aligned}
\mathcal{L} = \sum_{j=1}^{J}\Bigg\{&\int_0^l\int_0^T c_j(z_j, \alpha_j U, x, t)\,dxdt + \int_0^T w_j(z_j(l,t),t)\,dt +\\
&+\int_0^l q_j(z_j(x,T),x)\,dx + \int_0^l\int_0^T [(A_j z_j, y_j) + b_j(\alpha_j U, x, t)y_j -\\
&-\left(\frac{\partial z_j}{\partial t} + \xi_j\frac{\partial z_j}{\partial x}\right)y_j - p_j(z_j,\alpha_j U,x,t)\eta_j - d_j(z_j,\alpha_j U,x,t)\delta\Big]\,dxdt\Bigg\}.
\end{aligned}
$$

The marginal value is computed in the following form:

$$
\begin{aligned}
(g(0))' = \sum_{j=1}^{J}\Bigg\{&\int_0^l\int_0^T\left[\frac{\partial c_j}{\partial z_j}\frac{\partial \overset{\circ}{z}_j(x,t,0)}{\partial \rho} + \frac{\partial c_j}{\partial u_j}\frac{\partial \alpha_j(x,t,0)}{\partial \rho}\overset{\circ}{U}\right]dxdt +\\
&+\int_0^T \frac{\partial w_j}{\partial z_j(l,t)}\frac{\partial \overset{\circ}{z}_j(l,t,0)}{\partial \rho}\,dt + \int_0^l \frac{\partial q_j}{\partial z_j(x,T)}\frac{\partial \overset{\circ}{z}_j(x,T,0)}{\partial \rho}\,dx +\\
+\int_0^l\int_0^T\Bigg[&\left(A_j\frac{\partial \overset{\circ}{z}_j(x,t,0)}{\partial \rho}\overset{\circ}{y}_j\right) + \frac{\partial b_j}{\partial u_j}\frac{\partial \alpha_j(x,t,0)}{\partial \rho}\overset{\circ}{U}\,\overset{\circ}{y}_j -\\
&-\left(\frac{\partial}{\partial t}\left(\frac{\partial z_j(x,t,0)}{\partial \rho}\right) + \xi_j\frac{\partial}{\partial x}\left(\frac{\partial \overset{\circ}{z}_j(x,t,0)}{\partial \rho}\right)\right)\overset{\circ}{y}_j - \left(\frac{\partial p_j}{\partial z_j}\frac{\partial \overset{\circ}{z}_j(x,t,0)}{\partial \rho} +\right.
\end{aligned}
\tag{3.49}
$$

$$+\frac{\partial p_j}{\partial u_j}\frac{\partial \alpha_j(x,t,0)}{\partial \rho}\overset{\circ}{U}\Big)\overset{\circ}{\eta}_j - \left(\frac{\partial d_j}{\partial z_j}\frac{\partial \overset{\circ}{z}_j(x,t,0)}{\partial \rho} + \frac{\partial d_j}{\partial u_j}\frac{\partial \alpha_j(x,t,0)}{\partial \rho}\overset{\circ}{U}\right)\overset{\circ}{\delta}\Big] dxdt\Big\}.$$

We integrate by parts the third and the fourth terms in the fourth integral of the right-hand side of (3.49):

$$-\int_0^l\int_0^T \frac{\partial}{\partial t}\frac{\partial \overset{\circ}{z}_j(x,t,0)}{\partial \rho}\overset{\circ}{y}_j dxdt =$$

$$= -\int_0^l \frac{\partial \overset{\circ}{z}_j(x,t,0)}{\partial \rho}\overset{\circ}{y}_j\Big|_0^T dx + \int_0^l\int_0^T \frac{\partial \overset{\circ}{z}_j(x,t,0)}{\partial \rho}\frac{\partial \overset{\circ}{y}_j}{\partial t}dxdt; \qquad (3.50)$$

$$-\int_0^l\int_0^T \xi_j\frac{\partial \overset{\circ}{z}_j(x,t,0)}{\partial \rho}\overset{\circ}{y}_j dxdt =$$

$$= -\int_0^T \xi_j\frac{\partial \overset{\circ}{z}_j(x,t,0)}{\partial \rho}\overset{\circ}{y}_j\Big|_0^l dt + \int_0^l\int_0^T \frac{\partial \xi_j}{\partial x_j}\frac{\partial \overset{\circ}{z}_j(x,t,0)}{\partial \rho}\overset{\circ}{y}_j dxdt +$$

$$+\int_0^l\int_0^T \xi_j\frac{\partial \overset{\circ}{z}_j(x,t,0)}{\partial \rho}\frac{\partial \overset{\circ}{y}_j}{\partial x}dxdt. \qquad (3.51)$$

The lower limits in the first integrals in the right-hand sides of (3.50), (3.51) are equal to zero due to the initial and boundary conditions in (3.47). The upper limits of these integrals are cancelled with the second and third integrals in the right-hand side of (3.49) according to final and boundary conditions in (3.48). The first term in the first integral is cancelled as well as the first, fifth, seventh terms in the fourth integral of the right-hand side of (3.49) together with integrands in the remaining terms of the right-hand sides of (3.50), (3.51). This follows from the differential equations in (3.48). Thus, we obtain the relation analogous to (3.12), (3.40):

$$(\overset{\circ}{g}(0))' = \sum_{j=1}^{J}\left\{\int_0^l\int_0^T \frac{\partial}{\partial u_j}[c_j + b_j\overset{\circ}{y}_j - p_j\overset{\circ}{\eta}_j - d_j\overset{\circ}{\delta}](\hat{u}_j - \overset{\circ}{u}_j)dxdt\right\}. \qquad (3.52)$$

Moreover, the differential equations in (3.48) are multiplied by $\hat{z}_j(x,t) -$

$\overset{\circ}{z}_j(x,t)$, and the sum

$$\sum_{j=1}^{J}\left\{\int_0^l\int_0^T\left[\frac{\partial \overset{\circ}{y}_j}{\partial t}(\hat{z}_j-\overset{\circ}{z}_j)+\xi_j\frac{\partial \overset{\circ}{y}_j}{\partial x}(\hat{z}_j-\overset{\circ}{z}_j)+\right.\right.$$

$$+\frac{\partial\xi}{\partial x}\overset{\circ}{y}_j(\hat{z}_j-\overset{\circ}{z}_j)+(A_j^T\overset{\circ}{y}_j,\hat{z}_j-\overset{\circ}{z}_j)+$$

$$\left.\left.+\frac{\partial}{\partial z_j}(c_j-p_j\overset{\circ}{\eta}_j-d_j\overset{\circ}{\delta})(\hat{z}_j-\overset{\circ}{z}_j)\right]dxdt\right\}=0 \qquad (3.53)$$

equal to zero, is considered. We integrate by parts the first three terms of the left-hand side of (3.53). Here, we use the equality of the initial and boundary conditions for the functions $\hat{z}_j(x,t)$, $\overset{\circ}{z}_j(x,t)$ as well as the final and boundary conditions in (3.48). Finally, the left-hand side of (3.53) is transformed to the form analogous to (1.18):

$$\sum_{j=1}^{J}\left\{\int_0^l\int_0^T\frac{\partial}{\partial z_j}(c_j-p_j\overset{\circ}{\eta}_j-d_j\overset{\circ}{\delta})(\hat{z}_j-\overset{\circ}{z}_j)dxdt+\right.$$

$$+\int_0^T\frac{\partial w_j}{\partial z_j(l,t)}(\hat{z}_j(l,t)-\overset{\circ}{z}_j(l,t))dl++\int_0^l\frac{\partial q_j}{\partial z_j(x,T)}(\hat{z}_j(x,T)-\overset{\circ}{z}_j(x,T))dx+$$

$$\left.+\int_0^l\int_0^T(b_j(\overset{\circ}{u}_j,x,t)-b_j(\hat{u}_j,x,t))\,\overset{\circ}{y}_jdxdt\right\}=0.$$

The last relation and (3.52) imply the formula for the marginal value of the type (1.19). Here, it is worth mentioning that the functions $\overset{\circ}{y}_j(x,t)$, $j\in[1:J]$ are nonnegative. This follows from the differential equations in (3.48) as well as from the supposed monotonicity of entering functions and nonnegativeness of the components of matrices $A_j(x,t)$, $j\in[1:J]$. The reasoning is analogous to that of Section 1, but the interval of the characteristic curve lying in the rectangle $[0,l]\times[0,T]$ is taken instead of the time interval $[0,T]$.

§4. Linear-Quadratic Optimal Control Problems of Block Type

Consider the simplest class of block separable optimal control problems, where subsystems are described by linear differential equations, and functionals are quadratic with respect to phase variables and controls. Applying the Pontryagin maximum principle, the solution is reduced to systems of linear algebraic equations. In some cases, we can determine the specificity of these systems of equations and the lowering of dimension in a great number of subsystems is accomplished analytically. In other cases, it is efficient to use expansion by the method of iterative aggregation.

We consider the following simplest model of optimal control:

$$0,5x(c(T))^2 + 0,5\int_0^T (u(t))^2 dt \to \min,$$

$$dx(t)/dt = u(t), \qquad x(0) = k, \qquad k, c, \text{are constant values.}$$

We introduce the block separable statement based on this model. We have

$$0,5\sum_{j=1}^{J}\left[c_j(x_j(T))^2 + \gamma_j\int_0^T (u_j(t))^2 dt\right] \to \min,$$
$$\frac{dx_j(t)}{dt} = u_j(t), \qquad x_j(0) = k_j, \qquad j \in [1:J]; \qquad \sum_{j=1}^{J} u_j(t) \leq w, \tag{4.1}$$

where c_j, γ_j, $j \in [1:J]$, w are constant values and $\gamma_j \neq 0$, $j \in [1:J]$; $w > 0$.

We have a two-level system, where subsystems are described by differential equations, and there is a binding constraint on the control functions.

We will solve problem (4.1) by the direct use of the Pontryagin maximum principle. The Hamiltonian looks as follows:

$$H(t) = \sum_{j=1}^{J}\left[-0,5\gamma_j(u_j(t))^2 + \psi_j(t)u_j(t))\right],$$

and, for the conjugated variables $\psi_j(t)$, $j \in [1:J]$, we have the differential equations for $t = T$ of the form $d\psi_j(t)/dt = 0$, $\psi_j(T) = -c_j x_j(T)$, $j \in [1:J]$. From this, we obtain $\psi_j(t) = \text{const} = -c_j x_j(T)$, $t \in [0,T]$, where $x_j(T)$, $j \in [1:J]$ are not known yet. The Pontryagin maximum principle leads to the problem of quadratic programming with respect to $u_j(t)$, $j \in [1:J]$ for each $t \in [0, T]$

$$-\sum_{j=1}^{J} [0,5\gamma_j (u_j(t))^2 + c_j x_j(T) u_j(t)] \to \max, \qquad (4.2)$$
$$\sum_{j=1}^{J} u_j(t) \leq w.$$

Assume that the binding condition in (4.1), (4.2) is effective, then we have the problem of unconstrained maximization

$$-\sum_{j=1}^{J} [0,5\gamma_j (u_j)^2 + c_j x_j(T) u_j] + \lambda (w - \sum_{j=1}^{J} u_j) \to \max,$$

where λ is the Lagrangian multiplier for the binding constraint. The rule of Lagrangian multipliers leads to the following system of equations with respect to the variables u_j, $j \in [1 : J]$, λ:

$$\gamma_j u_j + c_j x_j(T) + \lambda = 0, \qquad j \in [1 : J]; \sum_{j=1}^{J} u_j = w. \qquad (4.3)$$

The system of linear equations (4.3) is solved as follows. From the first equations, the controls u_j, $j \in [1 : J]$ are expressed through the value λ, and the remaining relations are substituted in to the binding condition. We have $-\sum_{j=1}^{J} (c_j x_j(T) + \lambda)/\gamma_j = w$. From this equality, we find the unknown

$$\lambda = -\left[\sum_{j=1}^{J} (c_j x_j(T)/\gamma_j + w\right] \left(\sum_{j=1}^{J} \gamma_j^{-1}\right)^{-1}.$$

Then, using the first J equalities from (4.3), we obtain the formula for the equations

$$u_j = \left(-c_j x_j(T) \sum_{r=1}^{J} \gamma_r^{-1} + \sum_{r=1}^{J} c_r x_r(T) \gamma_r^{-1} + w\right) \left(\gamma_j \sum_{r=1}^{J} \gamma_r^{-1}\right)^{-1}, \qquad (4.4)$$
$$j \in [1 : J].$$

Thus, the optimal controls of problem (4.1) are constant in the interval $[0, T]$. It remains only to define the unknowns $x_j(T)$, $j \in [1 : J]$ and then the optimal controls of problem (4.1). To do this, we substitute (4.4) into differential equations (4.1) and integrate (4.1) from 0 to T. We obtain a set of linear algebraic equation in $x_j(T)$, $j \in [1:J]$

$$x_j(T)(1+c_j\gamma_j^{-1}T) - \left(\sum_{r=1}^{J} c_r x_r(T)\gamma_r^{-1}\right)\left(\gamma_j \sum_{r=1}^{J}\gamma_r^{-1}\right)^{-1} T = \\ = k_j + wT\left(\gamma_j\sum_{r=1}^{J}\gamma_r^{-1}\right)^{-1}, \qquad j \in [1:J]. \tag{4.5}$$

Using the standard notations, the set (4.5) is written by the following:

$$\bar{b}_j x_j(T) + \sum_{r=1}^{J} \bar{d}_r x_r(T) = \bar{m}_j, \qquad j \in [1:J]. \tag{4.6}$$

We denote

$$\bar{P} = \sum_{r=1}^{J} \bar{d}_r x_r(X). \tag{4.7}$$

Then, it follows from (4.6) that

$$x_j(T) = -\bar{b}_j^{-1}\bar{P} + \bar{b}_j^{-1}\bar{m}_j. \tag{4.8}$$

A substitution of (4.8) into (4.7) leads to an equation with respect to $\bar{P}$

$$\bar{P} = -\left(\sum_{r=1}^{J}\bar{b}_r^{-1}\bar{d}_r\right)\bar{P} + \sum_{r=1}^{J}\bar{b}_r^{-1}\bar{m}_r\bar{d}_r,$$

then $\bar{P} = \left(\sum_{r=1}^{J}\bar{b}_r^{-1}\bar{m}_r\bar{d}_r\right)\left(1+\sum_{s=1}^{J}\bar{b}_x^{-1}\bar{d}_s\right)^{-1}$ is obtained. At last, we deduce unknown values $x_j(T)$ from (4.8) as well as optimal controls of the problem (4.1).

Thus, for large J, the problem of reducing the dimension in problem (4.1) is solved analytically.

Below, we solve problem (4.1) by the iterative decomposition method described earlier using aggregation of variables. Consider the aggregated problem for fixed weights constant in the interval $[0, T]$, Let these weights be

α_j, $j \in [1:J]$, where $\sum\limits_{j=1}^{J} \alpha_j = 1$. We have

$$g = 0,5 \sum_{j=1}^{J} \left[c_j(x_j(T))^2 + \gamma_j(\alpha_j)^2 \int_0^T (U(t))^2 dt \right] \to \min, \tag{4.9}$$
$$dx_j(t)/dt = \alpha_j u(t), \qquad x_j(0) = k_j, \qquad j \in [1:J], U(t) \le w.$$

Let, for any weights, the condition $U(t) \le w$ of the aggregated problem (4.9) be effective. This is valid, for example, for $c_j, \gamma_j > 0$, $k_j < 0$, $j \in [1:J]$ when the absolute values of k_j, $j \in [1:J]$ are sufficiently large. Then, the optimal control $\overset{\circ}{U}(t)$ of the aggregated problem is always equal to w, and the conjugated variable $\overset{\circ}{\delta}$ that corresponds to the condition $U(t) \le w$ is computed in the form

$$\overset{\circ}{\delta} = \sum_{j=1}^{J} \left[-\alpha_j c_j (k_j + \alpha_j w T) - \gamma_j (\alpha_j)^2 w \right].$$

The local problems are formulated as follows:

$$0,5 c_j (x_j(T))^2 + 0,5 \gamma_j \int_0^T (u_j(t))^2 dt + \int_0^T \overset{\circ}{\delta}\, u_j(t) dt \to \min,$$
$$dx_j(t)/dt = u_j(t), \qquad x_j(0) = k_j,$$

and their solution by the maximum principle gives

$$\hat{u}_j = -(c_j \hat{x}_j(T) + \overset{\circ}{\delta})(\gamma_j^{-1}),$$

where $\hat{x}_j(T) = k_j + \hat{u}_j T$. In other words, the solution of local problems is reduced to linear equations with one unknown $\hat{x}_j(T)$.

In the particular case $c_j = c$, $\gamma_j = \gamma$, $j \in [1:J]$, we obtain the following simple formulas for optimal controls of problem (4.1) and local problems in the iterative process:

$$u_j^* = \bar{\mu} \left[\sum_{\substack{r=1 \\ r \ne j}} k_r - (J-1) k_j \right] [J(\bar{\mu}+1)T]^{-1} + wJ^{-1}, \qquad \bar{\mu} = cT\gamma^{-1},$$

$$\hat{u}_j = \left[\sum_{r=1}^{J} \alpha_r (k_r - k_j) + \gamma w(\bar{\mu} - 1) \sum_{r=1}^{J} (\alpha_r)^2 \right] [cT + \gamma]^{-1} .$$

These formulas allow numerical computation, where there is a possibility to compare solution obtained by the method of decomposition, with the finite precise solution of problem (4.1). We will consider this in Section 5 of the next chapter.

Note one more property of the solution to problem (4.1) by the method of iterative decomposition for the case $J = 2$. We consider the optimal value of the functional of aggregated problem as the function of weights α_1 and α_2:

$$\overset{\circ}{g}(\alpha_1, \alpha_2) = 0,5\, (c_1 k_1^2 - 2c_1 k_1 w T - c_1 \alpha_1^2 w^2 T_2 - \\ - \gamma_1 \alpha_1^2 w^2 T - c_2 k_2^2 - 2c_2 k_2 \alpha_2 w T - c_2 \alpha_2^2 w^2 T^2 - \gamma_2 \alpha_2^2 w^2 T) .$$

We put $\alpha_1 = \alpha$, $\alpha_2 = (1 - \alpha)$ and find the stationary point of the function $\overset{\circ}{g}(\alpha, (1 - \alpha))$:

$$\alpha^* = \left(\bar{\Lambda} w^{-1} + (c_2 T + \gamma_2)\right) \left[(c_1 T + \gamma_1) + (c_2 T + \gamma_2)\right]^{-1} ,$$

$$\bar{\Lambda} = c_2 k_2 - c_1 k_1 ,$$

which corresponds to the weight coefficient computed for the optimal solution of the initial problem. We will consider the weight coefficients α_1, α_2 as functions of parameters ρ in accordance with formula (1.11) for $\rho_1 = \rho_2 = \rho$. We try to find the point $\bar{\rho}$, where the equality $\alpha_1^* = \alpha_1(\bar{\rho})$ holds. After some cumbersome calculation, we obtain

$$\bar{\rho} = (\bar{A} - \bar{\Lambda}) \left[(\bar{A} - \bar{\Lambda}) + \bar{\Lambda}(-1 + (\alpha_2 \Lambda + \bar{\Omega})(\gamma_1 + c_1 T)^{-1} + \\ + (-\alpha_1 \bar{\Lambda} + \bar{\Omega})(\gamma_2 + c_2 T)^{-1})\right]^{-1} ,$$

where $\bar{A} = \alpha_1(c_1 T + \gamma_1) - \alpha_2(c_2 T + \gamma_2)$, $\Omega = \alpha_1^2(c_1 T + \gamma_1) + \alpha_2^2(c_2 T - \gamma_2)$.

If $\Lambda = 0$, then the maximum of the function $\overset{\circ}{g}(\rho)$ is attained for $\bar{\rho} = 1$, and the optimal solution of the main problem is always obtained in one iteration. By choosing the initial weights, we can make $\bar{A} = 0$ for definiteness. If the value $\bar{\Lambda}$ is positive, fairly small and moreover,

$$-1 + (\alpha_2 \bar{\Lambda} + \bar{\Omega})(\gamma_1 + c_1 T)^{-1} + (-\alpha_1 \bar{\Lambda} + \bar{\Omega})(\gamma_2 + c_2 T)^{-1} > 0,$$

then the value $\bar{\rho} < 1$, and it is close to one. In other words, the maximum of the function $\overset{\circ}{g}(\rho)$ is attained in the interval $[0, 1]$, and the solution of the main problem is obtained in one iteration. If the last inequality has the opposite meaning, then the value $\bar{\rho} > 1$, and it is close to one. Then, the solution of the initial problem is obtained in two iterations.

We consider a generalization of problem (4.1). We have

$$f = 0,5 \sum_{j=1}^{J} \left[\sum_{n-0}^{m} c_j^m (x_j^n(T))^2 + \int_0^T \sum_{i=1}^{I} \gamma_j^i (u_j^i(t))^2 dt \right] \to \min,$$

$$dx_j^m(t)/dt = \sum_{i=1}^{I} b_j^{in} u_j^i(t),$$

$$x_j^n(0) = k_k^n, \qquad j \in [1:J], \qquad n \in [0:m],$$

$$\sum_{j=1}^{J} \sum_{c=1}^{I} d_j^i u_j^i = w.$$

Here, phase coordinates and control in each system are vectors with dimensions $m+1$ and I, respectively, and the binding constraints are written in the form of equalities.

As for problem (4.1), upon the application of the maximum principle, we come to the system of equations for u_j^i, $j \in [1:J]$, $i \in [1:I]$, λ of the following form:

$$-\gamma_j^i u_j^i - \sum_{n=0}^{m} b_j^{in} c_j^n x_j^n(T) - \lambda d_j^i = 0, \qquad j \in [1:J], \qquad i \in [1:I],$$

$$\sum_{j=1}^{J} \sum_{i=1}^{I} d_j^i u_j^i = w.$$

By solving this system, we obtain the formula for optimal controls

$$u_j^i = d_j^i (\gamma_j^i)^{-1} w K^{-1} + d_j^i (\gamma_j^i)^{-1} K^{-1} \sum_{s=1}^{J} \sum_{p=1}^{I} d_s^p (\gamma_s^p)^{-1} \times$$

$$\times \sum_{n=0}^{m} b_s^{pn} c_s^n x_s^n(T) - (\gamma_j^i)^{-1} \sum_{n=0}^{m} b_j^{in} c_j^n x_j^n(T), \qquad j \in [1:J],$$

$$i \in [1:I],$$

where $K = \sum_{j=1}^{J} \sum_{i=1}^{I} (d_j^i)^2 (\gamma_j^i)^{-1}$.

Substituting the resulting formula in the problem's differential equations, we come to the system of linear algebraic equations with respect to unknowns $x_j^n(T)$, $j \in [1:J]$, $n \in [0,m]$:

$$x_j^n(T) = k_j^n + \sum_{i=1}^{I} b_j^{in} d_j^i (\gamma_j^i)^{-1} K^{-1} T + \sum_{i=1}^{I} b_j^{in} d_j^i (\gamma_j^i)^{-1} K^{-1} T +$$
$$\sum_{s=1}^{J} \sum_{p=1}^{I} d_s^p (\gamma_s^p)^{-1} \sum_{r=0}^{m} b_s^{pr} c_s^r x_s^r(T) -$$
$$- \sum_{i=1}^{I} b_j^{in} (\gamma_j^i)^{-1} \sum_{r=0}^{m} b_j^{ir} c_j^r x_j^r(T), \qquad j \in [1:J], \qquad n \in [0:m].$$

We introduce the notation $R = \sum_{s=1}^{J} \sum_{p=1}^{I} d_s^p (\gamma_s^p)^{-1} \sum_{r=0}^{m} b_s^{pr} c_s^r x_s^r(T)$ and assume that the value R is known. For each fixed $j \in [1:J]$, we will consider independent systems of equations with respect to $x_j^n(T)$, $n \in [0:m]$. Solving this system, we obtain

$$x_j^n(T) = \mu_j^n + \nu_j^n R, \qquad j \in [1:J], \qquad n \in [0:m].$$

Substituting the latter relations in the formula for the value R, we finally obtain an equation with one unknown.

Let the problem being considered contain a small amount L of effectively binding constraints, then the subsequent considerations hold only, when we have a system of linear algebraic equations for Lagrange multipliers λ^l, $l \in [1:L]$.

We consider one more generalization of problem (4.1), where we put

$$dx_j(t)/dt + \lambda_j x_j(t) = u_j(t), \qquad \lambda_j \neq 0, \qquad j \in [1:J].$$

Then, the terms $-\lambda_j x_j \psi_j$ are added to the Hamiltonian function, and the conjugated variables $\psi_j(t)$ satisfy equations and conditions for $t = T$ of the form

$$d\psi_j(t)/dt = \lambda_j \psi_j(t), \qquad \psi_j(T) = -c_j x_j(T), \qquad j \in [1:J].$$

From this, we obtain $\psi_j(t) = D_j \exp(\lambda_j t)$.

The connection of controls $u_j(t)$ and conjugated functions is the same as in formula (4.4), where we should replace $-c_j x_j(T)$ by $\psi_j(t)$. Using this formula in differential equations for phase variables and integrating it, we obtain

$$x_j(t) = C_j \exp(-\lambda_j t) + \Bigg[D_j \exp_t(\lambda_j t) \frac{1}{2\lambda_j} \sum_{r=1}^{J} \gamma_j^{-1} - $$
$$ - \sum_{r=1}^{J} \frac{D_r \exp(\lambda_r t)\gamma_r^{-1}}{(\lambda_j + \lambda_r)} + \exp(-\lambda_j t) \int_0^t w(\tau) \times $$
$$ \times \exp(\lambda_j \tau) d\tau \Bigg] \left[\gamma_j \sum_{s=1}^{J} \gamma_s^{-1} \right]^{-1}, \qquad j \in [1:J],$$

where C_j, D_j, $j \in [1 : J]$ are unknown constants. Further, we take into account the initial conditions for phase coordinates $x_j(t)$ and conditions for $t = T$ for dual functions. We come to the system of $2J$ linear algebraic equations with respect to the constants C_j, D_j, $j \in [1 : J]$. In this system, it is easy to express C_j through D_r, $r \in [1 : J]$. Finally, we obtain a system of more general form with respect to D_r, $r \in [1 : J]$. If we try to solve it by the Gauss elimination method, then it will be a hard problem for large J. If, for the considered problem, we use the iterative decomposition method, then, in local problems, everything is reduced to linear algebraic equations with one unknown.

Comments and References to Chapter 3

Statements of the problems of Boltz type can be found, for example, in the book of N.N. Moiseev, ed., [8]. Conditions of applicability of the duality theorem are taken from A.M. Ter-Krikorov [13,15]. The model of explosion afteraction is studied by L.G. Razdol'skii [11]. Problems of optimal control under uncertainty are widely discussed by A.B. Kurzhanskii [5]. Optimization of noise in dynamic systems under random perturbation is considered in the book by F.L. Chernous'ko and V.B. Kolmanovskii [3]. Block statements from Section 2 are formulated on the basis of three specified problem types. For systems with partial differential equations, the notations from J.-L. Lions [7] are used as well as the concept of Sobolev spaces (for example, see the book by O.A. Ladyzhenskaya [6]). In this book, inclusion theorems based on the Poincare-Friedrichs inequality are widely used. The Wolfe dual of control problems with partial differential equations is studied in J.-L. Lions [7]. The Kuhn-Tucker theory for distributed systems is developed in the work by A.M. Ter-Krikorov [14]. Properties of the Green function for the Dirichlet problem, with Laplace equations, can be found, for example, in the book by V.S. Vladimirov [21]. In constructing parabolic and hyperbolic equations, we use the formulation of V.I. Plotnikov [9] and V.A. Il'in [4]. Our formulation of the block separable optimal control problem, with systems of first order partial differential equations, is based on the book by B.L. Rozhdestvenskii and N.N. Yanenko [12]. Finally, for the maximum principle see the book by L.S. Pontryagin, V.G, Boltyanskii, R.V. Gamkrelidze, and E.F. Mishchenko [10]. Linear quadratic optimal control problems are studied, for example, in the monograph by D. Bryson and Ho Yu Shi [2]. For the Gauss elimination method see N.S. Bakhvalov [1]. The results of Chapter 3 are based on the work of V.I.Tsurkov [16–19].

References to Chapter 3

[1] Bakhvalov N.S., *Chislennye metody* (Numerical Methods), Moscow: Nauka, 1973.

[2] Bryson D. and Ho Yu Shi, *Prikladnaya teoriya optimal'nogo upravleniya* (Applied Theory of Optimal Control), Moscow: Mir, 1972 [Russian translation].

[3] Chernous'ko F.L. and Kolmanovskii V.B., *Optimal'noe upravlenie pri sluchainykh vozmushcheniyakh* (Optimal Control for Random Disturbance), Moscow: Nauka, 1978.

[4] Il'in V.A., *O razreshimosti smeshannykh zadach dlya giperbolicheskikh i parabolicheskikh uravnenii* (On Solvability of Mixed Problems for Hyperbolic and Parabolic Equations), *Uspekhi Mat. Nauk*, 1960, vol. XV, no. 2, pp. 97-154.

[5] Kurzhanskii A.B., *Upravlenie i nabludenie v usloviyakh neopredelennosti* (Control and Observation under Uncertainty), Moscow: Nauka, 1977.

[6] Ladyzhenskaya O.A., *Kraevye zadachi matematicheskoi fiziki* (Boundary Problems of Mathematical Physics), Moscow, Nauka: 1973.

[7] Lions J.-L., *Optimal'noe ipravlenie sistemami, opisyvaemymi uravneniyami s chastnymi proizvodnymi* (Optimal Control in Systems Described by Partial Differential Equations), Moscow: Mir, 1972 [in Russian].

[8] Moiseev N.N., ed., *Sovremennoe sostoyanie teorii issledovaniya operatsii* Operations Research: State of Art, Moscow: Nauka: 1979.

[9] Plotnikov V.I., *Energeticheskoe neravenstvo i svoistvo neopredelennosti sistemy sobstvennykh functsii* (Energetic Inequality and the Property of Overdefiniteness of the System of Eigenfunctions), *Izv. Akad. Nauk SSSR, Ser. Math.*, 1968, vol. 31, no. 4, pp. 743-755.

[10] Pontryagin L.S., Boltyanskii V.G., Gamkrelidze R.V., and Mishchenko E.F., *Matematicheskaya teoriya optimal'nykh protsessov* (Mathematical Theory of Optimal Processes), Moscow, Nauka: 1969. Moscow: Mir, 1972 [Russian translation].

[11] Razdol'skii L.G., *Optimal'noe upravlenie v teorii vzryva gazovozdushnykh smesey* (Optimal Control in the Explosion Theory for Gas-Air Mixtures), *Proc. III-d Winter School on Mathematical Programming and Related Problems*, 1970, vol. 2, pp. 468-474.

[12] Rozhdestvenskii B.L. and Yanenko N.N., *Sistemy kvazilineinykh uravnenii* (Systems of Quasilinear Equations), Moscow, Nauka: 1968.

[13] Ter-Krikorov A.M., *Vypukloe programmirovanie v prostranstve, sopryazhennom prostranstvu Banaha, i zadachi optimal'nogo upravleniya s fazovymi ogranicheniyami* (Convex Programming in the Space Conjugated to the Banach Space and the Optimal Control Problems with Phase Constraints), *Zh. Vych. Mat. Mat. Fiz.*, 1976, vol. 16, no. 2, pp. 351-358.

[14] Ter-Krikorov A.M., *Ob odnoi zadache optimal'nogo upravleniya so smeshannymi ogranicheniyami dlya sistem s raspredelennymi parametrami* (On a Problem of Optimal Control with Mixed Constraints for Systems with Distributed Parameters), *Zh. Vych. Mat. Mat. Fiz.*, 1981, vol. 21, no. 3, pp. 561-571.

[15] Ter-Krikorov A.M., *Optimal'noe upravlenie i matematicheskaya ekonomika* (Optimal Control and Mathematical Economics), Moscow, Nauka: 1977.

[16] Tsurkov V.I., *Dekompozitsiya v dinamicheskykh zadachakh s upravleniem* (Decomposition in Dynamical Problems with Control), *Prikl. Mat. Mekh.*, 1979, vol. 43, no.5, pp. 949-953.

[17] Tsurkov V.I., *Decompozitsiya v optimizatsii sistem s ellipticheskimi uravneniyami* (Decomposition in the Optimization of Systems with Elliptic Controls), *Zh. Vych. Mat. Mat. Fiz.*, 1980, vol. 20, no. 5, pp. 1310-1319.

[18] Tsurkov V.I., *Iterativnoe razlozhenie optimizatsionnykh zadach s uravneniyami* (Iterative Decomposition of Optimization Problems with Control in Partial Derivatives), *Dokl. Akad. Nauk, SSSR*, 1980, vol. 255, no. 3, pp. 535-537.

[19] Tsurkov V.I., *Razlozhenie v zadachakh upravleniya sistemami s uravneniyami parabolicheskogo tipa* (Decomposition in Problems of Control for Systems with Parabolic Equations), *Prikl. Mat. Mekh.*, 1981, vol. 45, no. 1, pp. 137-144.

[20] Tsurkov V.I., *Dekompozitsiya v upravlenii sistemami s raspredelennymi parametrami i giperbolicheskogo tipa* (Decompostion in Control of Systems with Distributed Parameters and Hyperbolic Type), Avtom. Telemekh., 1981, no. 3, pp. 14-26.

[21] Vladimirov V.S., *Uravneniya matematicheskoi ziziki* (Equations of Mathematical Physics), Moscow: Nauka, 1988.

Chapter 4

Effectiveness of Decomposition

In this chapter, we demonstrate the effective application of decomposition based on iterative aggregation as compared with direct methods of solution. We consider dynamical problems of optimal control. Their solution through the use of the maximum principle results in cumbersome computations. The application of two-level decomposition or the reduction method to linear quadratic problems amounts to solving simple problems of lower dimensions.

We present nonlinear hierarchical optimal control problems. The iterative algorithm effectively constructs both smooth and discontinuous control functions. We state separate block problems of optimization of systems with distributed parameters. These problems have a particular physical meaning: the distribution of the system's power resources over subsystems described by classical partial differential equations. We demonstrate the efficiency of iterative aggregation when we have multiple control functions constraints. We consider dimension reduction for hierarchical optimal control problems with nonseparable functionals and describe numerical computation by the decomposition method for test block separable optimal control problems. The results also testify to the efficiency of the proposed approach.

§1. Nonlinear Two-level Statements

We show that, for a number of block optimal control problems, iterative decomposition constructs smooth trajectories or switching points in discontinuous controls. Here, the direct solution of the problems is complicated for the large number of subsystems.

Consider the following block optimal control problem:

$$
\begin{aligned}
&f = \sum_{j=1}^{J} \gamma_j x_j(T) \to \min, \\
&dx_j(t)/dt = \bar{\mu}_j (u_j(t))^{-1} \exp(-\bar{\nu}_j/u_j(t)), \qquad x_j(0) = \kappa_j, \\
&j \in [1:J]; \qquad \sum_{j=1}^{J} u_j(t) \le w(t),
\end{aligned}
\tag{1.1}
$$

where γ_j, $\bar{\mu}_j$, $\bar{\nu}_j$ – are positive constants, $w(t) > 0$ is a given function. We will assume that, for the solutions, the conditions $u_j(t) > \bar{\nu}/(2-\sqrt{2})$, $j \in [1:J]$ hold everywhere. These inequalities provide the convexity of functions in the right-hand sides of differential equations in (1.1).

First, we try to solve problem (5.1) by using directly the Pontryagin maximum principle. The Hamiltonian takes the form

$$
H(t) = \sum_{j=1}^{J} \psi_j(t) \bar{\mu}_j (u_j(t))^{-1} \exp(-\bar{\nu}_j/u_j(t)),
$$

and the conjugated variables $\psi_j(t)$ satisfy the following equations and conditions for $t = T$:

$$
d\psi_j(t)/dt = 0, \qquad \psi_j(T) = -\gamma_j, \qquad j \in [1:J],
$$

from where we have $\psi_j(t) = -\gamma_j$, $j \in [1:J]$ for $t \in [0,T]$.

The maximum principle leads to the system

$$
\begin{aligned}
&-\gamma_j \bar{\mu}_j (\bar{\nu}_j - u_j(t))(u_j(t))^{-3} \exp(-\bar{\nu}_j(t)) - \lambda(t) = 0, \\
&j \in [1:J], \\
&\sum_{j=1}^{J} u_j(t) \le w(t),
\end{aligned}
\tag{1.2}
$$

where $\lambda(t)$ is the Lagrange multiplier for the binding constraint in (1.2). Assume that the binding constraint in (1.1), (1.2) is effective for all time instants $t \in [0,T]$. Then, (1.2) turns into a system of nonlinear equations with unknowns $u_j(t)$ $j \in [1:J]$, $\lambda(t)$ for each $t \in [0,T]$. This system is fairly complicated, and its solution becomes more complicated with the growth of J.

We will solve problem (1.1) by the direct decomposition method. If the aggregated problem for fixed weights $\alpha_j(t)$ has solution $\overset{\circ}{U}(t) = w(t)$, then the conjugated function $\delta(t)$ that corresponds to the condition $U(t) \leq w(t)$ is computed. The functionals of local problems have the form

$$\gamma_j x_j(T) + \int_0^T \overset{\circ}{\delta}(t) u_j(t)dt \to \min .$$

Applying the maximum principle to local problems, we obtain nonlinear equations

$$-\gamma_j \bar{\mu}_j(\bar{\nu}_j - u_j(t))(u_j(t))^{-3} \exp(-\bar{\nu}_j/u_j(t) - \overset{\circ}{\delta}(t) = 0 \qquad (1.3)$$

independent with respect to j. Optimal solutions are found graphically from these equations. It is clear that to solve the separate equations (1.3) is simpler than to solve the complicated system (1.2).

We discuss another block optimal control statement with nonlinear right-hand sides in differential equations:

$$\begin{aligned} & f = \sum_{j=1}^{J} \gamma_j x_j(T) \to \max, \\ & dx_j(t)/dt = -\mu_j(x_j(t) - u_j(t))^2, \qquad x_j(0) = \kappa_j, \\ & j \in [1:J]; \qquad \sum_{j=1}^{J} u_j(t) \leq w(t), \end{aligned} \qquad (1.4)$$

where γ_j, $\bar{\mu}_j$, $\bar{\nu}_j$ are positive constants, and $w(t) > 0$ is a given smooth function.

First, we try to solve the problem (1.4) by direct application of the Pontryagin maximum principle. The Hamiltonian takes the following form:

$$H(t) = -\sum_{j=1}^{J} \psi_j(t)\mu_j(x_j(t) - u_j(t))^2,$$

and the conjugated variables $\psi_j(t)$ satisfy equations and conditions for $t = T$ of the following form:

$$d\psi_j(t)/dt = 2\mu_j(x_j(t) - u_j(t))\psi_j(t), \qquad \psi_j(T) = \gamma_j, \qquad j \in [1:J]. \quad (1.5)$$

We will assume that the binding conditions are essential. This is true, for example, if the initial values κ_j, $j \in [1:J]$ exceed the maximum of the function $w(t)$. Applying the Pontryagin maximum principle, we obtain the system of equations with respect to unknowns $u_j(t)$, $j \in [1:J]$ and γ_j for $t \in [0,T]$:

$$
\begin{aligned}
&2\psi_j(t)\mu_j(x_j(t) - u_j(t)) - \lambda(t) = 0, \qquad j \in [1:J], \\
&\sum_{j=1}^{J} u_j(t) = w(t).
\end{aligned}
\tag{1.6}
$$

The solution of system (1.6) looks as follows:

$$
\begin{aligned}
&\lambda(t) = \left[\sum_{j=1}^{J} x_j(t) - w(t)\right]\left[\sum_{s=1}^{J}(2\mu_s\psi_s(t))^{-1}\right]^{-1}, \\
&u_j(t) = x_j(t) - \lambda(t)(2\mu_j\psi_j(t))^{-1}.
\end{aligned}
\tag{1.7}
$$

According to (1.6), the equations for conjugated variables (1.5) are transformed to the form

$$
d\psi_j(t)/dt = \lambda(t), \qquad \psi_j(T) = \gamma_j, \qquad j \in [1:J]. \tag{1.8}
$$

Thus, solution of the initial problem (1.4) is reduced to integration of the two-point problem, which is given by differential equations from (1.4) and equations from (1.8) with corresponding initial and final conditions. Here, the variable $\lambda(t)$ and controls $u_j(t)$, $j \in [1:J]$ are expressed according to (1.7). This nonlinear two-point problem becomes more complicated with the growth of the number of subsystems.

Now, consider the solution of problem (1.4) by decomposition based on variable aggregation. Here, we should note that, because the right-hand sides of differential equations from (1.4) do not have the form (1.2) from Chapter 3 but are functions concave with respect to x_j, u_j, the decomposition method is an iterative process monotone with respect to the functional. We give the normalized weight functions $\alpha_j(t)$, $j \in [1:J]$ and write the aggregated problem

$$g = \sum_{j=1}^{J} \gamma_j x_j(T) \to \max,$$
$$dx_j(t)/dt = -\mu_j(x_j(t) - \alpha_j(t)U(t))^2, \qquad x_j(0) = \kappa_j, j \in [1:J] \tag{1.9}$$
$$U(t) \le w(t).$$

If the optimal control $\overset{\circ}{U}(t)$ of this problem is equal to $w(t)$, then the conjugated function $\delta(t)$ for the condition $U(t) \le w(t)$ is computed in the form

$$\overset{\circ}{\delta}(t) = -\sum_{j=1}^{J} 2\alpha_j(t)\mu_j(\overset{\circ}{x}_j(t) - \alpha_j(t)w(t))\,\overset{\circ}{\psi}_j(t),$$

where the functions $\overset{\circ}{x}_j(t)$, $j \in [1:J]$ are obtained after integration of differential equations in (1.9), and variables $\overset{\circ}{\psi}_j(t)$, $j \in [1:J]$ are computed after the integration of equations

$$d\overset{\circ}{\psi}_j(t)/dt = 2\mu_j(\overset{\circ}{x}_j(t) - \alpha_j(t)w(t))\,\overset{\circ}{\psi}_j(t), \qquad \overset{\circ}{\psi}_j(T) = \gamma_j.$$

Consider local problems. Their functionals have the form

$$\gamma_j x_j(T) - \int_0^T \overset{\circ}{\delta}(t)u_j(t)dt \to \max,$$

and the conditions are given by local connections from (1.4) for each fixed $j \in [1:J]$.

The application of the maximum principle for local problems leads to the independent equations

$$2\psi_j(t)\mu_j(x_j(t) - u_j(t)) - \overset{\circ}{\delta}(t) = 0, \qquad j \in [1:J], \tag{1.10}$$

where the conjugated variables $\psi_j(t)$ satisfy equations (1.5). From equations (1.5), (1.10), we obtain the simple relation

$$d\psi_j(t)/dt = \overset{\circ}{\delta}(t), \qquad \psi_j(T) = \gamma_j, \qquad j \in [1:J]. \tag{1.11}$$

Having integrated (1.11) and substituted $\psi_j(t)$ in (1.10), we find the values $(x_j(t) - u_j(t))$. Then, by integrating differential equations in (1.4), we obtain phase variables and optimal controls of local problems. Thus, for large J, the solution of problem (1.4) by the decomposition method is essentially simpler than its direct solution.

We state one more block separable optimal control problem:

$$f = \sum_{j=1}^{J} \left[\int_{0}^{T} \gamma_j (1 - u_j(t)) x_j(t) dt \right] \to \max,$$
$$dx_j(t)/dt = \beta_j u_j(t), \quad x_j(0) = \kappa_j, \quad 0 \le u_j(t) \le w_j(t), \quad j \in [1:J], \tag{1.12}$$
$$\sum_{j=1}^{J} u_j(t) \le w(t),$$

where γ_j, β_j are positive constants and it is assumed that $0 < w(t) < 1$, $w_j(t) > w(t)$, $j \in [1:J]$.

We try to solve problem (1.12) by direct use of the Pontryagin maximum principle. The Hamiltonian takes the form

$$H(t) = \sum_{j=1}^{J} \left[\psi_j(t) \beta_j u_j + \gamma_j (1 - u_j(t)) x_j(t) \right],$$

and the conjugated variables $\psi_j(t)$, $j \in [1:J]$ satisfy the following equations and conditions for $t = T$:

$$d\psi_j(t)/dt = -\gamma_j (1 - u_j(t)), \qquad \psi_j(T) = 0. \tag{1.13}$$

The maximum principle leads to the following problem of linear programming for each $t \in [0, T]$ with respect to $u_j(t)$:

$$\sum_{j=1}^{J} \left[\beta_j \psi_j(t) - \gamma_j x_j(t) \right] u_j(t) \to \max,$$
$$\sum_{j=1}^{J} u_j(t) \le w(t); \qquad 0 \le u_j(t) \le w_j(t), \qquad j \in [1:J]. \tag{1.14}$$

If we assume that the binding constraints are effective, then switching points in controls $u_j(t)$, $j \in [1:J]$ are possible, depending on the relations for factors $(\beta_j \psi_j(t) - \gamma_j x_j(t))$ in the functional from (1.14). Finding these switching points for large J is fairly complicated, since it is also necessary to analyze two-point problems formulated in (1.12), (1.13).

We will consider problem (1.12) by the iterative method of decomposition. If the aggregated problems has optimal control $\overset{\circ}{U}(t) = w(t)$ for given weights $\alpha_j(t)$, then the conjugated variable $\overset{\circ}{\delta}(t)$ of the condition $U(t) \le w(t)$ is computed as follows:

$$\overset{\circ}{\delta}(t) = \sum_{j=1}^{J} \left[-\gamma_j \alpha_j(t)\, \overset{\circ}{x}_j(t) + \beta_j \alpha_j(t)\, \overset{\circ}{\psi}_j(t) \right],$$

where the functions $\overset{\circ}{x}_j$, $\overset{\circ}{\psi}_j$ are found by the well-known optimal control $\overset{\circ}{U}(t)$.

For fixed $j \in [1:J]$, the solution of each local problem by the maximum principle leads to independent two-point problems

$$\begin{aligned} dx_j(t)/dt &= \beta_j u_j(t), \qquad x_j(0) = \kappa_j, \\ d\psi_j(t)/dt &= -\gamma_j(1 - u_j(t)), \qquad \psi_j(T) = 0 \end{aligned}$$

as well as

$$\left[\beta_j \psi_j(t) - \gamma_j x_j(t) - \overset{\circ}{\delta}(t)\right] u_j(t) \to \max, \qquad 0 \le u_j(t) \le w_j(t).$$

It is clear that it is easier to find switching points of the written independent problems than to study (1.14). This fact is illustrated by the following particular case of problem (1.12), where we assume $\gamma_j = \beta_j = 1$, $w_j(t) = w -$ const, $j \in [1:J]$. Moreover, for definiteness, we assume that the following conditions hold for the entering parameters:

$$\begin{aligned} &\kappa_j \le \kappa_{j+1} < 0, \qquad j \in [1:J-1], \\ &wT - J\kappa_j + \sum_{s=1}^{J} \kappa_s > 0, \qquad j \in [1:J]. \end{aligned} \tag{1.15}$$

We will solve this problem by decomposition using variables aggregation. We give weights $\alpha_j = 1/J$, $j \in [1:J]$ constant on the interval $[0,T]$. Let the optimal control of the aggregated problem $\overset{\circ}{U}(t)$ be equal to w. Then, the dual estimate $\overset{\circ}{\delta}(t)$ of the condition $U(t) \le w$ is computed as follows:

$$\overset{\circ}{\delta}(t) = \sum_{j=1}^{J} \left[(1 - w/J)(T - t)/J - (\kappa_j + wt/J)/J\right].$$

Consider the first local problem

$$\int_0^T \left[(1-u_1(t))x_1(t) - \overset{\circ}{\delta}(t)u_1(t)\right] dt \to \max .$$

$$dx_1(t)/dt = u_1(t), \qquad x_1(0) = \kappa_1, \qquad 0 \le u_1(t) \le w.$$

We will search its solution in the form $\hat{u}_1(t) = w$, $t \in [0, t_1^*)$, $u_1(t) = 0$, $t \in (t_1^*, T]$. Then, $\hat{x}_1(t) = \kappa_1 + wt$, $t \in [0, t_1^*)$, $\hat{x}_1(t) = \kappa_1 + wt_1^*$, $t \in (t_1^*, T]$, $\hat{\psi}_1(t) = (1-w)(t_1^* - t) + (T - t_1^*)$, $t \in [0, t_1^*]$, $\hat{\psi}_1(t) = (T - t)$, $t \in [t_1^*, T]$.

The maximum principle gives

$$[\psi_1(t) - x_1(t) - \overset{\circ}{\delta}(t)]u_1(t) \to \max, \qquad 0 \le u_1(t) \le w,$$

and, for the switching point t_1^*, we obtain the equation

$$(T - t_1^*) - (\kappa_1 + wt_1^*) - \sum_{j=1}^{J} [(1-w)/J)(T - t_1^*)/J - (x_j + wt_1^*/J)/J] = 0,$$

from where we have

$$t_1^* = \left[-\kappa_1 J + \sum_{s=1}^{J} \kappa_s\right] /Jw + T/J.$$

Due to the first inequalities in (5.16), we obtain $t_1^* > T/J$. According to the second conditions in (1.15), the inequality $t_1^* < T$ holds.

Furthermore, consider the second local problem

$$\int_0^T \left[(1-u_2(t))x_2(t) - \overset{\circ}{\delta}(t)u_2(t)\right] dt \to \max .$$

$$dx_2(t)/dt = u_2(t), \qquad x_2(0) = \kappa_2, \qquad 0 \le u_2(t) \le w.$$

We look for its solution in the form: $\hat{u}_2(t) = 0$, $t \in [0, t_1^*)$, $\hat{u}_2(t) = w$, $t \in (t_1^*, t_2^*)$, $\hat{u}_2(t) = 0$, $t \in (t_2^*, T]$. Then, we have: $\hat{x}_2(t) = \kappa_2$, $t \in [0, t_1^*)$, $\hat{x}_2(t) = \kappa_2 + w(t - t_1^*)$, $t \in [t_1^*, t_2^*]$, $\hat{x}_2(t) = \kappa_2 + w(t_2^* - t_1^*)$, $t \in [t_2^*, T]$; $\hat{\psi}_2(t) = (1-w)(t_2^* - t_1^*) + (T - t_2^*)$, $t \in [0, t_1^*]$, $\hat{\psi}_2(t) = (1-w)(t_2^* - t) + (T - t_2^*)$, $t \in [t_1^*, t_2^*]$, $\hat{\psi}_2(t) = (T - t_2^*)$, $t \in [t_2^*, T]$. We obtain the following equation for the switching point t_2^*:

$$\hat{\psi}_2(t_2^*) - \hat{x}_2(t_2^*) - \overset{\circ}{\delta}(t_2^*) = 0. \tag{1.16}$$

which is rewritten in the form

$$(T - t_2^*) - [\kappa_2 + w(t_2^* - t_1^*)] - \\ - \sum_{j=1}^{J} [(1-w)/J)(T - t_2^*)/J - (\kappa_j + wt_2^*/J)/J] = 0,$$

from where we find

$$t_2^* = \left[-J(\kappa_1 + \kappa_2) + 2\sum_{s=1}^{J} \kappa_s \right] /Jw + 2T/J.$$

By continuing this process, we obtain the formula for the switching point with number j, $j \in [1 : J-1]$:

$$t_j^* = \left[-J\sum_{s=1}^{J} \kappa_s + j\sum_{s=1}^{J} \kappa_s \right] /Jw + jT/J.$$

Here, according to the first and second inequalities in (1.15), we have

$$t_j^* > jT/J; \qquad t_j^* < t_{j+1}^*, \qquad j \in [1 : J-1], \qquad t_J^* = T,$$

respectively. Furthermore, following the decomposition scheme, we deduce the aggregation weights by formula (1.11) from Chapter 3, where

$$\begin{aligned} &\hat{u}_j(t) = w, \qquad t \in (t_j^*, t_{j+1}^*), \\ &\hat{u}_j(t) = 0, \qquad t \in T \backslash (t_j^*, t_{j+1}^*), \qquad t_0^* = 0, \end{aligned} \tag{1.17}$$

and consider the optimal solution of the functional $\overset{\circ}{g}(\rho)$ as a function of parameter $\rho \in [0, 1]$. As in problem (2.1) from Chapter 3, we prove that the extremum of the function $\overset{\circ}{g}(\rho)$ is attained at the point $\overset{\circ}{\rho} = 1$. Thus, the optimal controls of problem (1.12) are found in one iteration and are expressed in form (1.17).

We can abandon the assumptions given by the second inequality in (1.15). The optimal controls of local problems are constructed according to the scheme described above, where the switching points are computed sequentially from relations analogous to (1.16). If, for an index j, we have $t_{j+1}^* < t_j^*$, then this means that $\hat{u}_{j+1}(t) = 0$, $t \in [0, T]$, and we need to pass to the next problem.

The decomposition method based on aggregation of variables is applied to block separable optimal control problems with delay arguments. Justification of the method for this class of problems is carried out along the scheme described in Section 1 of Chapter 3. Further, we consider application of the iterative algorithm to particular problems involving delay.

First, we consider the simplest linear model with two subsystems:

$$
\begin{aligned}
& f = c_1 x_1(T) + c_2 x_2(T) \to \max, \\
& dx_1(t)/dt = u_1(t-\tau_1), \qquad x_1(0) = \kappa_1; \qquad 0 \le u_1(t) \le w_1, \\
& dx_2(t)/dt = u_2(t-\tau_2), \qquad x_2(0) = \kappa_2; \qquad 0 \le u_2(t) \le w_2, \\
& u_1(t) + u_2(t) \le w, \qquad u_1(t) = e_1(t) > 0, \qquad t \in [-\tau_1, 0), \\
& u_2(t) = e_2(t), \qquad t \in [-\tau_2, 0),
\end{aligned}
\tag{1.18}
$$

where the following inequalities $w < w_1$, $w < w_2$, $c_2 > c_1$, $0 < \tau_1 < \tau_2 < T$ are assumed to hold for positive constants: c_1, c_2, τ_1, τ_2, w_1, w_2, w

We will solve problem (1.18) by the Pontryagin maximum principle for systems with a delay argument. The part of the Hamiltonian that depends on controls $u_1(t)$, $u_2(t)$ is described as follows:

$$
\begin{aligned}
& H(t) = c_1 u_1(t) + c_2 u_2(t), \qquad t \in [0, T-\tau_2). \\
& H(t) = c_1 u_1(t), \qquad t \in (T-\tau_2, T-\tau_1), \\
& H(t) = 0, \qquad t \in (T-\tau_1, T].
\end{aligned}
$$

To find optimal controls $u_1^*(t)$, $u_2^*(t)$, we solve the problem of linear programming for every $t \in [0, T-\tau_1]$:

$$
\begin{aligned}
& H(t) \to \max, \qquad 0 \le u_1(t) \le w_1, \\
& 0 \le u_2(t) \le w_2, \qquad u_1(t) + u_2(t) \le w.
\end{aligned}
$$

Solving this problem, we finally obtain

$$
\begin{aligned}
& u_1^*(t) = 0, \quad t \in [0, T-\tau_2), \quad u_1^*(t) = w, \quad t \in (T-\tau_2, T-\tau_1], \\
& u_2^*(t) = w, \quad t \in [0, T-\tau_2), \quad u_2^*(t) = 0, \quad t \in (T-\tau_2, T-\tau_1].
\end{aligned}
\tag{1.19}
$$

The control values for $\tau \in (T-\tau_2, T]$ do not affect the functional value. Thus, optimal controls of problems (1.18) have the switching point $t^* = (T-\tau_2)$.

Further, we apply the decomposition method based on aggregated variables to problem (5.18). We give weights α_1, α_2, $\alpha_1 + \alpha_2 = 1$ constant on the interval $[0, T]$ and introduce the aggregated problem

$$\begin{aligned} & g = c_1 x_1(T) + c_2 x_2(T) \to \max, \\ & dx_1(t)/dt = \alpha_1(t-\tau_1)U(t-\tau_1) \qquad x_1(0) = \kappa_1; \\ & dx_2(t)/dt = \alpha_2(t-\tau_2)U(t-\tau_2) \qquad x_2(0) = \kappa_2; \\ & 0 \le U(t) \le w, \qquad t \in [0, T]. \end{aligned} \tag{1.20}$$

Here, we assume that $U(t) = e_1(t) + e_2(t)$, $t \in [-\tau_1, 0)$, $U(t) = e_2(t)$, $t \in [-\tau_2, -\tau_1)$; $\alpha_1(t) = e_1(t)/[e_1(t) + e_2(t)]$, $\alpha_2(t) = e_2(t)/[e_1(t) + e_2(t)]$, $t \in [-\tau_2, 0)$, $\alpha_2(t) = 1$, $t \in [-\tau_2, -\tau_1]$. Use of the maximum principle for problem (5.20) gives its solution $\overset{\circ}{U}(t) = w$, $t \in [0, T-\tau_1]$. We do not need to find the control $\overset{\circ}{U}(t)$ for $t \in [T-\tau_1, t)$. The dual variable $\overset{\circ}{\delta}(t)$ for the condition $U(t) \le w$ in the aggregated problem is computed as follows:

$$\begin{aligned} & \overset{\circ}{\delta}(t) = \alpha_1 c_1 + \alpha_2 c_2, \qquad t \in [0, T-\tau_2); \\ & \overset{\circ}{\delta}(t) = \alpha_1 c_1, \qquad t \in (T-\tau_2, T-\tau_1], \\ & \overset{\circ}{\delta}(t) = 0, \qquad t \in [T-\tau_1, T]. \end{aligned}$$

The function $\overset{\circ}{\delta}(t)$ is discontinuous at points $(T-\tau_2)$, $(T-\tau_1)$.

We formulate the first local problem

$$\begin{aligned} & h_1 = c_1 x_1(T) - \int_0^T \overset{\circ}{\delta}(t) u_1(t)\,dt \to \max, \\ & dx_1(t)/dt = u_1(t-\tau_1), \qquad x_1(0) = \kappa_1, \qquad 0 \le u_1(t) \le w_1; \\ & u_1(t) = e_1(t), \qquad t \in [-\tau_1, 0), \end{aligned}$$

and will solve it by the Pontryagin maximum principle:

$$c_1u_1(t) - (c_1\alpha_1 + c_2\alpha_2)u_1(t) \to \max,$$

$$0 \le u_1(t) \le w_1, \qquad t \in [0, T - \tau_2),$$

$$c_1u_1(t) - (c_1\alpha_1)u_1(t) \to \max,$$

$$0 \le u_1(t) \le w_1, \qquad t \in (T - \tau_2, T - \tau_1).$$

After obvious transformations

$$\alpha_2(c_1 - c_2)u_1(t) \to \max,$$

$$0 \le u_1(t) \le w_1, \qquad t \in [0, T - \tau_2).$$

$$c_1\alpha_2u_1(t) \to \max, \qquad 0 \le u_1(t) \le w,$$

$$t \in (T - \tau_2, T - \tau_1),$$

we obtain

$$\hat{u}_1(t) = 0, \quad t \in (0, T - \tau_2); \qquad \hat{u}_1(t) = w, \quad t \in [T - \tau_2, T - \tau_1).$$

Analogous consideration of the second local problem gives us

$$\hat{u}_2(t) = w_2, \quad t \in (0, T - \tau_2), \qquad \hat{u}_2(t) = 0, \quad t \in [T - \tau_2, T - \tau_1).$$

Using the iterative algorithm, the aggregation weights are computed by the formula of (1.11) (Chapter 3), where we assume $\rho_1 = \rho_2 = \rho$. Then, the optimal value of the functional of the aggregated problem (1.20) is a function of parameter ρ. As in the first example from Section 2, we establish that $\overset{\circ}{g}(\rho)$ is a linear fractional function, and its maximum is attained at the point $\overset{\circ}{\rho} = 1$. Thus, the solution of problem (1.18) coincides with the optimal solution of local problems, where, instead of w_1 and w_2, we should assume w. The initial optimum (1.19) is attained in one iteration.

Consider the block separable problem of optimal control with delayed arguments under a quadratic functional with an arbitrary number of subsystems:

$$\begin{aligned}
&f = 0,5\sum_{j=1}^{J}\left[c_j(x_j(T))^2 + \gamma_j\int_0^T(u_j(t))^2dt\right] \to \min,\\
&dx_j(t)/dt = u_j(t-\tau_j), \qquad x_j(0) = \kappa_j; \qquad u_j(t) = e_j,\\
&t \in [-\tau_j, 0), \qquad j \in [1:J],\\
&\sum_{j=1}^{J} u_j(t) \le w,
\end{aligned} \tag{1.21}$$

where, for definiteness, it is assumed $0 < \tau_1 < \ldots < \tau_j < \ldots < \tau_J < T$.

We try to solve problem (1.21) by direct use of the maximum principle for systems with delay. The terms in the Hamiltonian that depend on u_j have the following form:

$$H = \sum_{j=1}^{J}\left(-0,5\gamma_j(u_j(t))^2 + \beta_j u_j(t-\tau_j)y_j(t)\Big|_{t+\tau_j}\right),$$

where $\beta_j = 1$ for $t \in [0, T-\tau_j)$, $\beta_j = 0$ for $t \in [T-\tau_j, T]$; $y_j(t)$, $j \in [1:J]$ are conjugated variables constant on the interval $[0,T]$ and equal to $-c_j x_j(T)$, $j \in [1:J]$. Assume that, in each interval $[0, T-\tau_J), \ldots, (T-\tau_{j+1}, T-\tau_j), \ldots, (T-\tau_2, T-\tau_1)$, the binding condition in problem (1.21) holds as a strict equality. Then, in the interval $[0, T-\tau_J)$, we obtain the formula for optimal controls, which is the same as in problem (4.1) from Chapter 3:

$$u_s = (-c_s x_s(T)\sum_{r=1}^{J}\gamma_r^{-1} + \sum_{r=1}^{J} c_r x_r(T)\gamma_r^{-1} + w)(\gamma_s\sum_{r=1}^{J}\gamma_r^{-1})^{-1},$$

$$s \in [1:J], \qquad t \in [0, T-\tau_J).$$

Consider the interval $(T-\tau_{j+1}, T-\tau_j)$. Here, the terms of the Hamiltonian depending on u_j are equal to

$$H = \sum_{s=1}^{J} 0,5(-\gamma_s(u_s(t))^2) + \sum_{s=1}^{j}(-u_s(t)c_s x_s(T)).$$

Use of the Pontryagin maximum principle leads to a system of algebraic equations with respect to u_s, $s \in [1:J]$ and λ, where λ is a Lagrange multiplier of the binding constraint. We have

$$
\begin{aligned}
&-\gamma_s u_s - c_s x_s(T) - \lambda = 0, \qquad s \in [1:J],\\
&-\gamma_s u_s - \lambda = 0, \qquad s \in [j+1, J]; \sum_{s=1}^{J} u_s = w.
\end{aligned}
\tag{1.22}
$$

System (1.22) is solved in the standard way. The controls u_s, $s \in [1 : J]$ derive from the value λ and are substituted in the binding condition. After that, we obtain

$$
\lambda = \left(-\sum_{s=1}^{J} c_s x_s(T)(\gamma_s)^{-1} - w \right) \left(\sum_{s=1}^{J} \gamma_s^{-1} \right)
$$

as well as

$$
\begin{aligned}
u_s &= \left[-c_s x_s(T) \sum_{r=1}^{J} \gamma_r^{-1} + \sum_{r=1}^{j} c_r x_r(T) \gamma_r^{-1} + w \right] \left[\gamma_s \sum_{s=1}^{J} \gamma_r^{-1} \right]^{-1},\\
s &\in [1:j],\\
u_s &= \left[\sum_{r=1}^{j} c_r x_r(T) \gamma_j^{-1} + w \right] \left[\gamma_s \sum_{r=1}^{J} \gamma_r^{-1} \right]^{-1}, \qquad s \in [j+1, J].
\end{aligned}
\tag{1.23}
$$

Further, controls expressed by formula (1.23) for all $j \in [1 : J]$ and $t \in (T - \tau_{j+1}, T - \tau_j)$ are used in the right-hand sides of differential equations in (1.21). By integrating these equations from 0 to T, we obtain a system of linear algebraic equations with respect to the unknowns $x_j(T)$, $j \in [1 : J]$. Here, it is impossible to obtain the analytical solution, as for system (4.5). For large J, a complicated problem can arise for a system of algebraic equations, solved by the Gauss method.

We will solve problem (1.21) by the iterative decomposition method. If the aggregated problem has a solution $\overset{\circ}{U}(t) = w$, $t \in [0, T]$, then the dual variable $\overset{\circ}{\delta}(t)$ is easily computed under condition $U(t) \le w$. The function $\overset{\circ}{\delta}(t)$ is discontinuous for $t = T - \tau_J, \ldots, T - \tau_j, \ldots, T - \tau_1$, as it was in the previous example.

Consider the local problem with number j:

$$
h_j = 0,5 c_j (x_j(T))^2 + \int_0^T (0,5 \gamma_j (u_j(t))^2 + \overset{\circ}{\delta} u_j(t)) dt \to \min,
$$

$$
dx_j(t)/dt = u_j(t - \tau_j), \qquad x_j(0) = \kappa_j; \qquad u_j(t) = e_j, \quad t \in [-\tau_j, 0).
$$

The terms of the Hamiltonian, depending on u_j, have the following form:

$$H_j = -0,5\gamma_j(u_j(t))^2 - \overset{\circ}{\delta}(t)u_j(t) - u_j(t)c_jx_j(T), \qquad t \in [0, T - \tau_j),$$
$$H_j = -0,5\gamma_j(u_j(t))^2 - \overset{\circ}{\delta}(t)u_j(t), \qquad t \in (T - \tau_j, t].$$

According to the maximum principle, we obtain the following relation:

$$u_j(t) = (-c_jx_j(T) - \overset{\circ}{\delta}(t))/\gamma_j, \qquad t \in [0, T - \tau_j). \tag{1.24}$$

The control for $t \in (T - \tau_j, T]$ is equal to zero.

By integrating the differential equation in the local problem, taking into account (1.24), we finally obtain an equation with one unknown $x_j(T)$. This testifies to the efficiency of the decomposition method based on aggregation of variables for problem (5.20), as compared with its direct solution.

Consider control model in chemical technological projects having delayed arguments with functional terms quadratic in phase coordinates. We state the block separable optimal control problem of the following form:

$$\begin{aligned}
&f = 0,5 \sum_{j=1}^{J} \int_0^T \left[\sigma_j(x_j(t))^2 + \gamma_j(u_j(t))^2\right] dt \to \min, \\
&dx_j(t)/dt = u_j(t - \tau_j), \qquad x_j(0) = \kappa_j; \qquad u_j(t) = e_j(t), \\
&t \in [-\tau_j, 0), \qquad j \in [1 : J], \\
&\sum_{j=1}^{J} u_j(t) \le w(t), \qquad t \in [0, T],
\end{aligned} \tag{1.25}$$

where we assume $0 < \tau_1 < \ldots < \tau_j < \ldots < \tau_J = T$ as before.

The terms of the Hamiltonian depending on controls $u_j(t)$ will be the same as in problem (1.21). The conjugated variables $y_j(t)$, $j \in [1 : J]$ satisfy the differential equations

$$dy_j(t)/dt = \sigma_j x_j(t), \qquad j \in [1 : J] \tag{1.26}$$

and conditions of the form

$$y_j(T) = 0, \qquad j \in [1 : J]$$

for $t = T$.

The maximum principle establishes the relation between the controls and the conjugated variables by analogy with formulas (1.23), where $c_r x_r(T)$ should be replaced by $y_r(t+\tau_r)$. The two-point problem based on differential equations in (1.25) and (1.26) is fairly difficult for large J.

We will solve problem (1.25) by the iterative decomposition method based on aggregated variables. If the aggregated problem has a solution $\overset{\circ}{U}(t) = w(t)$, then the dual variable $\overset{\circ}{\delta}(t)$ for the condition $U(t) \leq w$ is computed immediately. This function is discontinuous as well as in the previous two problems.

The local problem with number j is formulated as follows:

$$h_j = \int\limits_0^T \left[0,5\sigma_j(x_j(t))^2 + 0,5\gamma_j(u_j(t))^2 + \overset{\circ}{\delta}(t)u_j(t)\right] dt \rightarrow \min,$$

$$dx_j(t)/dt = u_j(t - \tau_j),$$

$$x_j(0) = \kappa_j, \qquad u_j(t) = e_j(t), \qquad t \in [-\tau_j, 0).$$

The maximum principle applied to this problem gives the constraint

$$u_j(t) = \left[y_j(t + \tau_j) - \overset{\circ}{\delta}(t)\right] / \gamma_j, \qquad t \in [0, T - \tau_j).$$

Consider the following two-point problem:

$$\begin{aligned} &dx_j(t)/dt = F_j(t), \quad x_j(0) = \kappa_j; \quad u_j(t) = e_j(t), \quad t \in [-\tau_j, 0), \\ &F_j(t) = u_j(t - \tau_j), \qquad t \in [0, \tau_j); \\ &F_j(t) = (y_j(t) - \overset{\circ}{\delta}(t - \tau_j))/\gamma_j, \qquad t \in [\tau_j, T]. \end{aligned} \tag{1.27}$$

To solve problem (1.27), we start by integrating the first equation in the interval $[0, \tau_j]$ with regard to the fixed function $e_j(t)$. Then, we solve the linear two-point boundary problem in the interval $[\tau_j, T]$. The characteristic roots of this problem are equal to $\pm\sqrt{\sigma_j/\gamma_j}$, and quadrature formulas for its solution are obtained easily. Everything is reduced to a system of two linear algebraic equations with respect to exponential coefficients. This shows the efficiency of iterative decomposition for problem (5.15).

§2. Models of Hierarchical Systems with Distributed Parameters

In this section, we state block separable optimal control problems, where subsystems are described by classical partial differential equations. These problems have concrete physical meaning. In particular, they formalize models of optimal power resource distribution over subsystems. We consider linear problems with quadratic functionals. Applying the Fourier method and the Pontryagin maximum principle, their final solution in some cases is reduced to systems of linear algebraic equations. For large number of subsystems, we use either reducing the specified systems of equations or the method of iterative aggregation.

Consider the problem about optimal heating of a thin membrane with width $l = 1$. The temperature of the membrane $z(x,t)$ along the width and time satisfies the parabolic heat conductivity equation

$$\partial z(x,t)/\partial t - \partial^2 z(x,t)/\partial x^2 = 0 \text{ for } 0 < x < l,\ t > 0$$

with initial (for definiteness) zero temperature

$$z(x,0) = 0$$

and boundary conditions

$$\partial z(0,t)/\partial x = 0, \qquad \partial z(1,t)/\partial x = a(u(t) - z(1,t)), \qquad a = \text{const}.$$

The latter relation means that there is no heat flow on the left end, and the heat exchange on the right end obeys Newton's law. The value $u(t)$ gives the temperature of the heating medium and plays the role of control. The quadratic functional

$$\int_0^1 (z(x,T) - \bar{z}^0(x))^2 dx + \gamma \int_0^T u^2(t)dt$$

is minimized, where $\bar{z}^0(x) \in W_2^1(0,1)$ is the given function, γ is a positive constant, and T is final time.

Consider a system that consists of J membranes with centralized heating and state the following optimization problem:

$$
\begin{gathered}
f=\sum_{j=1}^{J}\left[\sigma_j\int_0^1 (z_j(x,T)-z_j^0(x))^2dx+\gamma_j\int_0^T (u_j(t))^2dt\right]\to\min,\\
\partial z_j(x,t)\partial t-\partial^2 z_j(x,t)/\partial x^2=0,\\
z_j(x,0)=0,\\
\partial z_j(0,t)/\partial x=0,\qquad \partial z_j(1,t)/\partial x=a_j(u_j(t)-z_j(1,t)),\\
a_j=\text{const},\qquad j\in[1:J];\\
\sum_{j=1}^{J}u_j(t)\le w(t),
\end{gathered}
\tag{2.1}
$$

where $w(t)$ is a given function, which characterizes the energetic resource of the center.

For each $j\in[1:J]$, consider the approximation of solutions $z_j(x,t)$ of mixed problems in (6.1) by the final number of terms from Fourier series:

$$
z_j(x,t)=\sum_{n=0}^{m}z_j^n(t)\cos\lambda_j^n x,\qquad j\in[1:J],
$$

where the values λ_j^n are positive roots of equations $\lambda_j^n \operatorname{tg}\lambda_j^n=a_j$, $j\in[1:J]$. We obtain the following optimization problem for the functions $z_j^n(t)$:

$$
\begin{gathered}
f=\sum_{j=1}^{J}\left\{\sigma_j\sum_{n=0}^{m}\omega_j^n\left[(z_j^n(T))^2-2z_j^n(T)z_j^{0n}\right]+\gamma_j\int_0^T (u_j(t))^2dt\right\}\to\min,\\
dz_j^n(t)/dt+(\lambda_j^n)^2z_j^n(t)=a_j\cos\lambda_j^n u_j(t)/\omega_j^n,\qquad z_j^n(0)=0,\\
j\in[1:J],\qquad n\in[0:m];\qquad \sum_{j=1}^{J}u_j(t)\in w(t),
\end{gathered}
\tag{2.2}
$$

where

$$
w_j^n=\left[(\lambda_j^n)^2+(a_j)^2+a_j\right]\left[2(\lambda_j^n)^2+(a_j)^2\right]^{-1},
$$

$$
z_j^{0n}=(\omega_j^n)^{-1}\int_0^1 z_j^0(x)\cos\lambda_j^n x dx.
$$

Problem (2.2) is solved by the Pontryagin maximum principle. The Hamiltonian has the form

$$H = \sum_{j=1}^{J}\sum_{n=0}^{m}\left(\frac{a_j \cos\lambda_j^n}{\omega_j^n}u_j(t) - (\lambda_j^n)^2 z_j^n(t)\right)y_j^n(t) - \gamma_j(u_j(t))^2\Bigg],$$

where $y_j^n(t)$ are conjugated variables that satisfy the following equations and conditions for $t=T$:

$$\begin{aligned} &dy_j^n(t)/dt - (\lambda_j^n)^2 y_j^n(t) = 0, \quad y_j^n(T) = -2\sigma_j\omega_j^n\left(z_j^n(T) - z_j^{0n}\right), \\ &j \in [1:J], \quad n \in [0:m]. \end{aligned} \tag{2.3}$$

Assume that, for the optimal solution of problem (2.2), the binding condition holds as a strict equality. Physically, this means that the energy released by the "center" is totally consumed. Applying the maximum principle, we have a problem with unknown controls $u_j(t)$, $j \in [1:J]$ and Lagrange multiplier $\lambda(t)$ for the binding constraint

$$\sum_{n=0}^{m}\frac{a_j\cos\lambda_j^n}{\omega_j^n}y_j^n(t) - 2\gamma_j u_j(t) - \lambda(t) = 0, \quad j\in[1:J]; \quad \sum_{j=1}^{J}u_j(t) = w(t).$$

From this, we have

$$\begin{aligned} u_j(t) = \Bigg[a_j\left(\sum_{n=0}^{m}\frac{\cos\lambda_j^n}{\omega_j^n}y_j^n(t)\right)\sum_{s=1}^{J}\frac{1}{2\gamma_s}+ \\ +w - \sum_{s=1}^{J}a_s\sum_{n=0}^{m}\frac{\cos\lambda_j^n}{2\omega_s^n\gamma_s}y_s(t)\Bigg]\left[\gamma_j\sum_{s=1}^{J}\frac{1}{\gamma_s}\right]^{-1}. \end{aligned} \tag{2.4}$$

The latter equalities give the relation between the controls and the dual variables of problem (2.2). If we substitute these expressions in the right-hand sides of differential equations from (2.2) and unite them with (2.3) under the corresponding conditions for $t=0$ and $t=T$, we obtain the two-point problem for a system of $2J(m+1)$ differential equations. The solution of this problem is constructed as follows. First we integrate independent equations in (2.3). We have

$$y_j^n(t) = C_j^n\exp((\lambda_j^n)^2 t), \qquad j\in[1:J], \qquad n\in[0:m], \tag{2.5}$$

where C_j^n are unknown constants. We substitute (2.5) in the right-hand side of (2.4) and then substitute controls in the right-hand side of differential equations from (2.2). We integrate them and obtain

$$\begin{aligned}
z_j^n(t) = D_j^n \exp(-(\lambda_j^n)^2 t) + a_j(\cos\lambda_j^n)(\omega_j^n)^{-1}(\gamma_j^{-1})\times \\
\times\left[\sum_{r=0}^{m} \frac{\cos\lambda_j^r a_j C_j^r \exp(\lambda_j^r)^2 t}{\omega_j^r((\lambda_j^n)^2+(\lambda_j^r)^2)} \sum_{s=1}^{J}\frac{1}{2\gamma_s} + \right. \\
+\exp(-(\lambda_j^n)^2 t)\int_0^t w(\tau)\exp((\lambda_j^n)^2\tau)d\tau - \\
\left. -\sum_{s=1}^{J} a_s \sum_{r=0}^{m} \frac{(\cos\lambda_j^r)C_s^r \exp((\lambda_s^r)^2 t)}{2\omega_s^r\gamma_s((\lambda_j^n)^2+(\lambda_s^r)^2)}\right]\times \\
\times\left[\sum_{s=1}^{J}\frac{1}{\gamma_s}\right]^{-1}, \qquad j\in[1:J], \qquad n\in[0:m],
\end{aligned} \tag{2.6}$$

where D_j^n are unknown constants.

With due regard to the initial conditions for $z_j^n(t)$ and conditions for $y_j^n(t)$ under $t = T$, we obtain a system of $2J(m+1)$ linear algebraic equations with respect to C_j^n, D_j^n, $j \in [1:J]$, $n \in [0:m]$:

$$\begin{aligned}
D_j^n + \frac{a_j(\cos\lambda_j^n)\gamma_j^{-1}}{\omega_j^n}\left[\sum_{r=0}^{m}\frac{\cos\lambda_j^r a_j C_j^r}{\omega_j^r((\lambda_j^n)^2+(\lambda_j^r)^2)}\sum_{s=1}^{J}\frac{1}{2\gamma_s} - \right. \\
\left. -\sum_{s=1}^{J} a_s \sum_{r=0}^{m}\frac{(\cos\lambda_s^r)C_s^r}{2\omega_s^r\gamma_s((\lambda_j^n)^2+(\lambda_s^r)^2)}\right]\left[\sum_{s=1}^{J}\frac{1}{\gamma_s}\right]^{-1} = 0,
\end{aligned}$$

$$\begin{aligned}
C_j^n \exp((\lambda_j^n)^2 T) - 2\sigma_j\omega_j^n z_j^n = -2\sigma_j\omega_j^n D_j^n \exp(-(\lambda_j^n)^2 T) - \\
-2\sigma_j a_j \gamma_j^1(\cos\lambda_j^n)\left[\sum_{r=0}^{m}\frac{\cos\lambda_j^r a_j C_j^r \exp((\lambda_J^r)^2 T)}{\omega_j^r((\lambda_j^n)^2+(\lambda_j^r)^2)}\sum_{s=1}^{J}\frac{1}{2\gamma_s} + \right. \\
+\exp\left(-(\lambda_j^n)^2 T\int_0^T w(\tau)\exp((\lambda_j^n)^2\tau)d\tau\right) - \\
\left. -\sum_{s=1}^{J} a_s \sum_{r=0}^{m}\frac{(\cos\lambda_s^r)C_s^r \exp((\lambda_s^r)^2 T}{2\omega_s^r\gamma_s((\lambda_j^n)^2+(\lambda_s^r)^2)}\right]\left[\sum_{s=1}^{J}\frac{1}{\gamma_s}\right]^{-1}, \\
j\in[1:J], \qquad n\in[0:m].
\end{aligned}$$

We express the values D_j^n from the first set of system equations and substitute them in the second set of system equations. Thus, we arrive at a general system of linear algebraic equations for the values C_j^n with the incidence matrix of dimension $J(m+1) \times J(m+1)$. For large J, this is a difficult problem. If we apply the iterative decomposition method for (2.2) as it is done in Section 4 of Chapter 3, then we obtain systems of equations with $(m+1) \times (m+1)$-matrices in local problems.

Consider a particular case $a_j = a$, $j \in [1 : J]$, then we arrive at a system of the form

$$b_j^n C_j^n + \sum_{r=0}^{m} d_j^{nr} C_j^r + p_j^n \sum_{s=1}^{J} \sum_{r=0}^{m} g_s^{nr} C_s^r = m_j^n, \qquad j \in [1 : J], \qquad n \in [0 : m]. \tag{2.7}$$

The specificity of system (2.7) defines the following solution. We introduce the notation

$$P^n = \sum_{s=1}^{J} \sum_{r=0}^{m} g_s^{nr} C_s^r, \qquad n \in [0 : m] \tag{2.8}$$

and consider independent systems of equations with respect to unknowns C_j^n, $n \in [0 : m]$ for each fixed $j \in [1 : J]$. Here, the values P^n are considered to be given. Solving the systems that consist of $(m+1)$ equations, we obtain

$$C_j^n = \sum_{r=0}^{m} \kappa_j^{nr} P^r + \nu_j^n, \qquad j \in [1 : J], \qquad n \in [0 : m]. \tag{2.9}$$

Then, we substitute (2.9) in the right-hand side of (2.8) and arrive at the system of $(m+1)$ equations with respect to unknown values P^n, $n \in [0 : m]$. After solving this system from (2.9), we finally find the sought coefficients C_j^n. Thus, the solution of the initial problem (2.1) is reduced to the $(J+1)$-th system of $(m+1)$-order linear algebraic equations. This is the essence of the dimension reduction method, under large values of J, for the indicated class of problems.

Consider a hierarchical model with hyperbolic equations, representing the longitudinal oscillations of a system having J elastic rods with length l, a cross-section with constant area and rigidly fixed ends. The rod number j is described by the wave equation

$$\begin{gathered} m_j \partial^2 z_j(x,t)/\partial t^2 - E_j \partial^2 z_j(x,t)/\partial x^2 = u_j(x,t), \\ 0 < x < l, \qquad 0 < t < T \end{gathered} \tag{2.10}$$

with initial conditions

$$z_j(x,0) = z_j^{(0)}(x), \qquad z_{jt}(x,0) = z_j^{(1)}(x), \qquad x \in [0,l], \tag{2.11}$$

and boundary conditions

$$z_j(0,t) = z_j(l,t) = 0, \qquad t \in [0,T], \tag{2.12}$$

where $z_j(x,t)$ is the displacement value for the j-th rod, E_j is the Young's modulus, m_j is the mass density, $u_j(x,t)$ is the distributed control characterized by the density of environment.

We formulate the block separable optimal control problem for the vibrating equipment, to choose regulating parameters for dynamical damping of the oscillations.

We take the binding condition in the form

$$\sum_{j=1}^{J} u_j(x,t) \le w(x,t) \tag{2.13}$$

and state the problem of damping the system by the instant T in the sense of minimizing the following functional:

$$f = \sum_{j=1}^{J} \left[\sigma_j \int_0^1 (z_j(x,T) - z_j^0(x))^2 dx + \beta_j \int_0^1 (z_{jt}(x,T) - z_j^1(x))^2 dx + \right.$$
$$\left. + \gamma_j \int_0^1 \int_0^T (u_j(x,t))^2 dx dt \right]. \tag{2.14}$$

Here, $z_j^{(0)}$, $z_j^{(1)}(x)$, $z_j^0(x)$, $z_j^1(x)$, $w(x,t)$ and σ_j, β_j, γ_j, $j \in [1:J]$ are given functions and positive constants.

We solve the problem (2.10)–(2.14) by the Pontryagin maximum principle. The conjugated variables $y_j(x,t)$, $j \in [1:J]$ satisfy the equations

$$m_j \partial^2 y_i(x,t)/\partial t^2 - E_j \partial^2 y_j(x,t)/\partial x^2 = 0 \tag{2.15}$$

under conditions for $t = T$:

$$\begin{aligned} y_j(x,T) &= -(2\beta_j/m_j)(z_{jt}(x,T) - z_j^1(x)), \\ y_{jt}(x,T) &= (2\sigma_j/m_j)(z_j(x,T) - z_j^0(x)) \end{aligned} \tag{2.16}$$

and boundary conditions of (2.12) type. According to the maximum principle, we arrive at the following problem:

$$\sum_{j=1}^{J}(-\gamma_j(u_j)^2 + y_j u_j) \to \max, \qquad \sum_{j=1}^{J} u_j \le w.$$

If the binding condition (2.13) is effective, then, as before, we obtain the relation between controls with conjugated variables, which, in our case, have the following form:

$$u_j(x,t) = \left[y_j(x,t)\sum_{s=1}^{J}\frac{1}{2\gamma_s} + w(x,t) - \right.$$
$$\left. - \sum_{s=1}^{J}\frac{y_s(x,t)}{2\gamma_s}\right]\left[\gamma_j \sum_{s=1}^{J}\frac{1}{\gamma_s}\right]^{-1}, \qquad j \in [1:J]. \tag{2.17}$$

We solve the mixed problems (2.10)–(2.12) and (2.15)–(2.16) by the Fourier method:

$$z_j(x,t) = \sum_{n=1}^{\infty}(A_j^n \cos\lambda_j^n t + B_j^n \sin\lambda_j^n t)\sqrt{\frac{2}{l}}\sin\frac{n\pi x}{l} + \sum_{n=1}^{\infty}\sqrt{\frac{2}{l}}\times$$
$$\times \sin\frac{n\pi x}{l}(2m_j\lambda_j^n)^{-1}\int_0^l\int_0^t \sin\lambda_j^n(t-\tau)\sqrt{\frac{2}{l}}\sin\frac{n\pi}{l}\xi u_j(\xi,\tau)\,d\xi d\tau, \tag{2.18}$$

where

$$\lambda_j^n = \frac{n\pi}{l}\left(\frac{E_j}{m_j}\right)^{1/2},$$
$$A_j^n = \int_0^l \sqrt{\frac{2}{l}} z_j^{(0)}(\xi)\sin\frac{n\pi\xi}{l}\,d\xi,$$
$$B_j^n = \int_0^l (\lambda_j^n)^{-1}\sqrt{\frac{2}{l}} z_j^{(0)}(\xi)\sin\frac{n\pi}{l}\xi\,d\xi$$

as well as

$$y_j(x,t) = \sum_{n=1}^{\infty}(C_j^n \cos\lambda_j^n t + D_j^n \sin\lambda_j^n t)\sqrt{\frac{2}{l}}\sin\frac{n\pi x}{l}, \qquad j \in [1:J]. \tag{2.19}$$

Here, C_j^n, D_j^n are unknown constants.

Now, we substitute relations (2.19) into the right-hand sides of (2.17). Then, we substitute the obtained result into the integrands from (2.18). We obtain

$$z_j(x,t) = \sum_{n=1}^{\infty}\Bigg[A_j^n \cos\lambda_j^n t + B_j^n \sin\lambda_j^n t + \Lambda_j^n(t) - \\ - \sum_{s=1}^{J}(C_s^n \kappa_{js}^n(t) + D_s^n \mu_{js}^n(t))\Bigg]\frac{2}{l}\sin\frac{n\pi x}{l};$$

$$\Lambda_j^n(t) = \frac{1}{2m_j\lambda_j^n\gamma_j}\left(\sum_{r=1}^{J}\frac{1}{\gamma_r}\right)^{-1}\int_0^1\int_0^t \sin\lambda_j^n(t-\tau)\sqrt{\frac{2}{l}}\sin\frac{n\pi\xi}{l}w(\xi,\tau)d\xi d\tau;$$

$$\kappa_{js}^n(t) = \left(4m_j\lambda_j^n\gamma_s\gamma_j\sum_{r=1}^{J}\frac{1}{\gamma_r}\right)^{-1}\{\sin\lambda_j^n t\,[\lambda_j^n + \lambda_s^n)^{-1}\sin(\lambda_j^n+\lambda_s^n)t + \\ +(\lambda_j^n - \lambda_s^n)\sin(\lambda_j^n-\lambda_s^n)t] + \cos\lambda_j^n t\,[(\lambda_j^n+\lambda_s^n)^{-1}(\cos(\lambda_j^n+\lambda_s^n)t-1) + \\ +(\lambda_j^n-\lambda_s^n)^{-1}(\cos(\lambda_j^n-\lambda_s^n)t-1]\}, \qquad s \neq j;$$

$$\kappa_{jj}^n(t) = \left(2m_j\lambda_j^n\gamma_j\sum_{r=1}^{J}\frac{1}{\gamma_r}\right)^{-1}\left(\sum_{\substack{r=1\\ r\neq j}}\frac{1}{2\gamma_r}\right)\times \\ \{\sin\lambda_j^n t\,[(2\lambda_j^n)^{-1}\sin 2\lambda_j^n t + t] + \cos\lambda_j^n t\,[(2\lambda_j^n)^{-1}(\cos 2\lambda_j^n t - 1)]\};$$

$$\mu_{js}^n(t) = \left(4m_j\lambda_j^n\gamma_s\gamma_j\sum_{r=1}^{J}\frac{1}{\gamma_r}\right)^{-1}\{\sin(\lambda_j^n+\lambda_s^n)^{-1}(\cos(\lambda_j^n+\lambda_s^n)t-1) - \\ -(\lambda_j^n-\lambda_s^n)^{-1}(\cos(\lambda_j^n-\lambda_s^n)t-1) + \\ +\cos\lambda_j^n t\,[(\lambda_j^n+\lambda_s^n)^{-1}\sin(\lambda_j^n+\lambda_s^n)t - (\lambda_j^n-\lambda_s^n)^{-1}\sin(\lambda_j^n-\lambda_s^n)t]\}, \quad s\neq j;$$

$$\mu_{jj}^n(t) = -\sum_{r=1}^{J}\left(\frac{1}{2\gamma_r}\right)\left(16m_j\lambda_j^n\gamma_j\sum_{r=1}^{J}\frac{1}{\gamma_r}\right)^{-1}\times \\ \times\{\sin\lambda_j^n t\,[2\lambda_j^n)^{-1}(\cos 2\lambda_j^n t - 1)] + \cos\lambda_j^n t\,[(2\lambda_j^n)^{-1}\sin 2\lambda_j^n t - t]\}.$$

We substitute the written expressions for solutions $z_j(x,t)$ as well as $z_{jr}(x,t)$, into equalities (2.16) for $t = T$ and equate the coefficients for linear independent functions $\sqrt{2/l}\sin(n\pi x/l)$:

$$C_j^n \cos\lambda_j^n T + D_j^n \sin\lambda_j^n T =$$

$$= -(2\beta_j/m_j)\left(-A_j^n\lambda_j^n \sin\lambda_j^n T + B_j^n\lambda_j^n \cos\lambda_j^n T + \right.$$

$$\left. + \Lambda_j^{n\prime}(T) + \sum_{s=1}^{J}(C_s^n \kappa_{js}^{n\;\prime}(T) + D_s^n \mu_{js}^{n\;\prime}(T)) - z_j^{1n}\right); \qquad (2.20)$$

$$-C_j^n\lambda_j^n \sin\lambda_j^n T + D_j^n\lambda_j^n \cos\lambda_j^n T = (2\sigma_j/m_j)\left(A_j^n \cos\lambda_j^n T + \right.$$

$$\left. + B_j^n \sin\lambda_j^n T + \Lambda_j^n(T) + \sum_{s=1}^{J}(C_s^n \kappa_{js}^n(T) + D_s^n \mu_{js}^n(T)) - z_j^{0n}\right),$$

where

$$z_j^{0n} = \int_0^1 \sqrt{\frac{2}{l}}\, z_j^0(\xi) \sin\frac{n\pi\xi}{l}\, d\xi, \qquad z_j^{1n} = \int_0^1 \sqrt{\frac{2}{l}}\, z_j^1(\xi) \sin\frac{n\pi\xi}{l}\, d\xi.$$

We do the same with the initial values for $z_j(x,t)$, $z_{jt}(x,t)$, by adding them to (2.20). Then, we exclude A_j^n, B_j^n and obtain a system of general linear algebraic equations with respect to C_j^n, D_j^n, $j \in [1:J]$:

$$\sum_{s=1}^{J} q_{js}^{nk} C_s^n + \sum_{s=1}^{J} g_{js}^{-nk} D_s^n = b_j^{nk}, \qquad j \in [1:J], \qquad n \in [0:m], \qquad k = 1,2,$$

which has a $2J \times 2J$-matrix. For large J, a difficult problem arises. If we apply the iterative decomposition method for problem (2.10)–(2.14), then, in local problems, everything is reduced to linear algebraic equations with two unknowns.

Let E_j/m_j, $j \in [1:J]$, then $\lambda_j^n = \lambda^n$, $j \in [1:J]$. In this case, after the introduction of the corresponding notation, the latter system will take the following form:

$$d_j^{nk} C_j^n + \bar{d}_j^{nk} D_j^n + p_j^{nk} \sum_{s=1}^{J} g_s^{nk} C_s^n +$$

$$+ \bar{p}_j^{nk} \sum_{s=1}^{J} \bar{g}_s^{nk} D_s^n = e_j^{nk}, \qquad j \in [1:J], \qquad k = 1,2. \qquad (2.21)$$

The specificity of system (2.21) shows how it can be solved. We introduce the notation

$$P^{nk} = \sum_{s=1}^{J} g_s^{nk} C_s^n, \qquad \bar{P}^{nk} = \sum_{s=1}^{J} \bar{g}_s^{nk} D_s^n, \qquad n \in [0:m], \qquad k = 1,2. \qquad (2.22)$$

We solve (2.21) for each $j \in [1:J]$ as a system of two equations with respect to C_j^n, D_j^n and consider the values P^{nk}, $\bar{P}^{nk}$, $k = 1, 2$, to be given. We obtain

$$C_j^n = \sum_{k=1}^{2} \left(\pi_j^{nk} P^{nk} + \bar{\pi}_j^{nk} \bar{P}^{nk} \right) + \nu_j^n,$$
$$D_j^n = \sum_{k=1}^{2} \left(\varsigma_j^{nk} P^{nk} + \bar{\xi}^{nk} \bar{P}^{nk} \right) + \chi_j^n, \qquad j \in [1:J]. \tag{2.23}$$

We substitute (2.23) in the right-hand sides of (2.22) and determine the values P^{nk}, $\bar{P}^{nk}$, $k = 1, 2$ from the four linear equations. Then, we compute the required coefficients C_j^n, D_j^n, $j \in [1:J]$ from (2.23). Thus, the solution of the optimization problem (2.10)–(2.14) is reduced to linear algebraic equations with two and four unknowns, achieving the required reduction of dimensions.

Consider a problem obtained from (2.10)–(2.14) when functional (2.14) is replaced by the following integral functional:

$$f = \sum_{j=1}^{J} \left\{ \frac{\mu_j}{2} \int_0^l \int_0^T \left[m_j \left(\frac{\partial z_j(x,t)}{\partial t} \right)^2 + E_j \left(\frac{\partial z_j(x,t)}{\partial x} \right)^2 \right] dx dt + \right.$$
$$\left. + \gamma_j \int_0^l \int_0^T (u_j(x,t))^2 dx dt \right\}. \tag{2.24}$$

Here, the first integrals characterize the kinetic and potential energy of oscillations, and μ_j, $j \in [1:J]$ are positive preference coefficients.

We look for solutions of problem (2.10)–(2.13), (2.24) in the form of Fourier series

$$u_j(x,t) = \sum_{n=1}^{\infty} u_j^n(t) \sqrt{\frac{2}{l}} \sin \frac{n\pi x}{l},$$
$$z_j(x,t) = \sum_{n=1}^{\infty} z_j^n(t) \sqrt{\frac{2}{l}} \sin \frac{n\pi x}{l},$$

where, for each $n = 1, 2, \ldots$, we obtain the following optimization problem:

$$f = \sum_{j=1}^{J} \left\{ \frac{\mu_j}{2} \int_0^T \left[(w_j^n(t))^2 + (\lambda_j^n)^2 (z_j^n(t))^2 + (\gamma_j/m_j)(u_j^n(t))^2 \right] dt \right\} \to \min,$$

$$\begin{aligned}
&dw_j^n(t)/dt = -(\lambda_j^n)^2 z_j^n(t) + u_j^n(t)/m_j, \qquad dz_j^n(t)/dt = w_j^n(t),\\
&z_j^n(0) = z_j^{n(0)} = \int_0^l z_j^{(0)}(x)\frac{2}{l}\sin\frac{n\pi x}{l}dx,\\
&w_j^n(0) = z_j^{n(1)} = \int_0^l z_j^{(1)}(x)\frac{2}{l}\sin\frac{n\pi x}{l}dx,\\
&\sum_{j=1}^J u_j^n(t) \le w^n(t) = \int_0^l w(x,t)\frac{2}{l}\sin\frac{n\pi x}{l}dx.
\end{aligned} \tag{2.25}$$

We can try to solve problem (2.25), using the Pontryagin maximum principle. The conjugated variables $y_j^n(t)$, $v_j^n(t)$ satisfy the following equations and conditions for $t = T$:

$$\begin{aligned}
&dy_j^n(t)/dt = \mu_j^n w_j^n(t) - \nu_j^n(t),\\
&d\nu_j^n(t)/dt = \mu_j^n w_j(\lambda_j^n)^2 z_j^n(t) + (\lambda_j^n)^2 y_j^n(t),\\
&y_j^n(T) = \nu_j^n(T) = 0, \qquad j \in [1:J].
\end{aligned} \tag{2.26}$$

If the binding constraint is efficient, the maximum principle leads to the relation analogous to (2.17):

$$u_j^n(t) = \left[y_j^n(t)\sum_{s=1}^J \frac{m_s}{2\gamma_s}m_j + w^n(t) - m_j\sum_{s=1}^J\frac{y_s^n(t)}{2\gamma_s}\right]\left[\gamma_j\sum_{s=1}^J\frac{m_s}{2\gamma_s}\right]^{-1}, \tag{2.27}$$

$j \in [1:J]$.

We substitute (2.27) in the right-hand sides of differential equations in (2.25) and unite them with differential equations (2.26). We obtain a system of $4J$ equations, which form a two-point problem. As is known, its solution consists of the solution of the corresponding linear homogenous system and a partial solution. Here, the following problems arise: finding eigen values of $4J$-matrices, determination of the corresponding eigen vectors, finding a particular solution, and finally, finding a solution of a system of $4J$-order linear algebraic equations with respect to unknown exponent coefficients. The latter system arises when the conditions for $y_j^n(t)$, $\nu_j^n(t)$ under $t = T$ are used in general form for the solution $z_j^n(t)$, $w_j^n(t)$ of differential equations. Thus, for large J, we encounter fairly difficult computational problems. Here, we should not forget the possibility of a short eigenvalue, etc.

We solve the problem (2.10)–(2.13), (2.24) or, which is the same, problem (2.25) by the iterative decomposition method. If, for the normed weights $\alpha_1^n(t)$, the binding condition in (2.25) is essential, then we directly obtain an optimal solution of the aggregated problem: $\overset{\circ}{U}{}^n(t) = w^n(t)$. The conjugated function $\overset{\circ}{\delta}{}^n(t)$ that corresponds to the binding constraint is computed in the form

$$\overset{\circ}{\delta}{}^n(t) = \sum_{j=1}^{J} \left[\alpha_j^n(t)\, \overset{\circ}{y}{}_y^n(t) m_j - 2\gamma_j(\alpha_j(t))^2 w^n(t)/m_j\right],$$

where the conjugated variables $\overset{\circ}{y}{}_j^n$ are easily determined when the optimal control $\overset{\circ}{U}{}^n(t)$ is found. At each step of the iterative process, we also solve local problems independent of j, which are given by differential equations from (2.25), and the terms $\int_0^T \overset{\circ}{\delta}{}^n(t) u_j^n(t)dt$ are added to the functionals. Solving the local problems by means of the maximum principle, we obtain

$$u_j^n(t) = (y_j^n(t) - \overset{\circ}{\delta}{}^n(t) m_j)/2\gamma_j, \tag{2.28}$$

where $y_j^n(t)$ are conjugated variables that satisfy (2.26).

For each fixed j, we substitute the control expressed according to (2.28) in the right-hand sides of the differential equations from (2.25), (2.26) and obtain J fourth-order independent systems. The characteristic equations of these systems look as follows:

$$(\rho_j^n)^4 + (2(\lambda_j^n)^2 - \mu_j(2m_j\gamma_j)^{-1}(\rho_j^n)^2 + ((\lambda_j^n)^4 + \mu_j(\lambda_j^n)^2(2m_j\gamma_j)^{-1}) = 0.$$

If the inequality $\lambda_j^n < \mu_j(4m_j\gamma_j)^{-1}$ holds, then there are four different real roots, otherwise, we obtain two complex conjugated roots. Finding exponent coefficients in homogenous solutions of the systems considered is reduced to the solution of fourth-order algebraic equations. Thus, we reduce the dimension in the solution of problem (2.25) or, which is the same, of problem (2.10)–(2.13), (2.24).

We consider one more block separable statement. Consider the model problem about the heating of a thin membrane, which is moved with constant velocity a through a furnace with length l. The heat process in the furnace is described by the following first-order partial differential equation:

$$\partial z(x,t)/\partial t + a\partial z(x,t)/\partial x = B(u(x,t) - z(x,t)), \quad x \in (0,l), \quad t \in (0,T).$$

Here, $z(x,t)$ is the distribution of temperature over the membrane, $u(x,t)$ is the control function of the temperature in the furnace. The value B is called the Bio criterion and is equal to $\beta/c\gamma s$, where β is the heat exchange coefficient, c is specific heat, γ is specific mass, s is the width of the membrane.

The material at the entrance of the furnace has a definite temperature and a given initial distribution $z(0,t) = z(t)$, $z(0,x) = z^{(0)}(x)$. We state optimal control problem as minimizing integral functional

$$\int_0^l \int_0^T \left(\sigma(z(x,t))^2 + \gamma(u(x,t))^2\right) dxdt,$$

where σ, γ are fixed positive coefficients.

Consider the system consisting of J equal furnaces united by a common "center" that releases energy (heat) for heating furnaces, where each furnace has temperature $u_j(x,t)$ proportional to the energy obtained from the center. We state the following optimization problem:

$$\begin{gathered}
f = \sum_{j=1}^{J} \left[\int_0^l \int_0^T \left[\sigma_j(z_j(x,t))^2 + \gamma(u_j(x,t))^2\right] dxdt \right] \to \min, \\
\partial z_j(x,t)/\partial t + a\partial z_j(x,t)/\partial x = B_j(u_j(x,t) - z_j(x,t)), \\
x \in (0,l), \qquad t \in (0,T). \\
z_j(0,t) = z_j(t), \quad t \in [0,T], \quad z_j(x,0) = z_j^{(0)}(x), \quad x \in [0,l], \\
u_j(x,t) \geq 0, \qquad j \in [1:J]; \qquad \sum_{j=1}^{J} u_j(x,t) \leq w(x,t),
\end{gathered} \tag{2.29}$$

where $w(x,t)$ is a given function that characterizes the amount of energy released by the center.

First, we consider the direct solution of problem (2.29) using the Pontryagin maximum principle. The Hamiltonian is as follows:

$$H = -\sum_{j=1}^{J} \left[\sigma_j(z_j)^2 + \gamma_j(u_j)^2 - B_j(u_j - z_j)y_j\right],$$

where $y_j(x,t)$, $j \in [1:J]$ are conjugated variables that satisfy the following equations and conditions under $t = T$, $x = l$:

$$\partial y_j(x,t)/\partial t + a\partial y_j(x,t)/\partial x = (B_j y_j(x,t) + 2\sigma_j z_j(x,t)),$$
$$x \in (0,l), \qquad t \in (0,T), \qquad y_j(x,T) = 0, \qquad x \in [0,l], \tag{2.30}$$
$$y_j(l,t) = 0, \qquad t \in [0,T].$$

Using the maximum principle, we have the quadratic programming problem

$$\sum_{j=1}^{J} \left(\gamma_j (u_j)^2 - B_j u_j y_j\right) \to \min,$$
$$u_j \geq 0, \qquad j \in [1:J], \qquad \sum_{j=1}^{J} u_j(x,t) \leq w(x,t).$$

If we assume that the binding condition in (2.30) is effective, i.e., the energy released by the center is totally used and, moreover, the optimal controls are strictly positive, then we obtain the formula analogous to (2.17), (2.27):

$$u_j(x,t) = \left[B_j y_j(x,t) \sum_{s=1}^{J} \frac{1}{2\gamma_s} + \right.$$
$$\left. + w(x,t) - \sum_{s=1}^{J} \frac{B_s y_s(x,t)}{2\gamma_s}\right] \left[\gamma_j \sum_{s=1}^{J} \frac{1}{\gamma_s}\right]^{-1}. \tag{2.31}$$

By substituting (2.31) in the right-hand sides of (2.29) and uniting them with equations (2.30), we obtain $2J$ linear nonhomogenous differential equations with respect to z_j, y_j, $j \in [1:J]$, along the characteristics $t - x/a =$ const. Here, as in the previous case, the following problems arise: find eigenvalues of a matrix of order $2J$; determine the corresponding eigenvectors; find a particular solution; finally, compute exponential coefficients in the formula for the solution of a homogenous system. For the latter problem, we use the initial or boundary conditions at intersection points of characteristics with the rectangular $[0,l] \times [0,T]$. Thus, for large J, these are difficult problems, which were discussed during the analysis of model (2.10)–(2.13), (2.24).

Furthermore, consider the solution of problem (2.29) by means of the iterative decomposition method. If, for given normed weights $\alpha_j(x,t)$, the optimal solution of the aggregated problem has the form $\overset{\circ}{U}(x,t) = w(x,t)$, then the equations

$$\partial z_j(x,t)/\partial t + a\partial z_j(x,t)/\partial x = B_j(\alpha_j(x,t)\,\overset{\circ}{U}(x,t) - z_j(x,t))$$

are integrated immediately along the characteristics of the equation. The same is done for the conjugated variables. The conjugated function $\overset{\circ}{\delta}(x,t)$ that corresponds to the binding constraints is computed in the form $\overset{\circ}{\delta}(x,t) = \sum_{j=1}^{J}(B_j\alpha_j\overset{\circ}{y}_j - 2\gamma_j(\alpha_j)^2\overset{\circ}{U})$, and the functional of the local problem has the following form:

$$\int_0^l\int_0^T \left(\sigma_j(z_j)^2 + \gamma_j(u_j)^2 + \overset{\circ}{\delta}\,u_j\right)dxdt.$$

Independent local problems are solved for each $j \in [1:J]$ by means of the Pontryagin maximum principle. They are reduced to the system of two differential equations of the form

$$\partial z_j(x,t)/\partial t + a\partial z_j(x,t)/\partial x =$$
$$= B_j\left[B_jy_j(x,t) - \overset{\circ}{\delta}(x,t))(2\gamma_j)^{-1} - z_j(x,t)\right],$$
$$\partial y_j(x,t)/\partial t + a\partial y_j(x,t)/\partial x = (B_jy_j(x,t) + 2\sigma_jz_j(x,t)).$$

The characteristic roots of this system are $\lambda_j^{+,-} = \pm B_j(1+\sigma_j/\gamma_j)^{1/2}$. Solutions of these systems are reduced to pairs of independent linear algebraic equations with exponential coefficients. Thus, the dimension in problem (2.29) is reduced.

In the model of membrane heating, it is assumed that the parameter B changes randomly and N values $B^1,\ldots,B^N$ with probabilities $p^1,\ldots,p^N \geq 0$, $\sum_{q=1}^{J} p^q = 1$ occur. The mean value of this quadratic integral functional is minimized. In this case, for the block separable statement, the direct solution by the maximum principle results in the following system of equations:

$$\partial z_j^q(x,t)/\partial t + a\partial z_j^q(x,t)/\partial x = B_j^q(u_j(x,t) - z_j^q(x,t)),$$
$$\partial y_j^q(x,t)/\partial t + a\partial y_j^q(x,t)/\partial x = (B_j^qy_j^q + 2\sigma_jz_j^q),$$
$$j \in [1:J], \qquad q \in [1:N],$$

where we should substitute expression analogous to (2.31) for u_j. This expression has the form

$$u_j = \left[\sum_{q=1}^{N} B^q p^q y_j^q \sum_{s=1}^{J} \frac{1}{2\gamma_s} + w - \sum_{q=1}^{N}\sum_{s=1}^{J} \frac{B_s^q y_s^q}{2\gamma_s}\right] \left[\gamma_j \sum_{s=1}^{J} \frac{1}{\gamma_s}\right]^{-1}.$$

The order of the system in two-point problems increases up to $2NJ$ along the characteristics, and the utility of iterative aggregation becomes apparent.

§3. Block Separable Problems with Large Number of Binding Constraints

In previous sections of this and the third chapter, we considered the simplest statements of block separable optimal control problems with a single binding condition on the control functions. However, it is also interesting to consider the situation with a large number of binding constraints, as we saw, for example, in block linear programming. It turned out that, for linear quadratic problems, it is efficient to use the reduction method in systems of linear algebraic equations obtained by applying the Pontryagin maximum principle. In many other situations, it is expedient to use iterative decomposition based on aggregation of variables.

Consider the generalization of problem (4.1) of Chapter 3. We assume that, in each subsystem, the control is a I-dimensional vector, and the following binding constraints are imposed on the components of all vectors for $j \in [1 : J]$:

$$f = 0,5 \sum_{j=1}^{J} \left[c_j(x_j(T))^2 + \sum_{i=1}^{I} \int_0^T \gamma_j^i (u_j^i(t))^2 dt \right] \to \min,$$

$$dx_j(t)/dt = \sum_{i=1}^{I} b_j^i u_j^i(t), \qquad x_k(0) = \kappa_j, \qquad j \in [1 : J], \tag{3.1}$$

$$\sum_{j=1}^{J} u_j^i(t) = w^i, \qquad i \in [1 : I].$$

Here, we omitted the condition for the sign of control functions, and the binding constraints are taken as point-wise equalities.

We try to solve problem (3.1) by direct use of the Pontryagin maximum principle. The conjugated functions are $\psi_j(t) = -c_j x_j(T)$, $t \in [0, T]$, $j \in [1 :$

$J]$, and we come to the problem of unconstrained maximization

$$\sum_{j=1}^{J}\sum_{i=1}^{I}\left[0,5\gamma_j^i(u_j^i)^2+b_j^ic_jx_j(T)u_j^i\right]+\sum_{i=1}^{I}\lambda^i(w^i-\sum_{j=1}^{J}u_j^i)\to\max,$$

where λ^i, $i\in[1:I]$ are Lagrange multipliers of the binding constraints. After standard differentiation, we have the system of equations with respect to u_j^i, j, $i\in[1:I]$ and λ^i, $i\in[1:I]$:

$$-\gamma_j^iu_j^i-b_j^ic_jx_j(T)-\lambda^i=0,\qquad j\in[1:J],\qquad i\in[1:I],$$
$$\sum_{j=1}^{J}u_j^i=w^i,\qquad i\in[1:I].$$

This gives us

$$u_j^i=\frac{w^i}{\gamma_j^i}\left(\sum_{s=1}^{J}\frac{1}{\gamma_s^i}\right)+\left[\sum_{s=1}^{J}\frac{b_s^ic_sx_s(T)}{\gamma_s^i}\left(\sum_{r=1}^{J}\frac{1}{\gamma_r^i}\right)^{-1}\right]\frac{1}{\gamma_j^i}-\frac{b_j^ic_jx_j(T)}{\gamma_j}.$$

Using the written relation in differential equations of problem (5.1), we obtain the following system of linear algebraic equations with respect to unknowns $x_j(T)$, $j\in[1:J]$:

$$\left[1+c_jT\sum_{i=1}^{I}\frac{(b_j^i)^2}{\gamma_j^i}\right]x_j(T)=\kappa_j+\sum_{i=1}^{I}\frac{b_j^iw^i}{\gamma_j^i}\left(\sum_{r=1}^{J}\frac{1}{\gamma_r^i}\right)^{-1}T+$$
$$+\sum_{i=1}^{I}b_j^iT\left[\sum_{s=1}^{J}\frac{b_s^ic_sx_s(T)}{\gamma_s^i}\left(\sum_{r=1}^{J}\frac{1}{\gamma_r^i}\right)^{-1}\frac{1}{\gamma_j^i}\right],\qquad j\in[1:J].\tag{3.2}$$

We try to reduce system (3.2), as was done in Section 4 of Chapter 3. To this end, we introduce standard notation and write, according to (3.2), the following equalities:

$$x_j(T)=\pi_j+\sum_{i=1}^{I}\nu_j^iP^i,\qquad j\in[1:J],\tag{3.3}$$

where

$$P^i=\sum_{s=1}^{J}\frac{b_s^ic_sx_s(T)}{\gamma_s^i},\qquad i\in[1:I].\tag{3.4}$$

We substitute (3.3) in the right-hand side of (3.4) and obtain the following system of linear algebraic equations with respect to variables P^i:

$$P^i = \sum_{r=1}^{I} \sigma_r^i P^r + \omega^i; \qquad \omega^i = \sum_{s=1}^{J} \frac{b_s^i c_s \pi_s}{\gamma_s^i}; \qquad \sigma_r^i = \sum_{s=1}^{J} \frac{b_s^i c_s \nu_s^r}{\gamma_s^i}. \tag{3.5}$$

Thus, the solution to the initial optimal control problem (3.1) is reduced to the solution of a system of linear algebraic equations either of form (3.2) with respect to $x_j(T)$, $j \in [1:J]$ or of form (3.5) with respect to P^i, $i \in [1:I]$. It is clear that if $J < I$, then we need to consider the first system and vice versa. If the values J and I are approximately equal and sufficiently large, then large matrices are transformed.

We solve problem (3.1) by the decomposition method based on variable aggregation. For any fixed weights α_j^i, $j \in [1:J]$, $i \in [1:I]$ that satisfy the normalization condition, the aggregated problem has optimal controls $\overset{\circ}{U}{}^i = w$, $i \in [1:I]$. Therefore, we obtain

$$\overset{\circ}{x}_j(t) = \kappa_j + \sum_{i=1}^{I} b_j^i \alpha_j^i w^i t, \qquad \overset{\circ}{\psi}_j(T) = -c_j(\kappa_j + \sum_{i=1}^{I} b_j^i \alpha_j^i w^i T).$$

Optimal dual variables for binding constraints are computed in the form

$$\overset{\circ}{\delta}{}^i = -\sum_{j=1}^{J} c_j(\kappa_j + \sum_{r=1}^{I} b_j^r \alpha_j^r w^r T) b_j^i \alpha_j^i - \sum_{j=1}^{J} \gamma_j^i (\alpha_j^i)^2 w^i, \qquad i \in [1:I].$$

The functionals of local problems look as follows:

$$0,5 c_j (x_j(T))^2 + 0,5 \sum_{i=1}^{I} \int_0^T \gamma_j^i (u_j^i(t))^2 dt + \sum_{i=1}^{I} \int_0^T \overset{\circ}{\delta}{}^i u_j^i(t) dt \to \min,$$

and the use of the maximum principle for these problems results in the relation

$$u_j^i = -(b_j^i x_j x_j(T) + \overset{\circ}{\delta}{}^i)(\gamma_j^i)^{-1}.$$

This gives independent equations with respect to unique unknowns $x_j(T)$:

$$x_j(T) = \kappa_j - \sum_{i=1}^{I} b_j^i (\gamma_j^i)^{-1} c_j x_j(T) T - \sum_{i=1}^{I} \overset{\circ}{\delta}{}^i (\gamma_j^i)^{-1} T.$$

Thus, the dimension of problem (3.1) is reduced.

Consider the generalization of problem (3.1), where several phase coordinates occur in each subsystem:

$$f = 0,5\sum_{j=1}^{J}\left[\sum_{n=0}^{m} c_j^n (x_j^n(T))^2 + \sum_{i=1}^{I}\int_0^T \gamma_j^i (u_j^i(t))^2 dt\right] \to \min,$$

$$dx_j(t)/dt = \sum_{i=1}^{I} b_j^{in} u_j^i(t), \qquad x_j^n(0) = \kappa_j^n, \qquad j \in [1:J], \qquad n \in [0:m],$$

$$\sum_{j=1}^{J} u_j^i(t) = w^i, \qquad i \in [1:I]. \tag{3.6}$$

Direct solution of the problem by the Pontryagin maximum principle results in the following system of linear algebraic equations with respect to unknowns $x_j^n(T)$, $j \in [1:J]$, $n \in [0:m]$:

$$x_j^n(T) = \kappa_j^n + \sum_{i=1}^{I} \frac{w^i b_j^{in}}{\gamma_j^i}\left(\sum_{r=1}^{J}\frac{1}{\gamma_r^i}\right)^{-1} T +$$

$$+\sum_{i=1}^{I}\frac{b_j^{in}}{\gamma_j^i}\left(\sum_{s=1}^{J}\sum_{q=0}^{m}\frac{b_s^{iq} c_s^q x_s^q(T)}{\gamma_s^i}\left(\sum_{s=1}^{J}\frac{1}{\gamma_s^i}\right)^{-1}\right)T-$$

$$-\sum_{i=1}^{I} b_j^{in}\sum_{q=0}^{m}\frac{b_j^{iq} c_j^q(T)}{\gamma_j^i}T. \tag{3.7}$$

Let us try to reduce system (3.7). To this end, we introduce the notation

$$P^i = \sum_{s=1}^{J}\sum_{q=1}^{m}\frac{b_s^i c_s^q x_s^q(T)}{\gamma_s^i}, \qquad i \in [1:I]. \tag{3.8}$$

We assume provisionally that the values P^i, $i \in [1:I]$ are known and consider system (3.7) as a set of independent systems of equations with respect to j, decomposed with respect to $x_j^n(T)$, $n \in [0:m]$. By solving these systems of equations, we obtain

$$x_j^n(T) = \nu_j^n + \sum_{i=1}^{I}\pi_j^{ni} P^i. \tag{3.9}$$

Here, ν_j^n, π_j^{ni} are known coefficients. We substitute expressions (3.9) in the right-hand sides of (3.8), obtaining the general system of linear algebraic equations for the values

$$P^i = \sum_{s=1}^{I} \Lambda^{is} P^s + \Omega^i, \qquad i \in [1:I]. \tag{3.10}$$

Thus, solving the problem (3.6) reduces to the solution of the following system of linear algebraic equations: either (3.7) with respect to unknowns $x_j^n(T)$, $j \in [1:J]$, $n \in [0:m]$ or (3.10) with respect to P^i, $i \in [1:I]$ as well as to a series of linear algebraic equations of order $(m+1)$ with respect to $x_j^n(T)$, $n \in [0:m]$. If the value $J(m+1)$ is much less than I, then we should consider system (3.7). Otherwise, it is necessary to solve (3.10). In the case, where the values $J(m+1)$ and I are of the same order and are sufficiently large, we should use the iterative decomposition method. Then, the aggregation problem for (3.6) is solved directly, and the local problem finally reduces to a system of linear algebraic equations with respect to $(m+1)$ unknowns $x_j^n(T)$.

Consider the generalization of problem (3.6), when the equation for the phase coordinates has the following form:

$$dx_j^n(t)/dt + (\lambda^n)^2 x_j^n(t) = \sum_{i=1}^{I} b_j^{in} u_j^i(t).$$

This problem generalizes the model of optimal heating of a system of equal membranes (2.2) if the conduction of heat to the boundary of mediums is realized either by means of "some ingredients of the fuel" or from some "centers".

If we apply the previously mentioned technique that involves the Pontryagin maximum principle, then we finally come to the following system of linear algebraic equations with respect to unknown values C_j^n, $j \in [1:J]$, $n \in [0:m]$:

$$h_j^n C_j^n + \sum_{i=1}^{I} a_j^{in} \sum_{q=0}^{m} g_{jq}^{in} C_j^q + d_j^n \sum_{i=1}^{I} \nu_j^{in} \sum_{s=1}^{J} e_s^i \sum_{r=0}^{m} g_{sr}^{in} C_s^r = m_j^n, \tag{3.11}$$
$$j \in [1:J], \qquad n \in [0:m].$$

Here, the values C_j^n are exponential coefficients in expressions for the conjugated functions, and all other constants are expressed in terms of the known parameters of the considered problem. We try to reduce system (3.11) and parameters of the considered problem. We try to reduce system (3.11) and introduce the notation

$$P^{in} = \sum_{s=1}^{J} e_s^i \sum_{r=0}^{m} g_{sr}^{in} C_s^r, \qquad i \in [1:I], \qquad n \in [0:m]. \tag{3.12}$$

We assume the values P^{in} to be provisionally known and consider system (3.11) as a decomposed set of equations with respect to unknowns C_j^n, $n \in [0:m]$. Solving this system of order $(m+1)$ and substituting the obtained result from the right-hand sides of (3.12), we arrive at the general system of linear algebraic equations for unknowns P^{in}, $i \in [1:I]$, $n \in [0:m]$:

$$P^{in} = \sum_{i=1}^{I} \sum_{q=0}^{m} \Omega_{rq}^{in} P^{rq} + \Lambda^{in}. \tag{3.13}$$

Thus, solution of the problem is reduced either to the system of equations (3.11) of order $J(m+1)$ or to the system (3.13) of order $I(m+1)$. The system to be solved is determined by the relation of the values I and J ($I > J$ or $J > I$). In this case, if the values are approximately equal and sufficiently large, it is necessary to use the iterative decomposition method based on aggregated variables.

We consider some block separable statements of optimal control problems, for which the decomposition method based on aggregated variables is more efficient than the direct method based on the Pontryagin maximum principle. We have the problem

$$f = \sum_{j=1}^{J} \left[0,5 c_j (x_j(T))^2 + \sum_{i=1}^{I} \int_0^T \frac{\gamma_j^i}{u_j^i(t)} dt \right] \to \min,$$

$$dx_j(t)/dt = \sum_{i=1}^{I} b_j^i u_j^i(t), \quad x_j(0) = \kappa_j, \quad u_j^i(t) > 0, \quad j \in [1:J], \tag{3.14}$$

$$\sum_{j=1}^{J} u_j^i(t) = w^i, \qquad i \in [1:I].$$

Here, for definiteness, we assume that, for entering constant coefficient, the inequalities c_j, γ_j^i, b_j^i, κ_j, $w^i > 0$ hold. In this case, the sought controls are constant in the interval $[0, T]$.

Application of the maximum principle, gives us the unconstrained maximization of the function

$$-\sum_{j=1}^{J}\sum_{i=1}^{I}\frac{\gamma_j^i}{u_j^i}-\sum_{j=1}^{J}c_jx_j(T)\sum_{i=1}^{I}b_j^iu_j^i+\sum_{i=1}^{I}\lambda^iw^i-\sum_{j=1}^{J}u_j^i\to\max$$

with respect to u_j^i, where λ_j^i, $i\in[1:I]$ are Lagrange multipliers of binding constraints. Differentiating with respect to u_j^i, we obtain the following system of equations with respect to variables u_j^i, $j\in[1:J]$, $i\in[1:I]$, λ^i, $i\in[1:I]$:

$$\begin{aligned}&\gamma_j^i(u_j^i)^{-2}-c_jx_j(T)b_j^i-\lambda^i=0,\\&j\in[1:J],\qquad i\in[1:I],\\&\sum_{j=1}^{J}u_j^i=w^i,\qquad i\in[1:I].\end{aligned}\tag{3.15}$$

To solve this system, we express the controls from the first equations

$$u_j^i=\sqrt{\frac{\gamma_j^i}{\lambda^i+c_jx_j(T)b_j^i}},\qquad j\in[1:J],\qquad i\in[1:I],\tag{3.16}$$

and, then, the obtained result is substituted in the second relations in (3.15):

$$\sum_{j=1}^{J}\sqrt{\frac{\gamma_j^i}{\lambda^i+c_jx_j(T)b_j^i}}=w^i,\qquad i\in[1:I].\tag{3.17}$$

Unfortunately, we cannot deduce the analytical dependence of values λ^i, $i\in[1:I]$ on the entering parameters and unknown constants $x_j(T)$ in order to obtain, from (3.16) and differential equations in (3.14), the final system of equations with respect to $x_j(T)$, $j\in[1:J]$. This difficulty remains in the case of one binding condition in problem (3.14).

Further, we consider problem (3.14) by the iterative decomposition method based on the aggregated variable. For fixed weights α_j^i, we have the optimal solution of the aggregated problem $\overset{\circ}{U}{}^i=w^i$, $i\in[1:I]$ and the following formula for optimal dual variables $\overset{\circ}{\delta}{}^i$ that correspond to conditions $U^i=w^i$, $i\in[1:I]$:

$$\overset{\circ}{\delta}{}^i=-\sum_{j=1}^{J}\left[c_j\left(\kappa_j+\sum_{r=1}^{I}\frac{b_j^rT}{\alpha_j^rw^r}\right)b_j^i\alpha_j^i\right]+\sum_{j=1}^{J}\frac{\gamma_j^i}{\alpha_j(w^i)^2}.$$

The functionals of the local problems have the following form:

$$0,5(x_j(T))^2 + \int_0^T \left[\sum_{i=1}^{I} \left(\frac{\gamma_j^i}{u_j^i} + \overset{\circ}{\delta}{}^i u_j^i \right) \right] dt \to \min,$$

The application of the maximum principle to these problems results in the equalities

$$\gamma_j^i/(u_j^i)^2 - \overset{\circ}{\delta}{}^i - b_j^i c_j x_j(T) = 0.$$

From this, using differential equations in (3.14), we obtain the following one-dimensional equations in $x_j(T)$ independent of j:

$$x_j(T) = \kappa_j + T \sum_{i=1}^{I} b_j^i \sqrt{\frac{\gamma_j^i}{\overset{\circ}{\delta}{}^i + b_j^i c_j x_j(T)}}. \tag{3.18}$$

In the case of the binding condition, unique for the problem (3.14), equations (3.18) are cubic with respect to $(\overset{\circ}{\delta}{}^i + b_j^i x_j(T) c_j)^{1/2}$. Obviously, solving the independent equations (3.18) is much simpler than solving the complicated nonlinear system (3.17) and one more equation with respect to unknowns $x_j(T)$. This fact testifies to the efficiency of iterative decomposition for problem (3.14).

We state one more block separable optimal control problem:

$$\begin{gathered} f = \sum_{j=1}^{J} \left[0,5c_j(x_j(T))^2 + \sum_{i=1}^{I} \int_0^T \gamma_j^i u_j^i(t) dt \right] \to \min, \\ dx_j(t)/dt = 0,5 \sum_{i=1}^{I} b_j^i (u_j^i(t))^2, \quad x_j(0) = \kappa_j, \quad j \in [1:J], \\ \sum_{j=1}^{J} u_j^i(t) = w^i, \qquad i \in [1:I], \end{gathered} \tag{3.19}$$

where we consider for definiteness that the inequalities $c_j,\ \gamma_j^i,\ b_j^i,\ w^i > 0$, $\kappa_j < 0$ hold for the entering constants.

First, we try to solve problem (3.19) by direct application of the maximum principle. This gives the maximum of the function

$$-\sum_{j=1}^{J}\sum_{i=1}^{I} \gamma_j^i u_j^i - 0,5 \sum_{j=1}^{J} c_j x_j(T) \sum_{i=1}^{I} b_j^i (u_j^i)^2 + \sum_{i=1}^{I} \lambda^i (w^i - \sum_{j=1}^{J} u_j^i)$$

to unknown u^i_j, $j \in [1:J]$, $i \in [1:I]$; λ^i, $i \in [1:I]$:

$$
\begin{aligned}
&-\gamma^i_j - c_j x_j(T) u^i_j - \lambda^i = 0, \qquad j \in [1:J], \qquad i \in [1:I],\\
&\sum_{j=1}^{J} u^i_j = w^i, \qquad i \in [1:I].
\end{aligned}
\tag{3.20}
$$

We express the controls u^i_j from the first equalities (3.20) in terms of Lagrange multipliers λ^i and substitute them in the second relations from (3.20). Thus, we find the analytical dependence of λ^i, $i \in [1:I]$ on constants $x_j(T)$, $j \in [1:J]$:

$$
\lambda^i = -\left[w^i + \sum_{j=1}^{J} \frac{\gamma^i_j}{c_j x_j(T)}\right] \left[\sum_{s=1}^{J} \frac{1}{c_s x_s(T)}\right]^{-1}, \qquad i \in [1:I]. \tag{3.21}
$$

Then, using (3.21) in (3.20), we deduce the dependence of controls u^i_j on the values $x_j(T)$:

$$
u^i_j = \frac{1}{c_j x_j(T)} \left[w^i + \sum_{r=1}^{J} \frac{\gamma^i_s}{c_r x_r(T)}\right] \left[\sum_{s=1}^{J} \frac{1}{c_s x_s(T)}\right]^{-1} - \frac{\gamma^i_j}{c_j x_j(T)}. \tag{3.22}
$$

Then, relations (3.22) are used in the differential equations of problem (3.14). We obtain a system of equations with respect to unknowns $x_j(T)$, $j \in [1:J]$ of the following type:

$$
\begin{aligned}
x_j(T) = \kappa_j &+ \frac{1}{c_j x_j(T)} \sum_{i=1}^{I} b^i_j \left[w^i + \sum_{r=1}^{J} \frac{\gamma^i_j}{(c_r x_r(T))^2}\right] \times\\
&\times \left[\sum_{s=1}^{J} \frac{1}{c_s x_s(T)}\right]^{-2} T - \frac{1}{(c_j x_j(T))^2} \sum_{i=1}^{I} b^i_j (\gamma^i_j)^2 T.
\end{aligned}
\tag{3.23}
$$

Nonlinear system (3.23) is fairly complicated, even in the case of one binding constraint. System (3.23) becomes even more complicated with the growth of the values J and I.

Below, we will solve problem (3.19) by the decomposition method based on aggregation of variables. Conjugated variables $\overset{\circ}{\delta}{}^i$ for conditions in the aggregated problem are computed by the following:

$$
\overset{\circ}{\delta}{}^i = -\sum_{j=1}^{J} c_j (\kappa_j + 0,5 \sum_{q=1}^{I} b^q_j (\alpha^q_j)^2 (w^q)^2) b^i_j (\alpha^i_j)^2 - \sum_{j=1}^{J} \gamma^i_j \alpha^i_j, \qquad i \in [1:I].
$$

Here, $\overset{\circ}{U}{}^i = w^i$.

Independent local problems are formulated in the following way:

$$0,5c_j(x_j(T))^2 + \sum_{i=1}^{I} \int_0^T (\gamma_j^i + \overset{\circ}{\delta}{}^i)u_j^i dt \to \min,$$

$$dx_j(t)/dt = 0,5 \sum_{i=1}^{I} b_j(u_j^i)^2, \qquad x_j(0) = \kappa_j.$$

Their solution by the maximum principle leads to the maximization of the functions

$$-0,5c_jx_j(T)\left(\sum_{i=1}^{I} b_j^i(u_j^i)^2\right) - (\gamma_j^i + \overset{\circ}{\delta}{}^i)u_j^i \to \max$$

with respect to u_j^i, from whence, for each fixed $j \in [1:J]$, the controls u_j^i, $i \in [1:I]$ are expressed in terms of the unknown $x_j(T)$ by the formula

$$u_j^i = -\frac{\gamma_j^i + \overset{\circ}{\delta}{}^i}{b_j^i c_j x_j(T)}, \qquad i \in [1:I], j \in [1:J].$$

We use the latter equations in the differential equations of each local problem and finally arrive at the cubic equations with respect to $x_j(T)$ of the following form:

$$(x_j(T))^3 - \kappa_j(x_j(T))^2 - \frac{1}{c_j^2} \sum_{i=1}^{I} \frac{(\gamma_j^i + \overset{\circ}{\delta}{}^i)^2 b_j^i T}{(b_j^i)^2} = 0. \tag{3.24}$$

It is much simpler to analyze the cubic equations (3.24) independent of j than to solve the complex nonlinear system of equations (3.23). The efficiency of iterative decomposition in this case is apparent.

Consider another block separable statement of an optimal control problem. We have

$$f = \sum_{j=1}^{J} \left[0,5c_j(x_j(T))^2 + \sum_{i=1}^{I} \int_0^T \gamma_j^i u_j^i(t)dt\right] \to \min,$$

$$dx_j(t)/dt = \sum_{i=1}^{I} \frac{b_j^i}{u_j^i(t)}, \quad x_j(0) = \kappa_j, \quad u_j^i(t) > 0, \tag{3.25}$$

$$i \in [1:I], \qquad j \in [1:J]; \quad \sum_{j=1}^{J} u_j^i(t) = w^i, \qquad i \in [1:I],$$

where the following inequalities are assumed to hold for the coefficients c_j, γ_j^i, b_j, $w^i > 0$, $\kappa_j \geq 0$.

The direct solution of problem (3.25) by the maximum principle leads to maximization of the function

$$-\sum_{j=1}^{J}\sum_{i=1}^{I}\gamma_j^i u_j^i - \sum_{j=1}^{J} c_j x_j(T)\sum_{i=1}^{I}\frac{b_j^i}{u_j^i} + \sum_{i=1}^{I}\lambda^i(w^i - \sum_{j=1}^{J} u_j^i)$$

with respect to u_j^i. Upon differentiation, this gives a system of equations with unknowns u_j^i, $j \in [1:J]$, $i \in [1:I]$; λ^i, $i \in [1:I]$,

$$\begin{aligned} &c_j x_j(T) b_j^i/(u_j^i)^2 - \gamma_j^i - \lambda^i = 0, \qquad j \in [1:J], \qquad i \in [1:I], \\ &\sum_{j=1}^{J} u_j^i = w^i, \qquad i \in [1:I]. \end{aligned} \tag{3.26}$$

We establish the relation $u_j^i = \sqrt{c_j x_j(T) b_j^i/(\gamma_j^i + \lambda^i)}$, from the first inequalities of (3.26). If we use it in the second relation (3.26), then we obtain a complicated systems of equations with respect to λ^i of the form

$$\sum_{j=1}^{J}\sqrt{c_j x_j(T) b_j^i/(\gamma_j^i + \lambda^i)} = w^i, \qquad i \in [1:I],$$

where we cannot find an explicit dependence of the Lagrange multipliers λ^i, $i \in [1:I]$ on the values $x_j(T)$, $j \in [1:J]$.

Now, we consider constructions of the decomposition method based on the aggregated variables for problem (3.25). The dual estimates $\overset{\circ}{\delta}{}^i$ for conditions $U^i = w^i$, $i \in [1:I]$, in the aggregated problem are computed as follows:

$$\overset{\circ}{\delta}{}^i = \sum_{j=1}^{J} c_j \left(\kappa_j + \sum_{r=1}^{I}\frac{b_j^r}{\alpha_j^r w^r}\right)\frac{b_j^i}{\alpha_j^i (w^i)^2} - \sum_{j=1}^{J}\gamma_j^i \alpha_j^i, \qquad i \in [1:I].$$

The independent local problems have the following form:

$$\begin{aligned} &0,5c_j(x_j(T))^2 + \sum_{i=1}^{I}\int_0^T (\gamma_j^i + \overset{\circ}{\delta}{}^i) u_j^i dt \to \min, \\ &dx_j(t)/dt = \sum_{i=1}^{I}\frac{b_j^i}{(u_j^i(t))}, \quad x_j(0) = \kappa_j, \quad u_j^i(t) > 0, \quad i \in [1:I]. \end{aligned}$$

Use of the maximum principle for local problems leads to the unconstrained maximization of the function

$$-c_j x_j(T)\left(\sum_{i=1}^{I}\frac{b_j^i}{(u_j^i(t))}\right)-\sum_{i=1}^{I}(\gamma_j^i+\overset{\circ}{\delta}{}^i)u_j^i \to \max,$$

from whence we obtain the relation between optimal controls and the values $x_j(T)$: $u_j^i=\sqrt{c_j x_j(T) b_j^i(\gamma_j^i+\overset{\circ}{\delta}{}^i)}$.

We substitute the relations in differential equations of local problems, whence we obtain cubic equations, independent of j, with respect to the values $(x_j(T))^{1/2}$

$$\left[(x_j(T))^{1/2}\right]^3-\kappa_j(x_j(T))^{1/2}-\frac{1}{\sqrt{c_j}}\sum_{i=1}^{I}\sqrt{b_j^i(\gamma_j^o+\overset{\circ}{\delta}{}^i)}=0.$$

To solve this cubic equations is simpler than to solve the complicated system (3.26), which testifies to the efficiency of decomposition for problem (3.25).

In conclusion, we note that the results of this section can be generalized to the case, where coefficients γ_j^i, b_j^{in}, w^i and so on, are functions of time.

§4. Nonseparable Functionals

In previous sections of this chapter, we considered block separable optimal control problems. In this section, the functionals will characterize the alignment of phase coordinates for separate subsystems, the functions are no longer additive separable, and the decomposition method based on the aggregated variables is not applicable. Nevertheless, the reduction method for systems of linear algebraic equations described above, can be generalized to linear quadratic block statements with nonseparable functionals.

We describe the proposed approach for the simplest hierarchical problems, where subsystems are described by ordinary differential equations:

$$f = 0,5 \left[\sum_{j=1}^{J-1} d_j(x_j(T) - x_{j+1}(T))^2 + d_J(x_J(T) - \right.$$
$$\left. -x_1(T))^2 + \sum_{j=1}^{J} \gamma_j \int_0^T (u_j(t)^2) dt \right] \to \min, \tag{4.1}$$
$$dx_j(t)/dt = u_j(t), \qquad x_j(0) = \kappa_j, \qquad j = 1, 2, \ldots, J;$$
$$\sum_{j=1}^{J} u_j(t) = w(t).$$

Here, as before, c_j,γ_j are positive priority coefficients.

For the sake of convenience, we introduce the notation $x_{J+1}(T) = x_1(T)$, then, the functional of problem (4.1) is rewritten in the form

$$f = 0,5 \sum_{j=1}^{J} \left[d_j(x_j(T) - x_{j+1}(T))^2 + \gamma_j \int_0^T (u_j(t))^2 dt \right].$$

The solution of problem (8.1) by the maximum principle results in the formula similar to (4.4):

$$u_j(t) = \left[\psi_j \sum_{r=1}^{J} \gamma_r^{-1} + w(t) - \sum_{r=1}^{J} \psi_r \gamma_r^{-1} \right] \left[\gamma_j \sum_{r=1}^{J} \gamma_r^{-1} \right]^{-1}, \quad j = 1, 2, \ldots, J, \tag{4.2}$$

where the conjugated variables $\psi_j(t)$, $j \in [1 : J]$ satisfy the following equations and conditions for $t = T$:

$$d\psi_j(t)/dt = 0,$$
$$\psi_j(T) = d_{j-1}x_{j-1}(T) - (d_{j-1} + d_j)x_j(T) + d_j x_{j+1}(T), \tag{4.3}$$
$$j = 1, 2, \ldots, J.$$

Here, the notation $x_0(T) = x_J(T)$, $d_0 = d_J$ is introduced.

Relation (4.3) implies that the dual variables $\psi_j(t)$, $j \in [1 : J]$ are constant in the interval $[0, T]$. If we assume that $w(t) = w -$ const, then, due to (4.2), the same holds for the controls. We substitute the values $\psi_j(t)$ from (4.3) in (4.2) and integrate the differential equations in (4.1) from 0 to T. This gives a system of linear algebraic equations with respect to unknown values $x_j(T)$, $j \in [1 : J]$:

$$a_j x_{j-1}(T) - c_j x_j(T) + b_j x_{j+1}(T) = f_j \sum_{r=1}^{J} (d_{r-1} x_{r-1}(T) -$$
$$-(d_{r-1} + d_r) x_r(T) + d_r x_{r+1}(T)) \gamma_r^{-1} + h_j, \qquad j = 1, 2, \ldots, J, \tag{4.4}$$

where

$$a_j T d_{j-1} \gamma_j^{-1}, \qquad b_j = T d_j \gamma_j^{-1}, \qquad c_j = 1 + T(d_{j-1} + d_j) \gamma_j^{-1},$$

$$f_j = T \left[\gamma_j \sum_{r=1}^{J} \gamma_r^{-1} \right]^{-1}, \qquad h_j = -Tw \left[\gamma_j \sum_{r=1}^{J} \gamma_r^{-1} \right]^{-1} - \kappa_j, \qquad j = 1, 2, \ldots, J.$$

The specificity of system (4.4) dictates the following solution based on the reduction method proposed in Section 4. We introduce the notation

$$P = \sum_{r=1}^{J} (d_{r-1} x_{r-1}(T) - (d_{r-1} + d_r) x_r(T) + d_r x_{r+1}(T)) \gamma_r^{-1} \tag{4.5}$$

and rewrite (4.4) in the form

$$a_j x_{j-1}(T) - c_j x_j(T) + b_j x_{j+1}(T) = f_j P + h_j, \qquad j = 1, 2, \ldots, J. \tag{4.6}$$

Here, $x_0(T) = x_J(T)$, $x_1(T) = x_{J+1}(T)$. First, we assume that the value P is known, and the system (4.6) with a diagonal matrix is solved by the standard Gauss elimination method. The first line is subtracted from the second line multiplied by the coefficient that would exclude the variable $x_1(T)$. The result is written instead of the second line in the form

$$\bar{a}_2 x_2(T) + b_2 x_3(T) + \bar{c}_2 x_J(T) = \bar{f}_2 P + \bar{h}_2.$$

We repeat the analogous procedure for the third and second lines, thus obtaining the following lines instead of the third one:

$$\bar{a}_3 x_3(T) + b_3 x_4(T) + \bar{c}_3 x_J(T) = \bar{f}_3 P + \bar{h}_3.$$

We continue this process sequentially and arrive at the system

$$\begin{aligned} &\bar{a}_j x_j(T) + b_j x_j(T) + \bar{c}_j x_J(T) = \bar{f}_j P + \bar{h}_j, \qquad j = 1, 2, \ldots, J-2, \\ &\bar{a}_{J-1} x_{J-1}(T) + \bar{c}_{J-1} x_J(T) - \bar{f}_{J-1} P + h_{J-1}, \\ &b_J x_1(T) + \bar{c}_J x_j(T) = \bar{f}_J P + \bar{h}_J. \end{aligned} \tag{4.7}$$

New coefficients $\hat{c}_j$, $\hat{f}_j$, $\hat{h}_j$, $j \in [1 : J-1]$ are saved in the same computer storage locations, where a_j, c_j, f_j, h_j were before. Next, we transform system (4.7). We subtract the last row multiplied by a coefficient from the first row. The coefficient should be such that $x_1(T)$ is excluded. Then, we subtract the first row obtained at the previous step and $x_2(T)$ is excluded. The procedure is iterated, and, finally, we obtain a system of the form

$$\begin{aligned} b_J x_1(T) + \bar{c}_J x_J(T) &= \bar{f}_J P + \bar{h}_J, \\ b_j x_{j+1}(T) + \hat{c}_j x_J(T) &= \hat{f}_j P + \hat{h}_j, \qquad j = 1, 2, \ldots, J-2, \\ c_{J-1} x_J(T) &= \hat{f}_{J-1} P + \hat{h}_{J-1}. \end{aligned} \tag{4.8}$$

The new coefficients $\hat{x}_j$, $\hat{f}_j$, $\hat{h}_j$, $j = 1, 2, \ldots, J-1$ are saved in the same computer storage locations, where $\bar{c}_j$, $\bar{f}_j$, $\bar{h}_j$ were before. System (4.8) is solved, and, finally, we obtain

$$x_j(T) = \xi_j P + \theta_j, \qquad j = 1, 2, \ldots, J. \tag{4.9}$$

Thus, to compute coefficients ξ_j, θ_j, $j = 1, 2, \ldots, J$, we require $5J$ computer storage locations. We do not need to save the whole $J \times J$-matrix of system (4.4).

The recurrent formulas for the procedure look as follows:

$$\begin{aligned}
&\xi_j = p_j + \xi_J s_j, \qquad j - 1, 2, \ldots, J-1, \\
&\xi_J = (\beta_{J+1} + \alpha_{J+1} p_1)(1 - \alpha_{J+1} s_1 - \gamma_{J+1})^{-1}, \\
&\theta_j = g_j + \theta_J s_j, \qquad j = 1, 2, \ldots, J_1, \\
&\theta_J = (\delta_{J+1} + \alpha_{J+1} q_1)(1 - \alpha_{J+1} s_1 - \gamma_{J+1})^{-1}, \\
&p_j = \alpha_{j+1} p_{j+1} + \beta_{j+1}, \qquad j = J-1, \ldots, 1, \qquad p_J = 0, \\
&q_j = \alpha_{j+1} q_{j+1} + \delta_{j+1}, \qquad j = J-1, \ldots, 1, \qquad q_J = 0, \\
&s_j = \alpha_{j+1} s_{j+1} + \gamma_{j+1}, \qquad j = J-1, \ldots, 1, \qquad s_J = 0, \\
&\alpha_{j+1} = b_j (c_j - a_j \alpha_j)^{-1}, \qquad j = 2, \ldots, J, \qquad \alpha_2 = b_1 c_1^{-1}, \\
&\gamma_{j+1} = a_j \gamma_j (c_j - a_j \alpha_j)^{-1}, \qquad j = 2, \ldots, J, \qquad \gamma_2 = a_1 c_1^{-1}, \\
&\beta_{j+1} = (f_j + a_j \beta_j)(c_j - a_j \alpha_j)^{-1}, \qquad j = 2, \ldots, J, \qquad \beta_2 = f_1 c_1^{-1}, \\
&\delta_{j+1} = (h_j + a_j \delta_j)(c_j - a_j \alpha_j)^{-1}, \qquad j = 2, \ldots, J, \qquad \delta_2 = h_1 c_1^{-1}.
\end{aligned} \tag{4.10}$$

Upon finding the coefficients ξ_j, θ_j, $j = 1, 2, \ldots, J$ for $r = 1, 2, \ldots, J$, the right-hand sides from (4.9) are substituted in (4.5). We obtain a linear equation with respect to a single variable P and, then, we find the values $x_j(T)$, $j \in [1 : J]$ from (4.8).

Let the subsystems in (4.1) be described by simplest second-order equations

$$d^2x_j(t)/dt^2 = u_j(t), \quad x_j(0) = \kappa_j^0, \quad x_{jt}(0) = \kappa_j^1, \quad j = 1, 2, \ldots, J \tag{4.11}$$

and minimize the functional of the form

$$f = 0,5 \sum_{j=1}^{J} \left[\sigma_j (x_j(T) - x_{j+1}(T))^2 + \zeta_j (x_{jt}(T) - x_{j+1t}(T))^2 + \gamma_j \int_0^T (u_j(t))^2 dt \right],$$

where the notation $x_{J+1,t}(T) = x_{1t}(T)$ is added.

The conjugated variables $\psi_j(t)$, $j = 1, 2, \ldots J$ satisfy the following equations and conditions for $t = T$:

$$\begin{aligned} &d^2\psi_j(t)/dt^2 = 0, \\ &\psi_j(T) = \zeta_{j-1} x_{j-1t}(T) - (\zeta_{j-1} + \zeta_j) x_{jt}(T) + \zeta_j x_{j+1t}(T), \\ &\psi_{jt}(T) = \sigma_{j-1} x_{j-1}(T) - (\sigma_{j-1} + \sigma_j) x_j(T) + \sigma_j x_{j+1}(T), \qquad j = 1, 2, \ldots, J, \end{aligned} \tag{4.12}$$

from whence we obtain

$$\psi_j(t) = A_j t + B_j, \qquad j = 1, 2, \ldots, J, \tag{4.13}$$

where A_j, B_j, $j = 1, 2, \ldots, J$ are unknown constants.

The relation of constants and dual variables is the same as in (4.2). By integrating (4.11), we obtain

$$x_j(t) = \kappa_j^1 t + \kappa_j^0 + \int_0^t (t - \tau) u_j(\tau) d\tau, \qquad j = 1, 2, \ldots, J. \tag{4.14}$$

By substituting (4.13) in (4.2) and then the result in (4.14), we find the dependence of phase coordinates on the variables A_j, B_j:

$$\begin{aligned} x_j(t) &= \kappa_j^1 t + \kappa_j^0 + \Lambda_j^0(t) + (A_j t^3/6 + B_j t^2/2)\gamma_j^{-1} = \\ &= \left[\gamma_j \sum_{r=1}^{J} \gamma_r^{-1} \right]^{-1} \sum_{r=1}^{J} (A_r t^3/6 + B_j t^2/2)\gamma_r^{-1}, \end{aligned} \tag{4.15}$$

where

$$\Lambda_j^1(t) = \left[\gamma_j \sum_{r=1}^{J} \gamma_r^{-1}\right]^{-1} \int\limits_0^t w(\tau)d\tau, \qquad j = 1, 2, \ldots, J. \tag{4.16}$$

Upon the substitution of (4.13), (4.15), (4.16) for $t = T$ in the terminal conditions from (4.12), we obtain a system of linear algebraic equations with respect to A_j, B_j in the form

$$\begin{aligned}
&a_j^{11} A_j - c_j^{11} A_j + b_j^{11} A_{j+1} - a_j^{12} B_{j-1} - c_j^{12} B_j + b_j^{12} B_{j+1} = \\
&= f_j^{11} \sum_{r=1}^{J} A_r \gamma_r^{-1} + f_j^{12} \sum_{r=1}^{J} B_r \gamma_r^{-1} + h_j^1, \\
&a_j^{21} A_{j-1} - c_j^{21} A_j + b_j^{21} A_{j-1} + a_j^{22} B_j - c_j^{22} B_j + b_j^{22} B_j = \\
&= f_j^{21} \sum_{r=1}^{J} A_r \gamma_r^{-1} + f_j^{12} \sum_{r=1}^{J} B_r \gamma_r^{-1} + h_j^2, \qquad j = 1, 2, \ldots, J,
\end{aligned} \tag{4.17}$$

where the expressions for the coefficients are obvious and are not written here. For convenience, we put $A_0 = A_J$, $B_0 = B_J$, $A_1 = A_{J+1}$, $B_1 = B_{J+1}$.

To solve system (4.17), we compose the vector $X_j = [A_j, B_j]^T$ and introduce matrices for every j

$$a_j = \begin{bmatrix} a_j^{11} & a_j^{12} \\ a_j^{21} & a_j^{22} \end{bmatrix}, \qquad b_j = \begin{bmatrix} b_j^{11} & b_j^{12} \\ b_j^{21} & b_j^{22} \end{bmatrix}, \qquad c_j = \begin{bmatrix} c_j^{11} & c_j^{12} \\ c_j^{21} & c_j^{22} \end{bmatrix},$$
$$f_j = \begin{bmatrix} f_j^{11} & f_j^{12} \\ f_j^{21} & f_j^{22} \end{bmatrix}.$$

Consider also the vector $h_j = \left[h_j^1, h_j^2\right]^T$.

The sums in the right-hand sides (4.17) $\sum\limits_{r=1}^{J} A_r \gamma_r^{-1}$, $\sum\limits_{r=1}^{J} B_r \gamma_r^{-1}$ will be denoted by means of the vector

$$P = \begin{bmatrix} P_1 \\ P_2 \end{bmatrix} = \begin{bmatrix} \sum\limits_{r=1}^{J} A_r \gamma_r^{-1} \\ \sum\limits_{r=1}^{J} B_r \gamma_r^{-1} \end{bmatrix}. \tag{4.18}$$

In terms of this notation, system (4.17) is rewritten in the form

$$\begin{aligned}
&a_j X_{j-1} - c_j X_j + b_j X_{j+1} = f_j P + h_j, \qquad j = 1, 2, \ldots, J, \\
&X_{J+1} = X_1, \qquad X_J = X_0.
\end{aligned} \tag{4.19}$$

In solution of (4.19) by the Gauss method, the recurrent formulas (4.10) hold, where the scalar values are replaced by vector ones. At each step, a 2×2-matrix is inverted. Here, $18 \times J$ computer storage locations is used. Finally, we obtain

$$X_j = \xi_j P + \theta_j, \qquad j = 1, 2, \dots, J,$$

where ξ_j are 2×2-matrices, θ_j is a vector. We substitute the latter equalities in (4.18) and obtain a system of two linear algebraic equations for unknown P_1 and P_2. The problem can now be solved.

Consider the problem about the central heating of a system of similar plates from Section 2. Here, the functional denotes the leveling in the average temperature of the membranes by a given time instant. Thus, we have

$$\begin{aligned} & f = \sum_{j=1}^{J} \left[\sigma_j \int_0^T (z_j(x,T) - z_{j+1}(x,T))^2 dx + \gamma_j \int_0^T (u_j(t))^2 dt \right] \to \min, \\ & \partial z_j(x,t)/\partial t - \partial^2 z_j(x,t)/\partial x^2 = 0, \quad t \in (0,T), \quad x \in (0,l), \\ & z_j(x,0) = 0, \quad x \in [0,l], \\ & \partial z_j(0,t)/\partial x = 0, \\ & \partial z_j(1,t)/\partial x = a(u_j(t) - z_j(1,t)), \quad j = 1, 2, \dots, J, \\ & \sum_{j=1}^{J} u_j(t) = w(t). \end{aligned} \tag{4.20}$$

As in Section 2, we find an approximate solutions $z_j(x,t)$ to mixed problems by means of a finite number of terms of the Fourier series. For components z_j^n, $j \in [1:J]$, $n \in [0:m]$, we obtain the following optimal control problem:

$$\begin{aligned} & dz_j^n(t)/dt + \mu^n z_j^n(t) = \eta^n u_j(t), \\ & z_j^n(0) = 0, \qquad j = 1, 2, \dots, J, \quad n = 0, 1, \dots, m, \\ & \sum_{j=1}^{J} u_j(t) = w(t), \end{aligned} \tag{4.21}$$

where

$$\mu^n = (\lambda^n)^2, \qquad \textit{and} \qquad w^n = (a^n - a + \mu^n)(a^2 - \mu^n)^{-1},$$

$$\eta^n = 2a\lambda^n(a^2+\mu^n)^{1/2}(a^2-a+\mu^n)^{-1}.$$

The conjugated variables $y_j^n(t)$, $j=1,2,\dots,J$, $n=0,1,\dots,m$ satisfy the following differential equations and conditions for $t=T$:

$$\begin{aligned} &dy_j^n(t)/dt - \mu^n y_j^n(t) = 0, \qquad j=1,2,\dots,J, \qquad n=0,1,\dots,m \\ &y_j^n(T) = 2w^n(\sigma_{j-1}z_{j-1}^n(t) - (\sigma_{j-1}+\sigma_j)z_j^n(T) + \sigma_j z_{j+1}^n(T)). \end{aligned} \tag{4.22}$$

As before, on applying the Pontryagin maximum principle, we obtain the relation between controls and dual variables. We have

$$u_j(t) = \left[\sum_{n=0}^m \eta^n y_j^n(t)\sum_{r=1}^J \gamma_r^{-1} + w(t) - \frac{1}{2}\sum_{r=1}^J y_r^n(t)\gamma_r^{-1}\right]\left[\gamma_j\sum_{r=1}^J\gamma_r^{-1}\right]^{-1},$$
$$j=1,2,\dots,J. \tag{4.23}$$

The general solution of the primal and the conjugated problem looks as follows:

$$z_j^n(t) = B_j^n\exp(-\mu^n t) + \exp(-\mu^n t)\int_0^t \eta^n u(\tau)\exp(\mu^n\tau)d\tau, \tag{4.24}$$

$$y_j^n(t) = A_j^n\exp(\mu^n t), \qquad j=1,2,\dots,J, \qquad n-0,1,\dots,m, \tag{4.25}$$

where A_j^n, B_j^n are unknown constants.

We substitute (4.25) into (4.23) and substitute the result in (4.24). We obtain

$$\begin{aligned} z_j^n(t) = {} & B_j^n\exp(-\mu^n t) + \eta^n(2\gamma_j)^{-1}\sum_{s=0}^m \frac{\eta^s\exp(\mu^s t)}{\mu^s+\mu^n}A_j^s - \\ & -\eta^n\left[2\gamma_j\sum_{r=1}^J\gamma_r^{-1}\right]\sum_{s=0}^m\frac{\eta^s\exp(\mu^s t)}{\mu^s+\mu^n}\sum_{r=1}^J -A_r^s\gamma_r^{-1} \mid \\ & +\eta^n\left[\gamma_j\sum_{r=1}^J\gamma_r^{-1}\right]^{-1}\exp(-\mu^n t)\int_0^t w(\tau)\exp(\mu^n\tau)d\tau, \\ & j=1,2,\dots,J, \qquad n=0,1,\dots,m. \end{aligned} \tag{4.26}$$

We substitute (4.25), (4.26) in the terminal condition from (4.22) and take into account the initial condition from (4.21). We arrive at the system of linear algebraic equations with respect to variables A_j^n, B_j^n:

$$0 = B_j^n + \eta^n (2\gamma_j)^{-1} \sum_{s=0}^{m} \frac{\eta^s A_j^s}{\mu^s + \mu^n} - \eta^n \left[2\gamma_j \sum_{r=1}^{J} \gamma_r^{-1} \right]^{-1} \sum_{s=0}^{m} \frac{n^s}{\mu^s + \mu^n} \sum_{r=1}^{J} A_r^s \gamma_r^{-1},$$

$$A_j^n \exp(\mu^n T) = 2w^n \exp(-\mu^n T) \left(\sigma_{j-1} B_{j-1}^n - (\sigma_{j-1} + \sigma_j) B_j^n + \sigma_j B_{j+1}^n \right) -$$
$$-\omega^n \eta^n \sum_{s=0}^{m} \left(\sigma_{j-1} A_j^s \gamma_{j-1}^s - (\sigma_{j-1} + \sigma_j A_{j+1}^s \gamma_{j+1}^{-1} \right) \frac{\eta^s \exp(\mu^s T)}{\mu^s + \mu^n} +$$
$$\omega^n \eta^n \left[\sum_{r=1}^{J} \gamma_r^{-1} \right]^{-1} \left(\sigma_{j-1} \gamma_{j-1}^{-1} - (\sigma_{j-1} + \sigma_j) \gamma_j^{-1} + + \sigma_j \gamma_{j+1}^{-1} \right) \times$$
$$\sum_{s=0}^{m} \frac{\eta^s \exp(\mu^s T)}{\mu^s + \mu^n} \sum_{r=1}^{J} A_r^s \gamma_r^{-1} + h_j^n,$$

where

$$h_j^n = 2\omega^n \eta^n \exp(-\mu^n T)(\sigma_{j-1} \gamma_{j-1}^{-1} - (c_{j-1} + \sigma_j) \gamma_j^{-1} + \sigma_j \gamma_{j+1}^{-1} \times$$
$$\times \int_0^T w(\tau) \exp(\mu^n \tau) d\tau.$$

We express the values B_j^n from the first equations of the system and substitute the result in the second equations. We obtain a system with unknowns A_j^n, $j = 1, 2, \ldots, J$, $n = 0, 1, \ldots, m$ of the following form:

$$\sum_{s=0}^{m} a_j^{ns} A_{j-1}^s - \sum_{s=0}^{m} c_j^{ns} A_j^s + \sum_{s=0}^{m} b_j^{ns} A_{j+1}^s =$$
$$= f_j^n \sum_{s=0}^{m} g^{ns} \sum_{r=1}^{J} A_r^s \gamma_r^{-1} + h_j^n, \qquad j = 1, 2, \ldots, J, \qquad n = 0, 1, \ldots, m \tag{4.27}$$

with standard notation for the entering coefficients. We should not forget that $A_0 = A_J$, $A_1 = A_{J+1}$.

We reduce the dimension by Gauss elimination in (4.27), by analogy with the corresponding constructions for system (4.17). First, we introduce the notation

$$P^n = \sum_{s=0}^{m} g^{ns} \sum_{r=1}^{J} A_r^s \gamma_r^{-1}, \qquad n = 0, 1, \ldots, m \tag{4.28}$$

and consider the values P^n to be known. System (4.27) is rewritten in the vector form:

$$a_j A_{j-1} - c_j A_j + b_j A_{j+1} = f_j P + h_j, \qquad j = 1, 2, \ldots, J. \tag{4.29}$$

Here, a_j, c_j, b_j, f_j, $j = 1, 2, \ldots, J$ are square $(m+1) \times (m+1)$-matrices, $(m+1) \times (m+1)$; P, A_j, h_j are vectors with the $(m+1)$-th component.

The recurrent formulas (4.10), where scalar values are replaced by vectors, hold for the solution of system (4.29). At each step of the elimination, an $(m+1)\times(m+1)$-matrix is inverted. For each $j = 1, 2, \ldots, J$, we need memory for the storage of three $(m+1) \times (m+1)$-matrices and two $(m+1)$-vectors. Thus, instead of the $J(m+1) \times (m+1)$-matrices of system (4.27), the array of dimension $3J(m+1) \times (m+1)$ is transformed.

The described procedure gives the following relations:

$$A_j = D_j P + E_j, \qquad j \in [1 : J], \tag{4.30}$$

where D_j, E_j are known. We substitute (4.30) in (4.28) and arrive at a system of linear algebraic equations with respect to unknowns P^n, $n \in [0 : m]$. This completes the solution of problem (4.20).

Consider the optimal control problem for a system of equal elastic rods with fixed ends from Section 2. Let the functional here have mean leveling the average phase coordinates and velocities by a fixed time instant T. Then, we obtain

$$f = \sum_{j=1}^{J} \left[\sigma_j \int_0^l (z_j(x,T) - z_{j+1}(x,T))^2 dx + \zeta_j \int_0^l (z_{jt}(x,T) - z_{j+1,t}(x,T))^2 dx + \right.$$
$$\left. + \gamma_j \int_0^l \int_0^T (u_j(x,t))^2 dx dt \right] \to \min,$$

$$\partial^2 z_j(x,t)/\partial t^2 - \partial^2 z_j(x,t)/\partial x^2 = u_j(x,t). \quad x \in (0,t), \quad t \in (0,T),$$

$$z_j(x,0) = z_j^0(t), \quad z_{jt}(x,0) = z_j^1(x), \quad x \in [0,l], \quad z_j(0,t) = z_j(l,t),$$

$$t \in [0,T], \quad j \in [1 : J],$$

$$\sum_{j=1}^{J} u_j(x,t) = w(x,t), \tag{4.31}$$

where we have $z_{J+1} = z_1$, $z_{J+1t} = z_{1t}$.

The conjugated variables satisfy the following equations, conditions for $t = T$, and boundary conditions:

$$\begin{aligned}
&\partial^2 y_j(x,t)/\partial t^2 - \partial^2 y_j(x,t)/\partial x^2 = 0,\\
&y_j(x,T) = 2(\zeta_{j-1}\dot{z}_{j-1,t}(x,T) - (\zeta_{j-1}+\zeta_j)\dot{z}_{jt}(x,T) + \zeta_j\dot{z}_{j+1,t}(x,T)),\\
&\dot{y}_{jt}(x,T) = 2(\sigma_{j-1}z_{j-1}(x,T) - (\sigma_{j-1}+\sigma_j)z_j(x,T) + \sigma_j z_{j+1}(x,T)), \quad x \in [0,l],\\
&y_j(0,t) = y_j(l,t) = 0, \quad t \in [0,T], \quad j \in [1:J].
\end{aligned} \tag{4.32}$$

The relation of controls and dual variables is given by formula (2.17). As before, the relations (2.18), (2.19) are given. Finally, we obtain

$$\begin{aligned}
z_j(x,t) = &\sum_{n=0}^{\infty}(A_j^n \cos\lambda^n t + B_j^n \sin\lambda^n t))\sqrt{\frac{2}{l}}\sin\frac{n\pi x}{l}\times\\
&\times\sum_{n=0}^{\infty}\sqrt{\frac{2}{l}}\sin\frac{n\pi x}{l}\left[2\lambda^n\gamma_j\sum_{r=1}^{J}\gamma_r^{-1}\right]^{-1}\int_0^t\int_0^l\sqrt{\frac{2}{l}}\sin\frac{n\pi\xi}{l}\times\\
&\times\sin\lambda^n(t-\tau)w(\xi,\tau)d\xi d\tau + (4\lambda^n\gamma_j)^{-1}\int_0^l\int_0^t\sin\lambda^n(t-\tau)\sqrt{\frac{2}{l}}\times\\
&\times\sin\frac{n\pi\xi}{l}\gamma_j^n(\xi,\tau)d\xi d\tau - \left[4\lambda^n\gamma_j\sum_{r=1}^{J}\gamma_r^{-1}\right]^{-1}\times\\
&\times\int_0^l\int_0^T\sum_{l=1}^{J}\sin\lambda^n(t-T)\sqrt{\frac{2}{l}}\sin\frac{n\pi\xi}{l}y_r^{-1}d\xi d\tau.
\end{aligned}$$

We substitute the values $y_j(x,t)$, $j \in [1:J]$, from (2.19) into the right-hand side of the equality and arrive at the following expressions for $z_j(x,t)$, $j \in [1:J]$:

$$\begin{aligned}
z_j(x,t) = &\sum_{n=1}^{\infty}\Big[A_j^n\cos\lambda^n t - B_j^n\sin\lambda^n t + \Lambda_j^n(t) + C_j^n\mu_j^n(t) + D_j^n\eta_j^n(t)+\\
&+\theta_j^n(t)\sum_{r=1}^{J}C_r^n\gamma_r^{-1} + \kappa_j^n(t)\sum_{r=1}^{J}D_r^n\gamma_r^{-1}\Big]\sqrt{\frac{2}{l}}\sin\frac{n\pi x}{l},
\end{aligned} \tag{4.33}$$

where

$$\Lambda_j^n(t) = \left[2\lambda^n\gamma_j\sum_{r=1}^{J}\gamma_r^{-1}\right]^{-1}\int_0^l\int_0^t \sin\lambda^n(t-\tau)\sqrt{\frac{2}{l}}\sin\frac{n\pi\xi}{l}w(\xi,\tau)d\xi d\tau,$$

$$\mu^n(t) = [8\lambda^n\gamma_j]^{-1}t\sin\lambda^n t,$$

$$\eta_j^n(t) = [8\gamma_j]^{-1}[\sin\lambda^n t - \lambda^n t\cos\lambda^n t],$$

$$\theta_j^n(t) = \left[8\lambda^n\gamma_j\sum_{r=1}^{J}\gamma_r^{-1}\right]^{-1}\sin\lambda^n t,$$

$$\kappa_j^n(t) = \left[8\gamma_j\sum_{r=1}^{J}\gamma_t^{-1}\right]^{-1}[\sin\lambda^n t - \lambda^n t\cos\lambda^n t].$$

We use (2.19) and (4.33) in the terminal conditions of (4.32). For each harmonic (i.e., for fixed index n), we obtain a system of linear algebraic equations with respect to unknown constants C_j^n, D_j^n:

$$\begin{aligned}
&a_j^{11}C_{j-1}D - c_j^{11}C_j + b_j^{11}C_{j+1} + a_j^{12}D_{j-1} - c_j^{11}D_j + b_j^{12}D_{j+1} = \\
&= f_j^{11}\sum_{r=1}^{J}C_r\gamma_r^{-1} + f_j^{12}\sum_{r=1}^{J}D_r\gamma_r^{-1} + h_j^{-1}, \\
&a_j^{21}C_{j-1} + C_j^{21}C_j + b_j^{21}C_{j+1} + a_j^{22}D_{j-1} - c_j^{22}D_j + b_j^{22}D_{j+1} = \\
&= f_j^{21}\sum_{r=1}^{J}C_r\gamma_r^{-1} + f_j^{22}\sum_{r=1}^{J}D_r\gamma_r^{-1} + h_j^2, \qquad j\in[1:J],
\end{aligned} \tag{4.34}$$

where, for simplicity, the index n is omitted, and standard notation is used for the coefficients.

Solving system (4.34) is carried out using the Gauss elimination method by analogy with the corresponding constructions for (4.17). The following values are introduced:

$$P_1 = \sum_{r=1}^{J}C_r\gamma_r^{-1}, \qquad P_2 = \sum_{r=1}^{J}D_r\gamma_r^{-1}. \tag{4.35}$$

The vector $X = [C_j, D_j]^T$, $j\in[1:J]$ is composed. We also define the following matrices and vectors:

$$a_j = \begin{bmatrix} a_j^{11} & a_j^{12} \\ a_j^{21} & a_j^{22}\end{bmatrix}, \quad b_j = \begin{bmatrix} b_j^{11} & b_j^{12} \\ b_j^{21} & b_j^{22}\end{bmatrix}, \quad c_j = \begin{bmatrix} c_j^{11} & c_j^{12} \\ c_j^{21} & c_j^{22}\end{bmatrix}, f_j = \begin{bmatrix} f_j^{11} & f_j^{12} \\ f_j^{21} & f_j^{22}\end{bmatrix},$$

$$h_j = \left[h_j^1, h_j^2\right]^T, \qquad P = [P_1, P_2]^T.$$

System (4.34) is rewritten in the form

$$a_j X_{j-1} - c_j X_j + b_j X_{j+1} = f_j P + h_j, \qquad j \in [1:J], \tag{4.36}$$

where $X_{y+1} = X_1$, $X_J = X_0$.

In solution (4.36), the recurrent formulas (4.10), where scalar values are replaced by vector ones, are valid. At each step, 2×2-matrices are inverted. The main point here is that the array of dimension $12 \times J$ is transformed instead of the $2J \times 2J$-matrices of system (4.34). This denotes the reduction of the dimension for large values of J. Transformation by the Gauss method result in the equality

$$\begin{bmatrix} C_j \\ D_j \end{bmatrix} = \begin{bmatrix} \xi_j^1 \\ \xi_j^2 \end{bmatrix} \begin{bmatrix} P_1 \\ P_2 \end{bmatrix}^T + \begin{bmatrix} \theta_j^1 \\ \theta_j^2 \end{bmatrix}, \qquad j \in [1:J]. \tag{4.37}$$

We substitute (4.37) in (4.35) and obtain a system of two equations with two unknowns with respect to P_1 and P_2. Thus, problem (4.31) is solved.

§5. Results of Numerical Computation

In this section, we present results of numerical computation by iterative decomposition based on aggregation for some optimal control problems. These results show the convergence of the iterative algorithm and its efficiency. In a few iterations, we obtain good approximations of the optimum. In numerical tests, the difference between functional values of the conjugated and primal problems for intermediate feasible solutions becomes 5 – 6 orders of magnitude less at each iteration. This value is a measure of approximation to the optimum by the functional, and demonstrates the fast convergence of the iterative process.

Consider the problem (4.1) of Chapter 3, where we take $c_j = c = 1$, $\gamma_j = \gamma = 1$, $j \in [1:J]$; $w = 100$, $J = 501$. As was mentioned in Section 4, for equal coefficient c_j, γ_j, $j \in [1:J]$, the following simple formula for optimal control holds in the main problem. We have

$$u_j^* = 0,5 \sum_{r=1}^{J} \kappa_r J^{-1} - 0,5\kappa_j + wJ^{-1}.$$

In the same Section, we deduced relations for other variables used in the iterative process.

For this case, the corresponding formulas have the following form:

$$\overset{\circ}{\delta} = -\sum_{j=1}^{J} \alpha_j \kappa_j - 2w \sum_{j=1}^{J} (\alpha_j)^2,$$

$$\hat{u}_j = 0,5 \sum_{r=1}^{J} \alpha_r \kappa_r - 0,5\kappa_j + w \sum_{r=1}^{J} (\alpha_r)^2, \qquad j \in [1:J],$$

$$\pi = -\sum_{j=1}^{J} \left[(\hat{u}_j - \alpha_j w) 2\kappa_j + ((\hat{u}_j)^2 - (\alpha_j w)^2) + \overset{\circ}{\delta}\, \hat{u}_j \right] + \overset{\circ}{\delta}\, w,$$

$$g = -0,5 \sum_{j=1}^{J} ((\kappa_j)^2 + 2\kappa_j \alpha_j w + (\alpha_j w)^2) - 0,5w^2 \sum_{j=1}^{J} (\alpha_j)^2.$$

The values of phase coordinates for $t = 0$ are fixed as follows: $x_j = -50 + h_j$, $j \in [1 : J]$. The initial weight control coefficients are given in the form of an arithmetic progression with the sum equal to 0, 2, therefore $\alpha_j = 2j(501 \cdot 502)^{-1}$, $j \in [1 : J]$. To ensure that the sum of weights is equal to one, the terms $j = 101$, 301, 501 are added to the weights with numbers 0, 26, 0, 27, 0, 27, respectively. We could have levelled the weight coefficients at the initial iteration, i.e., set them equal to $\alpha_j = 1/J$, $j \in [1 : J]$. Here, we take the "spoiled" initial weights on, to see how the algorithm "couples" with this situation.

Going to the next iteration, the new weights are computed by formula (1.11), where we set $\rho_j = \rho$, $j \in [1 : J]$. To maximize the function $\overset{\circ}{g}(\rho)$ in the interval $[0, 1]$, we take its values at points $\rho_k = 0,05k$, $k \in [0 : 20]$ and find the extremum by exhaustive search. Every fiftieth optimal, disaggregated, and local optimal controls are printed. We obtained the following numerical values:

$\overset{*}{u}$ = 0,324601 0,299601 0,274601 0,249601 0,224601

0,199601

0,174601 0,149601 0,124601 0,099601 0,074601

Iteration 1:

$\overset{\circ}{u}$ = 0,000150 0,008111 26,016063 0,024016 0,031968

0,039920 27,047872 0,055825 0,063777 0,071729 27,079681

$\hat{u}$ = 21,582200 21,557200 21,532200 21,507200 21,482200

21,457200 21,432200 21,407200 21,482200 21,357200 21,332200

$\overset{\circ}{\delta}$ = 6,834600 $\pi = 228525,9625559,\ \overset{\circ}{\rho} = 1$

$-\overset{\circ}{g}(\overset{\circ}{\rho}) = 6,15028,482954.$

Iteration 2:

$\overset{\circ}{u}$ = 0,200764 0,200531 0,200298 0,200066 0,199833

0,199601 0,199368 0,199136 0,198903 0,198671 0,198436

$\hat{u}$ = 0,324350 0,299359 0,274359 0,249359 0,224359

0,199359 0,174359 0,299359 0,274359 0,249359 0,224359

$\overset{\circ}{\delta}$ = 49,350281 $\pi = 2,571327,\ \overset{\circ}{\rho} = 1$

$-\overset{\circ}{g}(\overset{\circ}{\rho}) = -615025,911656.$

Iteration 3:

$\overset{\circ}{u}$ = 0,324752 0,299722 0,274692 0,249661 0,224631

0,199601 0,174571 0,149540 0,124510 0,99480 0,074449

$\hat{u}$ = 0,324633 0,299633 0,274633 0,249358 0,224633

0,199359 0,174633 0,149633 0,124633 0,099633 0,074633

$\overset{\circ}{\delta}$ = 49,349735 $\pi = 0,000005,\ \overset{\circ}{\rho} = 0,15$

$-\overset{\circ}{g}(\overset{\circ}{\rho}) = -615025,911654.$

Iteration 4:

$\overset{\circ}{u}$ = 0,324726 0,299701 0,274676 0,249651 0,224626

0,199601 0,174576 0,149551 0,124525 0,099500 0,074475

$\hat{u}$ = 0,324627 0,299627 0,274627 0,249627 0,224627

0,199627 0,174627 0,147627 0,124627 0,099627 0,074627

$\overset{\circ}{\delta}$ $= 49{,}349746 \quad \pi = 0,000004, \; \overset{\circ}{\rho} = 0,05$

$-\overset{\circ}{g}(\overset{\circ}{\rho}) = -615025,911654.$

In iterations 5, 6, we obtained almost the same values as in iteration 4, however $\pi = 0,000003; \, 0,000002; \; \overset{\circ}{\rho} = 0,1; \, 0$. The convergence of the method in the first three iterations to optimal control with an accuracy up to the fourth decimal place is worth mentioning. In other tests, the initial weights were changed, however the result remains the same.

In another series of numerical experiments, we consider the simplest block separable model with quadratic integral functional, which reflects the main specificity of the latter two problems from Section 2. We have

$$\begin{gathered}\sum_{j=1}^{J} \int_0^T (\sigma_j(z_j(t))^2 + \gamma_j(u_j(t))^2)dt \to \min, \\ dz_j(t)/dt = u_j(t), \quad z_j(0) = \kappa_j, \quad u_j(t) \geq 0, \quad j \in [1:J], \\ \sum_{j=1}^{J} u_j(t) \leq w(t).\end{gathered} \tag{5.1}$$

The form of the function $w(t)$ can be made more precise after solving problem (5.1) for $j = 1$, which is written in the form

$$\begin{gathered}\int_0^T (\sigma(z(t))^2 + \gamma(u(t))^2 dt \to \min, \\ dz(t)/dt = u(t), \quad z(0) = \kappa.\end{gathered} \tag{5.2}$$

The Hamiltonian for problem (5.2) has the following form:

$$H = -\sigma z^2 - \gamma u^2 + uy,$$

and the conjugated variable y satisfies the following equation and condition for $t = T$:

$$dy(t)/dt = 2\sigma z, \qquad y(T) = 0. \tag{5.3}$$

Use of the maximum principle connects the control and the dual function $u = y/2\gamma$.

If we substitute the latter equality in the differential equation from (5.2) and unite it with (5.3), we obtain a two-point problem for a system of two differential equations. The characteristic roots of this system $\pm\sqrt{\sigma/\gamma}$ are written in the following form:

$$z(t) = -C^1(\sqrt{\sigma\gamma})^{-1}\exp(-\sqrt{\sigma/\gamma}t) + C^2(\sqrt{\sigma\gamma})^{-1}\exp(\sqrt{\sigma/\gamma}t),$$

$$y(t) = C^1\exp(-\sqrt{\sigma/\gamma}t) + C^2\exp(\sqrt{\sigma/\gamma}t),$$

where the constants C^1, C^2 are defined from the condition for $t = 0$ and $t = T$. Finally, we obtain

$$u(t) = -\kappa\sqrt{\sigma/\gamma}\left[\exp(\sqrt{\sigma/\gamma}t) - \exp(2\sqrt{\sigma/\gamma}T - \sqrt{\sigma/\gamma}t)\right] \times \\ \times \left[\exp(2\sqrt{\sigma/\gamma}T) + 1\right]^{-1}. \tag{5.4}$$

For $\kappa < 0$, the right-hand side of (5.4) is a strictly monotone decreasing nonnegative function equal to zero for $t = T$.

In numerical solution of problem (5.1) by the iterative decomposition method, we take $J = 46$, $T = 1$. The function $w(t)$ is given in the form (9.4), where $\kappa = -25$, $\sigma = \gamma = 1$. The constants $\sigma_j = \gamma_j$, $j \in [1 : J]$ are fixed by the formula $1 + 0,002(j-1)$, and the initial values are given as follows: $\kappa_j = -10 + 0,01(j-1)$, $j \in [1 : J]$.

The input data are selected so that the aggregated problem for (5.1) always has an optimal control $\overset{\circ}{U}(t) = w(t)$. For the conjugated variables $\overset{\circ}{y}_j(t)$ of this problem to be found, the interval $[0, T]$ is partitioned into 200 equal parts, and the equation for phase coordinates is numerically integrated with the use of approximation by the Euler formula. We have

$$\overset{\circ}{z}_j(t + \bar{h}) = \overset{\circ}{z}_j(t) + \bar{h}(\alpha_j(t)w(t) + \alpha_j(t + \bar{h})w(t + \bar{h})),$$

where $\bar{h} = 0,005$. Then, we integrate numerically the equations of (5.3) type by the same formula.

The dual functions of the binding constraint of the aggregated problem is computed in the form

$$\overset{\circ}{\delta}(t) = \sum_{j=1}^{J}(\alpha_j(t)\overset{\circ}{y}_j(t) - 2\gamma_j(\alpha_j(t))^2 w(t)).$$

For each fixed $j \in [1 : J]$, the local problems are formulated as follows:

$$\int_0^T (\sigma_j(z_j(t))^2 + \gamma_j(u_j(t))^2 + \overset{\circ}{\delta}(t)u_j(t))dt \to \min,$$

$$dz_j(t)/dt = u_j(t), \qquad z_j(0) = \kappa_j.$$

Use of the maximum principle for these problems results in the following two-point problems:

$$dz_j(t)/dt = (y_j(t) - \overset{\circ}{\delta}/(2\gamma_j), \qquad dy_j(t)/dt = 2\sigma_j z_j(t),$$

$$y_j(0) = \kappa_j, \qquad y_j(T) = 0,$$

whose solution has the form

$$\begin{aligned}\hat{z}_j(t) &= -\exp(-\sqrt{\sigma_j/\gamma_j t})\left[\hat{C}_j^1 + \int_0^t 0,25(\gamma_j)^{-1}\overset{\circ}{\delta}(\tau)\exp(\sqrt{\sigma_j/\gamma_j\tau})dt\right] + \\ &+ \exp(\sqrt{\sigma_j/\gamma_j t})\left[\hat{C}_j^2 - \int_0^T 0,25(\gamma_j)^{-1}\exp(-\sqrt{\sigma_j/\gamma_j\tau})d\tau\right], \\ \hat{y}_j(t) &= \exp(-\sqrt{\sigma_j/\gamma_j t})\left[\hat{C}_j^1 + \int_0^t 0,25\,\overset{\circ}{\delta}(\tau)\sqrt{\sigma_j/\gamma_j}\exp(\sqrt{\sigma_j/\gamma_j}d\tau\right] + \\ &+ \exp(\sqrt{\sigma_j/\gamma_j t})\left[\hat{C}_j^2 - \int_0^T 0,5\,\overset{\circ}{\delta}(\tau)(\sqrt{\sigma_j/\gamma_j}\exp(-\sqrt{\sigma_j/\gamma_j\tau})d\tau\right],\end{aligned} \tag{5.5}$$

where

$$\hat{u}_j(t) = (\hat{y}_j(t) - \overset{\circ}{\delta}(t))/2\gamma_j.$$

The constants $\hat{C}_j^1$, $\hat{C}_j^2$ are found from the pair of conditions for $t = 0$, $t = T$ and are not written here. To compute integrals in (5.5) and expressions for constants C_j^1, C_j^2, the known trapezoid formula is applied. The value of the left-hand side in (1.12) from Chapter 3 in the optimality criterion is written as follows:

$$\pi = -\int_0^T \left[\sum_{j=1}^J (\sigma_j(\hat{z}_j(t))^2 + \gamma_j(\hat{u}_j(t))^2) + \left(w(t) - \sum_{j=1}^J u_j(t) \right) \overset{\circ}{\delta}(t) - \right.$$
$$\left. - - \sum_{j=1}^J \left(\sigma_j(\overset{\circ}{z}_j(t))^2 + \gamma_j(\overset{\circ}{u}_j(t))^2 \right) \right] dt.$$

The function $-\overset{\circ}{g}(\rho)$ is maximized in the same way as in the previous example. Further, we represent disaggregated controls and optimal solutions of block problems with numbers $j = 1, 46$ at the time instants $t_n = 0,1 \times n$, $n \in [0 : 10]$. The initial weights for each t are given in the form of the arithmetic progression with the sum equal to 0, 1. Moreover, the terms 0, 1, 0, 25, 0, 55 are added to the weights with numbers 16, 31, 46, so that their sum is equal to zero. The results of the computation are as follows:

Iteration 1:

$\overset{\circ}{u}_1$ = 0,001761 0,001538 0,001331 0,001137 0,000954

0,000781 0,000616 0,000456 0,000302 0,000150 0,000000

$\overset{\circ}{u}_{46}$ = 10,552940 9,217800 7,974915 6,811845 5,716951

4,670923 3,688428 2,734497 1,807935 0,899466 0,000000

$\hat{u}_1$ = 7,615527 6,652024 5,755097 4,915768 4,125638

3,376798 2,661755 1,973351 1,304696 0,649100 0,000000

$\hat{u}_{46}$ = 7,272844 6,352697 5,496129 4,694569 3,939993

3,224840 2,541981 1,884554 1,245088 0,619892 0,000000

$\overset{\circ}{\delta}$ = 0,001087 0,000976 0,000866 0,000757 0,000648

0,000540 0,000431 0,000323 0,000215 0,000108 0,000000

π = 1032,714677 $\overset{\circ}{\rho} = 0,6$ $-\overset{\circ}{g}(\overset{\circ}{\rho}) = -4470,224367.$

Iteration 2:

$\overset{\circ}{u}_1$ = 0,408365 0,356699 0,308604 0,263596 0,221228

0,181073 0,142730 0,105816 0,069961 0,034806 0,000000

$\overset{\circ}{u}_{46}$ = 0,767122 0,670067 0,579718 0,495172 0,415581

0,340149 0,268122 0,198778 0,131424 0,065385 0,000000

$\hat{u}_1$ = 0,279021 0,243720 0,210859 0,180107 0,151158

0,123722 0,097524 0,072301 0,047803 0,023782 0,000000

$\hat{u}_{46}$ = 0,542105 0,473517 0,409672 0,349926 0,293681

0,240375 0,189476 0,140472 0,092874 0,046206 0,000000

$\overset{\circ}{\delta}$ = 19,267364 17,340627 15,413891 13,487154 11,560418

9,633682 7,706945 5,780209 3,853473 1,926736 0,000000

π = 0,153817 $\overset{\circ}{\rho} = 1$ $-\overset{\circ}{g}(\overset{\circ}{\rho}) = -4470,017572.$

Iteration 3:

$\overset{\circ}{u}_1$ = 0,275586 0,240720 0,208263 0,177889 0,249297

0,122198 0,096322 0,071411 0,047214 0,023480 0,000000

$\overset{\circ}{u}_{46}$ = 0,535431 0,467689 0,404628 0,345617 0,290064

0,237415 0,187142 0,138742 0,091730 0,045632 0,000000

$\hat{u}_1$ = 0,273430 0,238837 0,206634 0,176498 0,148129

0,121243 0,095560 0,070853 0,046845 0,023306 0,000000

$\hat{u}_{46}$ = 0,539975 0,469039 0,405796 0,346615 0,290902

0,238101 0,187683 0,139143 0,091955 0,045769 0,000000

$\overset{\circ}{\delta}$ = 19,282046 17,353841 15,425637 13,497432 11,569227

9,641023 7,712818 5,784614 3,856409 1,928205 0,000000

π = 0,006150 $\overset{\circ}{\rho} = 1$ $-\overset{\circ}{g}(\overset{\circ}{\rho}) = -4470,014984.$

Iteration 4:

$\overset{\circ}{u}_1$	=	0,273559	0,238949	0,206730	0,176581	0,148198
0,121299		0,095614	0,070885	0,046866	0,023317	0,000000
$\overset{\circ}{u}_{46}$	=	0,537228	0,469259	0,405986	0,346777	0,291038
0,238212		0,187770	0,139207	0,092038	0,045790	0,000000
$\hat{u}_1$	=	0,273678	0,239053	0,206821	0,176658	0,148264
0,121353		0,095656	0,070917	0,046887	0,023327	0,000000
$\hat{u}_{46}$	=	0,537203	0,469237	0,405968	0,346761	0,291025
0,238202		0,187762	0,139202	0,092034	0,045788	0,000000
$\overset{\circ}{\delta}$	=	19,281395	17,353256	15,425116	13,496976	11,568837
9,640697		7,712558	5,784418	3,856279	1,928139	0,000000
π	=	0,006134	$\overset{\circ}{\rho} = 0$	$-\overset{\circ}{g}(\overset{\circ}{\rho}) = -4470,014984.$		

It is clear that we have an approximation of disaggregated solutions and optimal controls for block problems, accurate to the third decimal place, which testifies to the good convergence of the decomposition method. The fast decrease of the value in the first three iterations is noticeable.

Direct solution of problem (6.1) by the Pontryagin maximum principle leads to difficult problems indicated in Section 2. In particular, we need to compute the roots of a polynomial of the 92nd order. Systems of linear algebraic equations of the 92nd order are also solved. Here, 92 eigen vectors should be found. This demonstrates the essential gain in computation time when we use the decomposition method based on aggregated variables.

We consider the following problem with many binding constraints of type (3.6) from Section 3 in the next series of numerical experiments:

$$f = 0,5\left\{\sum_{j=1}^{J}\left[(x_j^1(T)^2 + (x_j^2(T))^2 + (x_j^3)(T))^2 + \sum_{i=1}^{I}\int_0^T (u_j^i(t))^2 dt\right]\right\} \to \min,$$

$$dx_j^1(t)/dt = \sum_{i=1}^{I} b_j^{i1} u_j^i(t), \qquad x_j^1(0) = \kappa_j^1,$$

$$dx_j^2(t)/dt = \sum_{i=1}^{I} b_j^{i2} u_j^i(t), \qquad x_j^2(0) = \kappa_j^2,$$

$$dx_j^3(t)/dt = \sum_{i=1}^{I} b_j^{i3} u_j^i(t), \qquad x_j^3(0) = \kappa_j^3, \qquad j \in [1:J],$$

$$\sum_{j=1}^{J} u_j^i(t) = w^i, \qquad i \in [1:I].$$

Here, as in every subsystem, three phase variables are taken. The dimensions are as follows: $J = 61$, $I = 181$. As it was shown in Section 7, the direct solution of a problem of this type by the maximum principle is reduced to a system of linear algebraic equations with a matrix of order 181×181 or of order $(61 \times 3) \times (61 \times 3)$. The solution of this problem by the iterative decomposition method based on the aggregated variables reduces to J independent linear algebraic equations with respect to unknowns $x_j^1(T)$, $x_j^2(T)$, $x_j^3(T)$, $j \in [1:J]$. The indicated systems of equations appear in the consideration of block problems at each iteration step. They have the following form:

$$\sum_{r=1}^{3} A_j^{nr} x_j^r = C_j^n, \qquad n = 1,2,3,$$

where

$$A_j^{nn} = 1 + \sum_{i=1}^{I} b_j^{ni} b_j^{ni}, \qquad A_j^{nr} = \sum_{i=1}^{I} b_j^{ni} b_j^{ri}, \qquad n \neq r, C_j^n = \kappa_j^n - \sum_{i=1}^{I} b_j^{in} \overset{\circ}{\delta}{}^i.$$

The input parameters are given as follows: $T = 1$; $\kappa_j^1 = -250 + h_1 j$, $\kappa_j^2 = -230 + h_2 j$, $\kappa_j^3 = -210 + h_1 j$, $h_1 = 0,1$, $j \in [1:J]$; $w^i = 58$, $i \in [1:I]$; $b_j^{i1} = 0,9 + h_{2j} + h_3 i$, $b_j^{i2} = 1,02 + h_2 j + h_3 i$, $b_j^{i3} = 1,1 + h_2 j + h_3 i$, $h_2 = -0,01$, $h_3 = -0,0005$, $j \in [1:J]$, $i \in [1:I]$. In one test, the initial weight coefficients are given in the form $\alpha_j^i = 0,1/61$, $j \in [1:J]$, $i \in [1:I]$, and the term $0,3$ is added to the weights with numbers $j,i = (1,37), (31,37), (61,37)$;

(1, 109), (31, 109), (61, 109); (1, 81), (31, 181), (61, 181). Thus, only for weights for $i = 37, 109, 181$, the normalization condition is satisfied. Since the binding constraints of the considered problem are written in the form of equalities, the algorithm begins from an infeasible disaggregated solution. Nevertheless, we have $\overset{\circ}{\rho} = 1$ after the iteration, and the new weights are constructed by optimal solutions of the local problems. Therefore, in what follows, the normalization condition always holds.

Disaggregated controls $\overset{\circ}{u}{}_j^i$ and optimal controls $\hat{u}_j^i$of block problems with numbers $j = 1, 61$, $i = 1, 37, 73, 109, 145, 181$ are printed. We have dual variables for the binding constraints in the aggregated problem with numbers $i = 1, 37, 109, 145, 181$ as well as the values π, $\overset{\circ}{\rho}$, $\overset{\circ}{g}(\overset{\circ}{\rho})$. In the maximization of the function $-\overset{\circ}{g}(\rho)$, its values are computed at points $\rho_k = 0,1k$, $k \in [0 : 10]$. We obtain the following numerical results:

Iteration 1:

$\overset{\circ}{u}{}_j^1$	$\overset{\circ}{u}{}_j^{37}$	$\overset{\circ}{u}{}_j^{73}$	$\overset{\circ}{u}{}_j^{109}$	$\overset{\circ}{u}{}_j^{145}$	$\overset{\circ}{u}{}_j^{181}$
0,095082	17,495082	0,095082	17,495082	0,095082	17,495082
0,095082	17,495082	0,095082	17,495082	0,095082	17,495082
$\hat{u}_j^1$	$\hat{u}_j^{37}$	$\hat{u}_j^{73}$	$\hat{u}_j^{109}$	$\hat{u}_j^{145}$	$\hat{u}_j^{181}$
8,103595	-392, 247624	7,606880	-376,919491	7,110164	-361,591358
8,360730	-392, 084226	7,676540	-376,943568	6,992349	-361,802910
$\overset{\circ}{\delta}{}_j^1$	$\overset{\circ}{\delta}{}_j^{37}$	$\overset{\circ}{\delta}{}_j^{73}$	$\overset{\circ}{\delta}{}_j^{109}$	$\overset{\circ}{\delta}{}_j^{145}$	$\overset{\circ}{\delta}{}_j^{181}$
0,605623	4,595391	0,583102	4,414621	0,560581	4,233852

$\pi = 169857, 860268 \qquad \overset{\circ}{\rho} = 1 \qquad -\overset{\circ}{g}(\overset{\circ}{\rho}) = -8575, 616040.$

Iteration 2:

$\overset{\circ}{u}{}_j^1$	$\overset{\circ}{u}{}_j^{37}$	$\overset{\circ}{u}{}_j^{73}$	$\overset{\circ}{u}{}_j^{109}$	$\overset{\circ}{u}{}_j^{145}$	$\overset{\circ}{u}{}_j^{181}$
0,036264	0,951014	0,946559	0,950790	0,958571	09505482
0,065973	0,950618	0,955277	0,950851	0,942688	0,951104

$\hat{u}_j^1$	$\hat{u}_j^{37}$	$\hat{u}_j^{73}$	$\hat{u}_j^{109}$	$\hat{u}_j^{145}$	$\hat{u}_j^{181}$
0,673012	0,784824	0,899175	1,011895	0,123360	1,239007
1,243872	1,123608	1,005883	0,886527	0,765916	0,649486
1,869859	1,833792	1,797700	1,761624	1,725561	1,689456

$\pi = 0,477308 \qquad \overset{\circ}{\rho} = 1 \qquad -\overset{\circ}{g}(\overset{\circ}{\rho}) = -8573,544221.$

Iteration 3:

$\overset{\circ}{u}{}_j^1$	$\overset{\circ}{u}{}_j^{37}$	$\overset{\circ}{u}{}_j^{73}$	$\overset{\circ}{u}{}_j^{109}$	$\overset{\circ}{u}{}_j^{145}$	$\overset{\circ}{u}{}_j^{181}$
0,671556	0,784620	0,898257	1,011825	1,125841	1,239104
1,241182	1,123316	1,004856	0,886466	0,767607	0,649537

$\hat{u}_j^1$	$\hat{u}_j^{37}$	$\hat{u}_j^{73}$	$\hat{u}_j^{109}$	$\hat{u}_j^{145}$	$\hat{u}_j^{181}$
0,654565	0,775135	0,896434	1,015484	1,135337	1,254854
1,225377	1,113914	1,002180	0,890196	0,778016	0,665500

$\overset{\circ}{\delta}{}_j^1$	$\overset{\circ}{\delta}{}_j^{37}$	$\overset{\circ}{\delta}{}_j^{73}$	$\overset{\circ}{\delta}{}_j^{109}$	$\overset{\circ}{\delta}{}_j^{145}$	$\overset{\circ}{\delta}{}_j^{181}$
1,869518	1,833435	1,797351	1,761268	1,725185	1,689103

$\pi = 0,035361 \qquad \overset{\circ}{\rho} = 0 \qquad -\overset{\circ}{g}(\overset{\circ}{\rho}) = -8573,544220.$

The drastic decrease of the difference between the functional values of a conjugated pair of problems in the first iteration (here in 6 orders of magnitude) is characteristic of this test as well as of the previous two series. We also note that the values of the disaggregated controls and optimal solutions of block problems differ no more than 0,17 after the third iteration. We can attain better approximation if, for example, we reduce the step by the parameter ρ or use the multidimensional optimization function $\overset{\circ}{g}(\rho_j)$ with respect to ρ_j, $j \in [1 : J]$. However, we do not really need this, since the final disaggregated controls in this test and in the test, where all initial weights are taken to be 1/61, differ very little.

Comments and References to Chapter 4

Block problems of optimal control are formulated in Section 1 on the basis of statements from the books of A.G. Butkovskii, S.A. Malyi, and Yu.N. Andreev [1], Yu.M. Svirezhev and E.Ya. Elizarov [7] Yu.P. Ivanilov and A.V. Lotov [5]. These problems were also studied for the case with delayed arguments introduced in the monographs of T.K. Sirazetdinov [7], R.T. Yanushevskii [11]. The iterative aggregation for block optimal control problems with delayed arguments is studied in detail in the thesis by O.A. Fed'ko [4]. A hierarchical mathematical-physical model of heat propagation is constructed, based on the problem of heating a thin membrane, from the paper by A.I. Egorov [2]. A model with hyperbolic equations from the work of A.I. Egorov and G.B. Shenfel'd [3] is used in the consideration of the corresponding two-level system. The problem about the heating of membranes that move through furnaces is introduced, based on statements from the book of T.K. Sirazetdinov [6]. Chapter 4 is based on the works of V.I. Tsurkov [8–10]

References to Chapter 4

[1] Butkovskii A.G., Malyi S.A., and Andreev Yu.N., *Optimal'noe upravlenie nagrevom metalla*(Optimal Control in Metal Heating), Moscow: Metallurgiya, 1972.

[2] Egorov A.I. *Ob usloviyakh optimal'nosti v odnoy zadache upravleniya protsessom teploperedachi* (On Optimality Conditions in a Problem of Controlling the Heat Exchange Process), *Zh. Vych. Mat. Mat. Fiz.*, 1972, vol. 12, no. 4, pp. 791-799.

[3] Egorov A.I. and Shenfel'd G.B., *Ob odnoi zadache optimal'nogo upravleniya izgibnymi kolebanijami balki* (On a Problem of Optimal Control of Bend Oscillations of a Beam), *Trudy Frunze Politekhn. Inst.*, Mashinostroenie, 1971, vol. 45, pp. 77-89.

[4] Fed'ko O.A., *Decompozitsija v zadachakh optimal'nogo upravlenija s zapazdyvanijami*(Decomposition in Optimal Control Problems with Delays, Dissertation Thesis, Moscow Physical-Technical Institute), 1984.

[5] Ivanilov Yu.P. and Lotov A.V., *Matematicheskie modeli v economike* (Mathematical Models in Economics), Moscow: Nauka, 1979.

[6] Sirazetdinov T.K., *Optimizatsija sistem s raspredelennymi parametrami* (Optimization of Systems with Distributed Parameters), Moscow: Nauka, 1977.

[7] Svirezhev Yu.M., and Elizarov E.Ya., *Matematicheskoe modelirovanie biologicheskykh sistem*(Mathematical Modeling of Biological Systems), Moscow: Nauka, 1972.

[8] Tsurkov V.I., *Razlozhenie v optimizatsii sistem s uravneniyami v chastnykh proizvodnykh pervogo poryadka* (Decomposition in the Optimization of Systems with Differential Equations of the First Order), Kibernetika, 1984, no. 1, pp. 60-64.

[9] Tsurkov V.I., Chislennye experimanty po iterativnomy metody decompozitsii dlja zadach optimal'nogo upravlenija (Numerical Experiments in the Iterative Decomposition Method for Optimal Control Problems), *Vychisl. Prikl. Mat.*, Kiev, 1983, no. 51, pp. 112-117.

[10] Tsurkov V.I., *Dvuhurovnevye sistemy optimal'njgj upravlenija* (Two-level Optimal Control Systems), *Izv. Akad. Nauk SSSR*, Tekh. Kibern., 1984, no. 4, pp. 45-59.

[11] Yanushevskii R.T., Upravlenije objectami s zapazdyvaniem (Control of Objects with Delays), Moscow: Nauka, 1978.

Chapter 5

Appendix. The Main Approaches in Hierarchical Optimization

Below, we present the main types of optimization problems with block constraint structure. We describe the Dantzig-Wolfe decomposition. This approach is fairly well covered in the literature, therefore, we present it briefly here, just to introduce the main ideas. The Kornai-Liptak decomposition principle is presented more comprehensively. Along with the main published constructions, we consider various schemes of reducing dimension based on the iterative resource redistribution. Special attention is given to the method of parametric decomposition, which is the standard method for optimizing large-scale systems and generalizes many well-known decomposition approaches. Here, the foundations of iterative aggregation are described. This method was initially applied in particular economic models. However, it was later modified and generalized for a wide class of problems of hierarchical optimization. Finally, the use of the Lagrange functional for the organization of the two-level iterative process with respect to primal and dual variables allows us to formulate a fairly universal approach to reducing dimensions in block problems of optimal control.

§1. Dantzig-Wolfe Principle

Consider the linear programming problem

$$(c, x) \to \max, \tag{1.1}$$

$$A^0 x = b^0, \tag{1.2}$$

$$A^1 x = b^1, \tag{1.3}$$

$$x \geq 0, \tag{1.4}$$

where $c = (c_1, \ldots, c_n)$ is n-dimensional row vector, $x = (x_1, \ldots, x_n)$ is n-dimensional column vector, $b^0 = (b_1, \ldots, b_m)$ and $b^1 = (b_{m+1}, \ldots, b_{m+m_1})$ are

m and m_1-vector columns, respectively. A^0 and A^1 are matrices of dimensions $m \times n$ and $m_1 \times n$, respectively.

Any problem of linear programming is written in the form (1.1)–(1.4) if the matrix of conditions is partitioned in two blocks. This partition is done to show the possibility of reducing a problem's number of constraints.

It is assumed that the set M_x given by conditions (1.3), (1.4) is, for simplicity, a limited polyhedron. Let x^ν, $\nu \in [1 : N]$ be its vertices. Then, each point $x \in M_x$ is described in the form

$$x = \sum_{\nu=1}^{N} z_\nu x^\nu, \tag{1.5}$$

$$\sum_{\nu=1}^{N} z_\nu = 1, z_\nu \geq 0, \nu \in [1 : N]. \tag{1.6}$$

The substitution of (1.5) in (1.1), (1.2), under the assumption that all x^ν, $\nu \in [1 : N]$ are known, provides the transfer to the so-called z-problem:

$$\sum_{\nu=1}^{N} \sigma_\nu z_\nu \to \max, \tag{1.7}$$

$$\sum_{\nu=1}^{N} p_\nu z_\nu = b^0, \tag{1.8}$$

$$\sum_{\nu=1}^{N} z_\nu = 1, \tag{1.9}$$

$$z_\nu \geq 0, \qquad \nu \in [1 : N], \tag{1.10}$$

where $\sigma_\nu = (c, x^\nu)$, $p_\nu = A^0 x^\nu$, $\nu \in [1 : N]$. Problem (1.7)–(1.10) has $m + 1$ conditions and a huge number (equal to N) of unknowns. At first glance, the analysis of this problem seems fairly difficult. However, it turns out that, in the solution of the z-problem by means of the simplex method, we do not need to keep in computer memory all the variables σ_ν and vectors p_ν, $\nu \in [1 : N]$. At each step, it suffices to have $m + 1$ vectors that make the current basis. The vector that is included in the basis is determined by the solution of an auxiliary linear programming problem with m_1 constraints.

It is easy to test that if z_ν^* is an optimal solution of the problem (1.7)–(1.10), then the solution x^* of the initial problem (1.1)–(1.4) has the following form:

$$x^* = \sum_{\nu=1}^{N} z_\nu^* x^\nu .$$

We solve the z-problem by means of the so-called second algorithm of the plan improving method (see, e.g., the book of D.B. Yudin and E.G. Gol'shtein [70]). We briefly recall the main steps of this method.

We introduce the vector of dual variables $(\Lambda; \lambda_0) = (\lambda_1, \ldots, \lambda_m; \lambda_0)$ that correspond to the conditions (1.8), (1.9). Let the indices $s \in S_z$ be the numbers of vectors of the current basis, which is assumed to be nondegenerate. Then, the vector $\Lambda; \lambda_0)$ is determined as a solution of the system of linear equations

$$(\Lambda, p_s) + \lambda_0 = \sigma_s, \qquad s \in S_z.$$

We introduce the values Δ_ν, which are called *the estimates of conditions with respect to the current basis*: $\Delta_\nu = (\lambda, p_\nu) + \lambda_0 - \sigma_\nu$, $\nu \in [1 : N]$. The criterion for optimality of the current basis is the nonnegativity of all Δ_ν. Failing this, the vector that corresponds to the minimum estimate Δ_ν is introduced to the basis. Thus, the step of the simplex method for problem (1.7)–(1.10) would be reduced to the search through a large number of values Δ_ν. The main point of the Dantzig-Wolfe method is the replacement of this procedure by finding an optimal solution of the following problem:

$$f_\Lambda(x) = (c_\Lambda, x) \to \max, \tag{1.11}$$

$$A^1 x = b^1, \tag{1.12}$$

$$x \geq 0, \tag{1.13}$$

where $c_\Lambda = c - \Lambda A^0$.

We compute the value of the linear form (1.11) for the support solution x^ν of the problem (1.11)–(1.13). By means of the introduced notation, we can easily obtain that

$$f_\Lambda(x^\nu) = \lambda_0 - \Delta_\nu.$$

Hence, if $x^{\nu,*}$ is an optimal solution of problem (1.11)–(1.13), then the inequalities $f_\Lambda(x^{\nu,*}) \geq f_\Lambda(x^\nu)$, $\nu \in [1 : N]$ result in the following relation:

$$\Delta_{\nu,*} \leq \Delta_\nu, \nu \in [1 : N]. \tag{1.14}$$

Relation (1.14) implies that the optimality condition for the current basis of the z-problem is the equality to zero of the value $\Delta_{\nu,*}$. If $\Delta_{\nu,*} > 0$, then the vector $p = A^0 x^{\nu,*}$ is introduced in the basis of the z-problem, since this vector corresponds to the minimal estimate $\Delta_{\nu,*}$. In other words, a step of the simplex method reduces to the comparison of the values $f_\Lambda(x^{\nu,*})$ and λ_0. If these values are equal, then the support solution of z-problem is optimal. If $f_\Lambda(x^{\nu,*} > \lambda_0$, then we pass to the basis with the largest value of the functional.

Earlier, we supposed that all vertices x^ν, $\nu \in [1:N]$ of the polyhedron M_x are known. However, the latter assumption seems to be fairly conditional. At each step of the solution of the z-problem, it suffices to have $m+1$ vertices that correspond to the vectors p_s, $s \in S_z$ constituting the current basis. The vector introduced into the basis is determined by the solution of problem (1.11)–(1.13).

Thus, the initial problem (1.1)–(1.4) with $m + m_1$ constraints is reduced to the solution of problems that contain $m + 1$ and m_1 constraints.

The efficiency of the Dantzig-Wolfe decomposition method becomes more obvious when the A^1 matrix has the block diagonal structure:

$$A^1 = \begin{bmatrix} A_1 & 0 & \dots & 0 \\ 0 & A_2 & \dots & 0 \\ \vdots & \vdots & \ddots & \vdots \\ 0 & 0 & \dots & A_j \end{bmatrix},$$

where A_j, $\nu \in [1:N]$ are $m_j \times n_j$-matrices. The initial problem is written in the form

$$\sum_{j=1}^{J} (c_j, x_j) \to \max, \tag{1.15}$$

$$\sum_{j=1}^{J} A_j^0 x_j = b^0, \tag{1.16}$$

$$A_j x_j = b_j, \tag{1.17}$$

$$x_j \geq 0, \qquad j \in [1:J]. \tag{1.18}$$

Here, for all $j \in [1:J]$, x_j is an n_j-dimensional vector column, A_j^0 is a $m \times n_j$-matrix, b_j is an m_j-dimensional vector column, and c_j is an n_j-dimensional vector row.

Let M_x^j, $j \in [1:J]$ be limited polyhedrons given by conditions (1.17), (1.18). Then, for all points $x_j \in M_x^j$, we have the representation

$$x_j = \sum_{\nu=1}^{N_j} z_\nu^j x_j^\nu, \qquad j \in [1:J],$$

$$\sum_{\nu=1}^{N_j} z_\nu^j = 1, \quad z_\nu^j \geq 0, \quad j \in [1:J], \quad \nu \in [1:N_j],$$

where x_j^ν, $\nu \in [1:N_j]$ are vertices of the polyhedrons M_x^j, $j \in [1:J]$.

In this case, the z-problem has the form

$$\sum_{j=1}^{J}\sum_{\nu=1}^{N_j} \sigma_\nu^j z_\nu^j \to \max, \tag{1.19}$$

$$\sum_{j=1}^{J}\sum_{\nu=1}^{N_j} p_\nu^j z_\nu^j = b^0, \tag{1.20}$$

$$\sum_{\nu=1}^{N_j} z_\nu^j = 1, \qquad j \in [1:J], \tag{1.21}$$

$$z_\nu^j \geq 0, \qquad j \in [1:J], \qquad \nu \in [1:N_j], \tag{1.22}$$

where $\sigma_\nu^j = (c_j, x_j^\nu)$, $p_\nu^j = A_j^0 x_j^\nu$.

Solution of the z-problem by the scheme presented above assumes consideration of the vector of dual variables. In accordance with (1.20) and (1.21), this vector is partitioned in two parts: $\Lambda^1 = (\lambda_1^1, \ldots, \lambda_m^1)$ and $\Lambda^2 = (\lambda_1^2, \ldots, \lambda_J^2)$. The current basis consists of $m + J$ vectors. Let S_z^J be the set of pairs of indices (j, ν) that correspond to the given base. Then, the components Λ^1 and Λ^2 satisfy the linear equations $(\Lambda^1, p_\nu^j) + \lambda_j^2 = \sigma_\nu^j$, $(j, \nu) \in S_z^J$. The vector Λ^1 forms the functional of the auxiliary problem of the type (1.11)–(1.13), which, in this case, is partitioned into J independent block problems:

$$f_\Lambda^j(x_j) = (c_j - \Lambda^1 A_j^0, x_j) \to \max, \tag{1.23}$$

$$A_j x_j = b_j, \tag{1.24}$$

$$x_j \geq 0. \tag{1.25}$$

There is the following relation between the estimates Δ_ν^j with respect to the current basis of the z-problem and the functional values of block problems (1.23)–(1.25) for feasible values:

$$\Delta_\nu^j = (\Lambda^1, p_\nu^j) + \lambda_j^2 - \sigma_\nu^j,$$

$$f_\Lambda^j(x_j^\nu) = \Delta_\nu^j - \lambda_j^2.$$

Let $x_j^{\nu,*}$, $j \in [1:J]$ be optimal solutions of problems (1.23)–(1.25). Then, the minimal estimate $\Delta_{\nu,*}^j$ corresponds to each fixed j. If

$$\min_{j\in[1:J]} \Delta_{\nu,*}^j = 0, \tag{1.26}$$

then the feasible solution of the z-problem is optimal. If the minimum in (1.26) is less than zero and is attained for some $r \in [1:J]$, then a new vector is introduced in the basis. The first m components of this vector are equal to $A_r^0 x_r^{\nu,*}$, the $(m+1)$th component is equal to unity; and all other are equal to zero. The subsequent computation follows the second algorithm for improving the plan.

In the Dantzig-Wolfe decomposition method, the z-problem is called *the coordinating one*. Problem (1.11)–(1.13) and (1.23)–(1.25) are called *local*. In the block diagonal case, the local problem is partitioned into J independent problems. This explains the efficiency of the expansion.

We study the Dantzig-Wolfe method in more detail. We come to problem (1.1)–(1.4) and consider the dual problem for the coordinating one (1.7)–(1.10):

$$(b^0, \Lambda) + \lambda \to \min, \tag{1.27}$$

$$(\Lambda, p_\nu) + \lambda \geq \sigma_\nu, \qquad \nu \in [1:N]. \tag{1.28}$$

If we fix Λ, then solution of problem (1.27), (1.28) takes the following form:

$$\lambda = \max_{\nu\in[1:N]} \left[\sigma_\nu - (\Lambda, p_\nu)\right].$$

Definition of the values σ_ν, p_ν and of the last relation imply that problem (1.27), (1.28) is equivalent to minimization of the function of Λ of the following form:

$$\varphi(\Lambda) = \max_{\nu\in[1:N]} (c - \Lambda A^0, x^\nu) + (b^0, \Lambda),$$

or, which is the same,

$$\varphi(\Lambda) = \max_{x \in M_x} (x - \Lambda A^0, x) + (b^0, \Lambda). \tag{1.29}$$

The right-hand side of (1.29) is a Lagrange function of problem (1.1)–(1.4) constructed for the constraints (1.2).

It is easy to see that the function $\varphi(\Lambda)$ is a convex piecewise linear function. The difficulty in finding the minimum of this function is because $\varphi(\Lambda)$ is given algorithmically (to find its value at the point Λ, we need to solve a local problem). Moreover, this function is not differentiable, thus the methods of minimization of smooth functions are not applicable.

In minimizing the $\varphi(\Lambda)$ function, procedures based on finite methods of linear programming are applied. Whereas the constructions of J. Dantzig and P. Wolfe [9–11] are based on improving the plan in linear programming, other decomposition procedures that involve the Dantzig-Wolfe method are based on the corresponding finite methods for linear problems. Thus, in the paper [34] by S.M. Movshovitch, the method of decreasing discrepancies is used. In a paper by E.J. Bell [5], the method of simultaneous solution of the primal and dual problem is applied. The constructions in paper [1] by J.M. Abadie and A.C. Williams are based on the method of refining estimates, etc. All these approaches are given in detail in [14, 22].

In other approaches, $\varphi(\Lambda)$ is minimized by applying approximate iterative procedures. Thus, the paper [49] by N.Z. Shor uses the descent along the generalized gradient. The direction of descent for a fixed point Λ is determined by the vector $-[b^0 - A^0 x(\lambda)]$, where $x^*(\Lambda)$ maximizes the right-hand side of (1.29) for a fixed vector Λ.

In the paper by B.T. Polyak and N.V.Tret'yakov [38], smoothing of the function $\varphi(\Lambda)$ is accomplished by the modified Lagrange function

$$L(x, \Lambda, Q) = (c - \Lambda A^0, x) + (b^0, \Lambda) + Q||A^0 x - b^0)||^2,$$

where Q is a positive constant. In this case, the function $\varphi(\Lambda, Q) = \max_{x \in M_x} L(x, \Lambda, Q)$ is smooth, and we can expect fast convergence. On the other hand, the function $L(x, \lambda, Q)$ is such that the local problem is not partitioned in blocks. Here, we use the dual iterative process.

In the paper by V.A. Volkonskii [66], minimization of the function $\varphi(\Lambda)$ is interpreted as a search for the saddle point or, which is the same, a solution of the minimax problem:

$$\min_{\Lambda} \max_{x \in M_x} (c - \Lambda A^0, x) + (b^0, x).$$

An iterative process based on a Brown game is used. At each iteration, minimization over Λ for fixed $x \in M_x$ and maximization over x for fixed Λ are carried out. In this second case, we have an independent solution in separate blocks. This method will be considered in more detail in the next section as applied to the Kornai-Liptak decomposition.

The Dantzig-Wolfe approach is generalized to nonlinear problems. Here, ideas related to linearization on the grid are used (see, e.g., the book of L. Lasdon [22]). We consider the problem of mathematical programming:

$$f(x) \to \max, \tag{1.30}$$

$$g_k(x) \le 0, \qquad k \in [1:m], \tag{1.31}$$

where $x \in \mathbf{R}^n$, and functions $-f(x)$, $g_k(x)$ are assumed to be convex. We construct the iterative process of its solution.

Let the l-th iteration correspond to the points $x_1^{(l)}, \dots, x_\Pi^{(l)}$ that satisfy (1.31). The functions $f(x)$, $g_k(x)$ are replaced by its linearization on the grid, and the following problem of linear programming is introduced:

$$\begin{aligned} &\sum_{\pi=1}^{\Pi} z_\pi f(x_\pi^{(l)}) \to \max, \\ &\sum_{\pi=1}^{\Pi} z_\pi g_k(x_\pi^{(l)}) \le 0, \qquad k \in [1:m], \\ &\sum_{\pi=1}^{\Pi} z_\pi = 1, \\ &z_\pi \ge 0, \qquad \pi \in [1:\Pi]. \end{aligned} \tag{1.32}$$

Let $z_\pi^{(l)}$ be an optimal solution of problem (1.32), and $\lambda_k^{(l)}$, $k \in [1:m]$, $\lambda_0^{(l)}$ are dual estimates. The auxiliary problem

$$F(x) = f(x) - \sum_{k=1}^{m} \lambda_k^{(l)} g_k(x) \to \max, \qquad x \in \mathbf{R}^n \tag{1.33}$$

is introduced.

Assume that $\bar{x}^{(l)}$ is its solution, and $F^{(l)}$ is the optimal value of functional (1.33). We introduce the convex combination

$$x^{(l)} = \sum_{\pi=1}^{\Pi} z_{\pi}^{(l)} x_{\pi}^{(l)}. \tag{1.34}$$

The point $x^{(l)}$ is admissible to the original point (1.30), (1.31). This follows from the convexity of the functions $g_k(x)$ and relations

$$g_k(x^{(l)}) = g\left(\sum_{\pi=1}^{\Pi} z_{\pi}^{(l)} x_{\pi}^{(l)}\right) \leq \sum_{\pi=1}^{\Pi} z_{\pi}^{(l)} g_k(x_{\pi}^{(l)}) \leq 0, \qquad k \in [1:m].$$

Due to the convexity of $f(x)$, the value of the functional (1.30) for $x^{(l)}$ satisfies the inequality

$$f(x^{(l)}) = f\left(\sum_{\pi=1}^{\Pi} z_{\pi}^{(l)} x_{\pi}^{(l)}\right) \geq \sum_{\pi=1}^{\Pi} z_{\pi}^{(l)} f(x_{\pi}^{(l)}).$$

By the First Duality Theorem 1.3 from Chapter 1, we have

$$\sum_{\pi=1}^{\Pi} z_{\pi}^{(l)} f(x_{\pi}^{(l)}) = \lambda_0^{(l)},$$

for problem (1.32), from whence we deduce the inequality

$$f(x^{(l)}) \geq \lambda_0^{(l)}.$$

According to (1.33), $\bar{x}^{(l)}$ and $\lambda_k^{(l)}$, $k \in [1:m]$ are an admissible solution to the dual problem for the initial (1.30), (1.31). Therefore, we have the inequality $F^{(l)} \geq f(x^{l)})$. The last two relations imply that the equality

$$F^{(l)} = \lambda_0^{(l)} \tag{1.35}$$

is the criterion of optimality of $x^{(l)}$ for the initial problem (1.30), (1.31) or, which is the same, the condition for termination of the iterative process.

In the case, where $F^{(l)} \geq \lambda_0^{(l)}$, the new column $(g_1(x^{(l)}), \dots, g_m(x^{(l)}), 1)$ is introduced in problem (1.32), and the sum $z_{\Pi+1} f(x^{(l)})$ is added to its functional. This corresponds to transfer to the $(l+1)$th iteration or, in other terms, to a more precise approximation of the functions $f(x)$, $g_k(x)$. We have the process monotonic with respect to the functional. In the book of

L. Lasdon [22], it is established that the algorithm converges to the optimal solution of problem (1.30), (1.31), as l tends to infinity.

The proposed method allows us to generalize the Dantzig-Wolfe decomposition to nonlinear problems. In the problem of mathematical programming (1.30), (1.31), let the vector of conditions be partitioned into two parts. We have

$$f(x) \to \max, \tag{1.36}$$

$$g^0(x) \leq 0, \tag{1.37}$$

$$g^1(x) \leq 0, \tag{1.38}$$

where $g^0(x)$ and $g^1(x)$ are vector functions with m and m_1 components, respectively. It is assumed that the set M_x given by (1.38) is limited. Let the points from $x_1^{(l)}, \ldots, x_\Pi^{(l)}$ be known at the l-th iteration. We introduce the coordinating problem of linear programming (1.32), where all $g_k(x)$ are replaced by $g_k^0(x)$, $k \in [1 : m]$. Upon solution of this problem, we obtain an approximate solution $x^{(l)}$ (1.34) for the original problem (1.36)–(1.38). In this case, the local problem (1.33) has the form

$$F(x) = f(x) - \sum_{k=1}^{m} \lambda_k^{(l)} g_k^0(x) \to \max,$$

$$g^1(x) \leq 0.$$

The criterion of optimality of $x^{(l)}$ for problem (1.36)–(1.38) is the satisfaction of equality (1.35). If the strict inequality $F^{(l)} > \lambda_0^{(l)}$ holds, then the point $x_{\Pi+1}^{(l)}$ is added to the tuple $x_1^{(l)}, \ldots, x_\Pi^{(l)}$ (according to (1.34)), and the process continues.

§2. Kornai-Liptak Principle

This well-known approach in block programming is based on iterative redistribution of the general resource over subsystems. We present the main construction following the paper [20] by I. Kornai and T. Liptak.

Consider the following linear programming problem:

$$
\begin{aligned}
&(c, x) \to \max, \\
&A^0 x \le b^0, \\
&x \ge 0,
\end{aligned}
\tag{2.1}
$$

where $c = (c_1, \dots, c_n)$ is n-dimensional row vector, $x = (x_1, \dots, x_n)$ is an n-dimensional column vector, b^0 is m-dimensional column vector, A^0 is an $m \times n$-matrix.

Let us partition the matrix A^0 into submatrices $A_1^0, \dots, A_j^0, \dots, A_J^0$, where, for each $j \in [1 : J]$, the matrix A_j^0 is of dimension $m \times n_j$. Then, the vectors c and x are partitioned into $c_1, \dots, c_j, \dots, c_J$ and $x_1, \dots, x_j, \dots, x_J$, respectively, where the dimensions c_j, x_j for $j \in [1 : J]$ are equal to n_j. Problem (2.1) is transformed to the form

$$
\begin{aligned}
&\sum_{j=1}^{J} (c_j, x_j) \to \max, \\
&\sum_{j=1}^{J} A_j^0 x_j \ge b^0, \\
&x_j \ge 0, \qquad j \in [1 : J].
\end{aligned}
\tag{2.2}
$$

We introduce m-dimensional vector columns y_i, $j \in [1 : J]$ that satisfy the condition

$$
\sum_{j=1}^{J} y_j \le b^0 \tag{2.3}
$$

and formulate J problems of linear programming:

$$
\begin{aligned}
&(c_j, x_j) \to \max, \\
&A_j^0 x_j \le y_j, \\
&x_j \ge 0.
\end{aligned}
\tag{2.4}
$$

We introduce the vector y, whose components are y_j. By M_y, we denote the set of vectors y such that (2.3) holds, and problems (2.4) have solutions. The optimal values of the functionals of problems (2.4) are functions of the values y_j. The dependence is represented in the form $f_j = f_j(y_j)$. We introduce the notation

$$F(y) = \sum_{j=1}^{J} f_j(y_j).$$

Then, problem (2.4) is reduced to the following one:

$$\begin{aligned} &F(y) \to \max, \\ &\sum_{j=1}^{J} y_j \leq b^0. \end{aligned} \tag{2.5}$$

If we interpret decomposition of the matrix A^0 as partition into J subsystems and consider the vector b^0 as the general system resource, then problem (2.5) consists in finding an optimal distribution of the general resource. Note that, here, we do not assume that the signs of the values y_j, $j \in [1 : J]$ are constrained, since the subsystem can both consume and produce resources. In further constructions, problem (2.5) will play the role of the coordinating problem. Problems (2.4) will be local.

Consider problem (2.5). Its analysis is complicated since the function $F(y)$ is given algorithmically. To find the values of this function, we need to solve the linear programming problems (2.4). Application of one or another maximization scheme for the function $F(y)$ generates the corresponding method of decomposition based on the Kornai-Liptak principle. In the initial paper [20], reduction to the maximin problem solvable by the game-theoretic techniques is demonstrated. We study the proposed scheme in more detail.

We write dual problems for (2.4) for each $j \in [1 : J]$:

$$\begin{aligned} &(\lambda_j, y_j) \to \min, \\ &\lambda_j A_j^0 \geq c_j, \\ &\lambda_j \geq 0, \end{aligned} \tag{2.6}$$

where the vector columns λ_j, $j \in [1 : J]$ have m components $(\lambda_j^1, \ldots, \lambda_j^s, \ldots, \lambda_j^m)$. Let Ω_λ^j, $j \in [1 : J]$ denote the limited polyhedrons in the space $\mathbf{R}_+^m$ given by conditions of problems (2.6). We introduce vector λ with components $(\lambda_1, \ldots, \lambda_J)$ and the set

$$\Omega_\lambda = \prod_{j=1}^{J} \Omega_\lambda^j.$$

By the First Duality Theorem 1.3 of Chapter 1, for conjugated problems (2.4), (2.6), we have the equality of optimal functional values: $(c_j, x_j^*) = (\lambda_j^*, y_j) = f_j(y_j)$. In other words, we obtain

$$f_j(y_j) = \min_{x_j \in \Omega_\lambda^j} (\lambda_j, y_j),$$

and problem (2.5) is finally reduced to the following one:

$$\min_{\lambda \in \Omega_\lambda} (\lambda, y) \to \max, \qquad y \in M_j, \tag{2.7}$$

where $(\lambda, y) = \sum_{j=1}^{J} (\lambda_j, y_j)$.

The relation of the linear programming problem as well as maximin problems with matrix games is well known. It is proposed to solve problem (2.7) by the method of the Brown game (see, for example, the book by V.A. Volkonskii, V.Z. Belen'kii, S.A. Ivankov, A.B. Pomanskii, and A.D. Shapiro [66]). The method is an iterative process, where, in terms of game theory, each iteration is a game and corresponds to the choice of some strategies for the opposite players. These strategies for our problem are vectors λ and y. Strategies for each player are to be the most profitable, taking of the second player move and the information stored about the behavior of the second player in previous games. Optimal strategies for problem (2.7) are defined in the form

$$(\lambda^*(y), y) = \min_{\lambda \in \Omega_\lambda} (\lambda, y), \tag{2.8}$$

$$(\lambda, y^*(\lambda)) = \min_{y \in M_j} (\lambda, y). \tag{2.9}$$

According to the Brown method, the iterative process gives the following computation rules:

Initial iteration ($l = 1$).

1. An arbitrary strategy $y^{(1)} \in M_y$ is chosen.
2. $y^*[1] = y^{(1)}$ is set.
3. $\lambda^{(1)} = \lambda^*(y[1])$ is defined according to (2.8).
4. $\lambda^*[1] = \lambda^{(1)}$ is set.

Iteration number l ($l \geq 2$).

1. $y^{(l)} = y^*(\lambda^*[l-1])$ is determined according to (2.9).
2. $y^*[l] = \frac{l-1}{l} y^*[l-1] + \frac{1}{l} y^{(l)}$ is computed.
3. $\lambda^{(l)} = \lambda^*(y^*[l])$ is computed according to (2.8).

4. $\lambda^*[l] = \frac{l-1}{l}\lambda^*[l-1] + \frac{1}{l}\lambda^{(l)}$ is computed.

Due to the Robinson theorem on convergence of the Brown method (see, for example, the book [14] by E.G. Gol'shtein and D.B. Yudin), the sequence $\{y^*[l], \lambda^*[l]\}$ tends to the saddle point of problem (2.7) as $t \to \infty$.

The efficiency of the Kornai-Liptak decomposition method becomes obvious when some of the constraints have block diagonal structure. We introduce additional block constraints in problem (2.2). We have

$$\begin{aligned} &\sum_{j=1}^{J} (c_j, x_j) \to \max, \\ &\sum_{j=1}^{J} A_j^0 x_j \geq b^0, \\ &A_j x_j \leq b_j, \\ &x_j \geq 0, \qquad j \in [1:J], \end{aligned} \tag{2.10}$$

where, for each $j \in [1:J]$, b_j is a m_j-vector column, and the matrix A_j has dimension $m_j \times n_j$. By M_x^j, $j \in [1:J]$, we denote limited polyhedrons in spaces $\mathbf{R}^{n_j}$, which are given by the two last conditions in (2.10).

In accordance with (2.4), we introduce the Lagrange function

$$L(x, y, \lambda) = \sum_{j=1}^{J} (c_j, x_j) + \sum_{j=1}^{J} (\lambda_j, y_j - A_j^0 x_j),$$

where $x = (x_1, \dots, x_J)$, $x_j \in M_x^j$, $\lambda_j \in \mathbf{R}_+^m$, $j \in [1:J]$; $\lambda \in \Omega_\lambda$. We choose fairly large range of variables λ_j and y_j, $j \in [1:J]$: $0 \leq \lambda_j \leq \bar{\lambda}$, $-\bar{y} \leq y_j \leq \bar{y}$, $j \in [1:J]$, where $\bar{\lambda}$, $\bar{y} \in \mathbf{R}_+^m$.

We can show that problem (2.5) for (2.10) is reduced to the game of two players with strategies $y \in \bar{M}_y$ and $\lambda \in \bar{\Omega}_\lambda$, where

$$\bar{M}_y = \left\{ y \,\middle|\, -\bar{y} \leq y \leq \bar{y}, \qquad \sum_{j=1}^{J} y_j \leq b^0 \right\},$$

$$\bar{\Omega}_\lambda = \left\{ \lambda \,\middle|\, 0 \leq \lambda_j \leq \bar{\lambda}, \qquad j \in [1:J] \right\},$$

and the functional $\varphi(y, \lambda)$ has the form

$$\varphi(y, \lambda) = \max_{x_j \in M_x^j} L(x, y, \lambda) =$$

$$= \sum_{j=1}^{J} (\lambda_j, y_j) + \sum_{j=1}^{J} \max_{x_j \in M_x^j} (c_j - \lambda_j A_j^0, x_j).$$

Finally, we have the problem about finding the saddle point of the function $\varphi(y, \lambda)$:

$$\min_{\lambda \in \bar{\Omega}_\lambda} \varphi(y, \lambda) \to \max, \qquad y \in \bar{M}_y. \tag{2.11}$$

Problem (2.11) is solved by the Brown game method. Then, at each step of the iterative process, the first (minimizing) player solves the problem

$$\sum_{j=1}^{J} (\lambda_j, y_j) \to \min, \qquad y \in \bar{M}_y,$$

where λ_j, $j \in [1 : J]$ are fixed. Note that the written problem is solved separately for each of the s-th resource, therefore, the first player can be represented as m independent players. The first player can be interpreted as a center that distributes general resources. Larger resources being sent to the subsystem imply greater values for λ_j.

In the iteration process, the second (maximizing) player solves the problem of finding optimal dual variables that correspond to the first group of constraints of the problems

$$\begin{aligned} &(c_j, x_j) \to \max, \\ &A_j^0 x_j \geq y_j, \\ &A_j x_j \leq b_j, \\ &x_j \geq 0, \end{aligned} \tag{2.12}$$

where the values y_j are fixed. The second player is considered as J independent players (subsystems). Thus, the Kornai-Liptak method is decomposed into $J + m$ subproblems.

In the paper by T.N .Pervozvanskaya and A.A. Pervozvanskii [36], the slow convergency of the iterative Brown process in practical problems was mentioned. To maximize function $F(y)$, they proposed to apply the method of feasible directions.

We return to problem (2.5) when the initial problem has the form (2.10). We can show that the functions $f_j(y_j)$, $j \in [1 : J]$ are concave and piecewise linear. The breakpoints correspond to the change of the basis in problems (2.12). See, for example, the book by G. Zoutendijk [72].

Consider the scheme of feasible directions in the simplest case, where some given values y_{j0}, $j \in [1 : J]$ correspond to unique optimal dual variables $\overset{\circ}{\lambda}_j$, $j \in [1 : J]$. By the First Duality Theorem 1.3 from Chapter 1, we have the equality

$$f_j(y_{j0}) = \sum_{s=1}^{m} \overset{\circ}{\lambda}{}_j^s y_{j0}^s + \psi_j. \tag{2.13}$$

The values y_j are taken in the form

$$y_{j1} = y_j(\rho) = y_{j0} + \rho z_j, \qquad j \in [1 : J], \tag{2.14}$$

where, for every $j \in [1 : J]$, z_j is an m-dimensional vector of feasible directions that satisfies the normalization condition, for example, of the form $-1 \leq z_j^s \leq 1$, $j \in [1 : J]$, $s \in [1 : m]$, and ρ is a nonnegative parameter. Due to the nondegeneracy, we can state that relation (2.13) holds in a neighborhood of the point $\rho = 0$.

We substitute (2.14) in (2.5) taking account of (2.13) and arrive at the following linear programming problem with respect to unknowns of feasible directions:

$$\begin{aligned} &\sum_{j=1}^{J}\sum_{s=1}^{m} \overset{\circ}{\lambda}{}_j^s z_j^s \to \max, \\ &\sum_{j=1}^{J} z_j^s \leq 0, \qquad s \in [1 : m], \\ &-1 \leq z_j^s \leq 1, \qquad j \in [1 : J], \qquad s \in [1 : m]. \end{aligned} \tag{2.15}$$

Problem (2.15) splits in m independent subproblems. For each s, let us have equalities

$$\overset{\circ}{\lambda}{}_{j_1}^s = \max_j \overset{\circ}{\lambda}{}_j^s, \qquad \overset{\circ}{\lambda}{}_{j_2}^s = \min_j \overset{\circ}{\lambda}{}_j^s.$$

Then, the optimal solution of problem (2.15) has the form

$$z_j^s = \begin{cases} -1 \text{ for } j = j_2, \\ 1 \text{ for } j = j_1, \\ 0 \text{ for } j \neq j_1, j_2, \end{cases} \qquad s \in [1 : m].$$

This optimal redistribution of resources has an obvious economic meaning: resources are taken from the subsystems, where they are least valuable and transferred to the subsystems, where they are most valuable.

Parameter ρ that determines the length of the step in the chosen direction is found in the form $\rho = \min\{\rho_1, \rho_2, \rho_3\}$, where ρ_1 is the maximal value of ρ that satisfies the constraint

$$\sum_{j=1}^{J} y_{10} + \rho \sum_{j=1}^{J} z_j \leq b^0,$$

the value ρ_2 is determined by maximization of the function

$$F(\rho) = \sum_{j=1}^{J} f_j(y_j(\rho))$$

under conditions

$$A_j^0 x_j \leq y_{j0} + \rho z_j, \qquad A_j x_j \leq b_j, \qquad x_j \geq 0, \qquad j \in [1:J].$$

Finally, ρ_3 is the maximal value of the parameter ρ, for which all the initial bases in problem (2.12) are retained.

We can generalize the described approach to the case of nonuniqueness of $\overset{\circ}{\lambda}_j$ and establish eventually that the sequential application of the scheme of feasible directions guarantees monotonic convergence to the extremum in a finite number of steps.

The method of feasible directions in the Kornai-Liptak scheme was applied for block-separable nonlinear problems in papers [13, 50, 71] and is presented in the monograph of L.S.Lasdon [22].

In [30], I. Mauer considers a convex block separable problem, and the distribution of resources is achieved by the method of penalty functions.

We have the following mathematical programming problem:

$$\begin{aligned} & f(x) = \sum_{j=1}^{J} f_j(x_j) \to \max, \\ & \sum_{j=1}^{J} g_j(x_j) \leq b_0, \\ & x_j \in M_x^j, \qquad j \in [1:J], \end{aligned} \tag{2.16}$$

where, for each $j \in [1:J]$, $x_j = (x_j^1, \dots, x_j^{n_j})$ is a point in Euclidean space $\mathbf{R}^{n_j}$, $g_j(x_j)$ is an m-vector function, b^0 is an m-vector, and M_x^j denote sets in $\mathbf{R}^{n_j}$. Moreover, we introduce the vector $x = (x_1, \dots, x_J)$ such that $x \in \prod_{j=1}^{J} \mathbf{R}^{n_j}$. We assume that M_x^j, $j \in [1:J]$ are convex closed bounded sets, and $f_j(x_j)$, $g_j(x_j)$, $j \in [1:J]$ are convex functions. We introduce two tuples that consist of J m-dimensional vectors $y = (y_1, \dots, y_J)$, and $w = (w_1, \dots, w_J)$.

The following problem is equivalent to (2.16):

$$
\begin{aligned}
& f(x) \to \max, \\
& g_j(x_j) \le y_j, \\
& x_j \in M_x^j, \qquad j \in [1:J], \\
& \|y - w\|^2 = 0, \qquad \sum_{j=1}^{J} w_j \le b^0, \qquad \|w\| \le C,
\end{aligned}
\tag{2.17}
$$

where the number C satisfies the condition

$$
C \ge \sum_{j=1}^{J} \max_{x_j \in M_x^j} \|g_j(x_j)\|^2.
$$

We introduce the following notation:

$$
M_{x,y}^j = \left\{ (x_j, y_j) | g_j(x_j) - y_j \le 0,\ x_j \in M_x^j, \|y_j\|_{R^m}^2 \le C \right\},
$$

$$
j \in [1:J],
$$

$$
M_w = \left\{ w \middle| \sum_{j=1}^{J} w_j \le b^0,\ (w_j^s)^2 \le C,\ j \in [1:J], s \in [1:m] \right\}.
$$

We choose a strictly increasing sequence of positive penalty constants $\{Q_l\}$ that tends to zero as $l \to \infty$. Consider the sequence of problems

$$
\begin{aligned}
& -f(x) + Q_l \|y - w\|^2 \to \min, \\
& (x, y, w) \in M_w \times \prod_{j=1}^{J} M_{x,y}^j.
\end{aligned}
\tag{2.18}
$$

To solve problems (2.18) for each fixed l, we need to apply the method of componentwise ascent, which, in our case, looks as follows.

By $(x^{(l)}, y^{(l)}, w^{(l)})$, denote optimal solutions of problem (2.18) for all l. We fix $w^{(l-1)}$ (for $l = 1$, w^0 is chosen to satisfy the last constraints (2.17)). For a given $w^{(l-1)}$, we minimize the functional in (2.18), and the solution $(x^{(1,l)}, y^{(1,l)}$ is sought on the product $\prod_{j=1}^{J} M^j_{x,y}$. In this case, problem (2.18) is partitioned into J local problems:

$$\begin{aligned} &-f_j(x_j) + Q_l||y_j - w_j^{l-1)}||^2 \to \min, \\ &(x_j, y_j) \in M'_{x,y}, \end{aligned} \tag{2.19}$$

which are convex programming problems. After that, we fix $x^{(1,l)}, y^{(1,l)}$ and look for solution $w^{(1,l)} \in M_w$, which minimizes the functional in (2.18). The obtained problem is partitioned into m problems:

$$\begin{aligned} &\sum_{j=1}^{J} ||y_j^{s(1,l)} - w_j^s||^2_{R^m} \to \min, \\ &\sum_{j=1}^{J} w_j^s \le b_s^0, \qquad (s_j^w)^2 \le C, \qquad j \in [1:J], \end{aligned} \tag{2.20}$$

which are convex quadratic programming problems (2.20). Then, we fix $w^{(1,l)}$ and define $x^{(2,l)}$, $y^{(2,l)}$, and so on. In [30], it was shown that, from the obtained sequence, we can select a subsequence $(x^{(l)}, y^{(l)}, w^{(l)})$ that converges to the optimal solution of problem (2.18). Then, it turns out that the sequence $R(x^{(l)}, \bar{M}_x)$ tends to zero as $l \to \infty$. Here, $R(x^{(l)}, \bar{M}_x)$, as before, means the distance from the point $x^{(l)}$ to the set $\bar{M}_x$. Another approach connected to the application of penalty functions is studied in [61].

In [43], J. Sanders proposes to use Lagrange multipliers for the organization of the descent by gradient method in the solution of the coordinating problem by the Kornai-Liptak decomposition principle. The following minimization problem is considered in this paper:

$$\begin{aligned} &f_1(x_1) + f_2(x_2) \to \min, \\ &g_k^1(x_1) \ge b_k^1, \qquad k \in [1:K_1], \\ &g_k^2(x_2) \ge b_k^2, \qquad k \in [1:K_2], \\ &g_1^{(1)}(x_1) + g_2^{(1)}(x_2) \ge b^{(1)}, \\ &g_1^{(2)}(x_1) + g_2^{(2)}(x_2) \ge b^{(2)}, \end{aligned} \tag{2.21}$$

where x_1 and x_2 are vectors of dimensions n_1 and n_2 lying in Euclidean spaces $\mathbf{R}^{n_1}$ and $\mathbf{R}^{n_2}$, respectively, and all functions have first continuous partial derivatives. According to the Kornai-Liptak principle, the local problems are written as follows:

$$\begin{aligned}
&f_1(x_1) + \lambda_1^{(1)}\left[y_1^{(1)} - g_1^{(1)}(x_1)\right] + \lambda_1^{(2)}\left[y_1^{(2)} - g_1^{(2)}(x_1)\right] \to \min,\\
&g_k^1(x_1) \geq b_k^1, \qquad k \in [1:K_1],\\
&g_1^{(1)}(x_1) = y_1^{(1)},\\
&g_1^{(2)}(x_1) = y_1^{(2)};
\end{aligned} \tag{2.22}$$

$$\begin{aligned}
&f_2(x_2) + \lambda_2^{(1)}\left[y_2^{(1)} - g_2^{(1)}(x_2)\right] + \lambda_2^{(2)}\left[y_2^{(2)} - g_2^{(2)}(x_2)\right] \to \min,\\
&g_k^2(x_2) \geq b_k^2, \qquad k \in [1:K_2],\\
&g_2^{(1)}(x_2) = y_2^{(1)},\\
&g_2^{(2)}(x_2) = y_2^{(2)}.
\end{aligned} \tag{2.23}$$

Let $x_1^*(y_1^{(1)}, y_1^{(2)})$, $f_1(x_1^*)$ and $x_2^*(y_2^{(1)}, y_2^{(2)})$, $f_2(x_2^*)$ be optimal solutions of problems (2.22) and (2.23), respectively. If we set $f_1^*(y_1^{(1)}, y_1^{(2)}) = f(x_1^*)$, $f_2^*(y_2^{(1)}, y_2^{(2)}) = f_2(x_2^*)$, then the coordinating problem takes the form

$$\begin{aligned}
&f_1^*(y_1^{(1)}, y_1^{(2)}) + f_2^*(y_2^{(1)}, y_2^{(2)}) \to \min,\\
&y_1^{(1)} + y_2^{(1)} \geq b^{(1)},\\
&y_1^{(2)} + y_2^{(2)} \geq b^{(2)}.
\end{aligned} \tag{2.24}$$

The following statements establish some properties of the initial problem as well as coordinating and local problems.

Lemma 2.1. *The following relations hold for partial derivatives:* $\partial f_1^*(y_1^{(1)}, y_1^{(2)})/\partial y_1^{(i)} = \lambda_1^{(i)}$, $\partial f_2^*(y_2^{(1)}, y_2^{(2)})/\partial y_2^{(i)} = \lambda_2^{(i)}$, $i = 1, 2,$, *if these derivatives exist.*

Lemma 2.2. *Let $f(\bar{x})$ be a convex monotone nondecreasing function and let $\bar{x}$ be a vector function, each component of which is a convex function of y_1 and y_2. Then, $f(\bar{x}(y_1, y_2))$ is a convex function of y_1 and y_2.*

Lemma 2.3. *Let $\bar{\varphi}(\bar{x}) = \bar{y}$ for every $\bar{y} \in \mathbf{R}^n$ and $\bar{x}$ such that $\bar{\varphi}$ is a vector function of the vector $\bar{x} \in \mathbf{R}^n$. Assume that the Jacobian is nonzero:*

$\partial(\bar\varphi_1, \ldots, \bar\varphi_n)/\partial(x_1, \ldots, x_n)$. Then, if $\bar\varphi$ is a convex function monotone nondecreasing with respect to $\bar x$, the inverse function $\bar\varphi^{-1}$ is convex and monotone nondecreasing with respect to $\bar y$.

Theorem 2.1. *The minimal value of the functional of initial problem (2.21) coincides with the minimal value of the functional of the coordinating problem (2.24).*

Theorem 2.2. *If $f_i(x_i)$, $i = 1, 2$ are convex monotone nondecreasing functions, and $g_k^i(x_i)$, $g_i^{(i)}(x_1)$, $g_2^{(i)}(x_2)$, $i = 1, 2$ are convex nondecreasing and nonnegative functions, then the coordinating problem (2.24) is the problem of convex programming.*

As before, to solve problem (2.24), we need to know the optimal solutions of local problems. In contrast, according to Lemma 2.1, we can use Lagrange multipliers that give gradients of the functions f_1^* and f_2^* in order to organize iterations in minimizing the functional in (2.24). A possible iterative procedure consists of the following stages.

1. Problems (2.22), (2.23) are solved for some fixed values $y_1^{(1)}[0]$, $y_1^{(2)}[0]$, $y_2^{(1)}[0]$, $y_2^{(2)}[0]$ that satisfy inequalities of problem (2.24). As a result, we find Lagrange multipliers $\lambda_1^{(1)}[0]$, $\lambda_1^{(2)}[0]$, $\lambda_2^{(1)}[0]$, $\lambda_2^{(2)}[0]$.

2. For values $\lambda_1^{(1)}[l]$, $\lambda_1^{(2)}[l]$, $\lambda_2^{(1)}[l]$, $\lambda_2^{(2)}[l]$, we assume

$$y_1^{(i)}[l+1] = y_1^{(i)}[l] - \beta(\lambda_1^{(i)}[l] - \lambda_2^{(i)}[l]),$$

$$y_2^{(i)}[l+1] = y_2^{(i)}[l] + \beta(\lambda_i^{(i)}[l] - \lambda_2^{(i)}[l]), \qquad i = 1, 2,$$

where β is the number that determines the step size.

3. Problems (2.22), (2.23) are solved for the values $y_1^{(i)}[l+1]$, $y_2^{(i)}[l+1]$, $i = 1, 2$. Thereafter, the values $\lambda_1^{(i)}[l+1]$, $\lambda_2^{(i)}[l+1]$ are determined.

4. l is replaced by $l+1$, and we go to stage 2.

Iterations are made until the inequalities $\left|\lambda_1^{(i)}[l] - \lambda_2^{(i)}[l]\right| < \varepsilon^i$ are satisfied, where ε^i, $i = 1, 2$ is a given precision. It is clear that the considered scheme is applicable when the number of subsystems is greater than two.

In the paper [39] by V.S. Razumikhin, decomposition by the Kornai-Liptak principle is based on the method of physical modelling. Alignment of dual estimates is also used.

The idea about redistribution of resources is fruitful in constructing of special decomposition algorithms for the solution of some problems with block constraints. The specificity of particular problems results in techniques of distribution of resources that differ from schemes described earlier. The problems considered formalize the models of decentralized control. On the one hand, the proposed techniques of distribution of general resources correspond to planning practice. On the other hand, these problems are related to possibly simple formulation of concrete control systems. These methods are often heuristic, since, for example, the convergence of the intermediary solutions to the real optimum of the original problem is not proved. However, they are surely valuable due to their numerical support of convergence and their relevance to particular control processes. Further, we consider one such model from the paper [29] by V.D. Marshak.

In problem (2.10), we introduce a special criterion, adding new constraints and variables. We obtain

$$
\begin{aligned}
&\theta \to \max, \\
&\sum_{j=1}^{J} A_j^0 x_j \le b^0, \\
&A_j x_j \le b_j, \qquad x_j \ge 0, \\
&H_j x_j - d_j \theta \ge 0, \qquad \theta \ge 0, \qquad j \in [1:J],
\end{aligned}
\tag{2.25}
$$

where, for each $j \in [1:J]$, d_j is an m_j^1-dimensional column vector, H_j is an $m_j^1 \times n_j$-matrix, and θ is scalar. All other dimensions remain the same as in (2.10). For definiteness, we assume that the matrix entries are nonnegative, and vector components are strictly positive.

Problem (2.25) describes the branch planning model that consists of J subbranches with specific products. The vectors d_j, $j \in [1:J]$ denote finite products of subbranches, and the value θ is the proportionality coefficient, which characterizes the amount of product of all subbranches. The first condition (2.25) corresponds to the consumption of general branch resources. The second condition in (2.25) is related to the presence of subbranch resources. Coefficients of matrices H_j give technological means of subbranches. Optimal solution of problem (2.25), from the standpoint of the branch, corresponds to the distribution of general resources that ensures the highest level of the final product of that branch.

In real models of branch planning, the original system consists of a large number of subbranches that include large amounts of variables, so that the direct solution of problem (2.25) is often impossible.

At each iteration, the proposed decomposition results in solving the following local problems:

$$\begin{aligned} &\theta_j \to \max, \\ &A_j^0 x_j \le y_j, \\ &A_j x_j \le b_j, \qquad x_j \ge 0, \\ &H_j x_j - d_j\theta_j \ge 0, \qquad \theta_j \ge 0 \end{aligned} \tag{2.26}$$

under conditions

$$\sum_{j=1}^{J} y_j \le b^0. \tag{2.27}$$

Here, the levels θ_j, $j \in [1 : J]$ that correspond to each subbranch are introduced.

At each step, the iterative method redistributes the general resource b^0. Here, all optimal values tend to be aligned at the maximal level of the original problem (2.5) in the iterative process. In this case, the coordination corresponds to some rules of resource redistribution, which are given below.

The algorithm consists of the following main steps:

1. Solving problems (2.26) begins for some fixed resources $y_i[1]$, $j \in [1 : J]$ that satisfy constraint (2.27).

2. Let $\theta_j[1]$, $j \in [1 : J]$ be obtained optimal levels, where the minimal one is obtained for some p. The values $\theta_p[1]$ that correspond to solutions $x_j[1]$ of problems (2.26) are admissible to the original problem (2.25). To obtain a higher minimal level and to align the levels, the resources from subbranches $j \in [1 : J]$, $j \neq p$ are allocated in accordance with the excess of $\theta_j[1]$ over $\theta_p[1]$.

3. The allocated resources are determined over all subbranches taking into account resource importance.

4. The process is repeated until the criterion

$$\left(\max_{j\in[1:J]} \theta_j[l] - \min_{j\in[1:J]} \theta_j[l]\right) \Big/ \min_{j\in[1:J]} \theta_j[l] \le \varepsilon \tag{2.28}$$

is satisfied, where l is the number of iteration, ε is a given precision for the dispersion of maximal and minimal values of the levels.

Redistribution of the allocated resources results in a strictly monotone increasing sequence of values $\min\limits_{j\in[1:J]} \theta_j[l]$.

We describe the solution in more detail.

Consider the first iteration step ($l = 1$). Assume that the maximal possible amount of consumption of general resources over all branches and for all s, $s \in [1 : m]$ are known. These resources w_j^s correspond to the complete load of industrial capacities of the subbranch. The first constraints (2.26) are satisfied as strict inequalities. The initial distribution of resources is determined in the form

$$y_j^s[1] = b_s^0 \left(w_j^s \Big/ \sum_{j=1}^{J} w_j^s \right), \qquad j \in [1 : J], \qquad s \in [1 : m].$$

To obtain the values $y_j^s[1]$, we need to solve all problems (2.26).

Consider step number l ($l \geq 2$). We have $y_j^s[l]$, $\theta_j[l]$, $\lambda_j^s[l]$, where λ_j^s are optimal dual variables that correspond to the first constraints (2.26). We test criterion (2.28). If the termination condition for iterations is not satisfied, the new vector of general resources is introduced by the formula

$$\bar{y}_j^s[l] = y_j^s[l] \left(\theta_p[l]/\theta_j[l]\right), \qquad j \in [1 : J], \qquad s \in [1 : m]. \tag{2.29}$$

Here, as before, $\theta_p[l]$ is the minimal level attained at the l-th iteration.

Since $\theta_p[l]/\theta_j[l] \leq 1$, the vectors $\bar{y}_j[l]$ satisfy condition (2.27).

The resources are redistributed for a certain r that corresponds to the largest dispersion of dual estimates λ_j^s. If we solve problems (2.26) for the fixed values $y_j^r = \bar{y}_j^r[l]$, $y_j^s = y_j^s[l]$, $s \neq r$, then the inequality $\bar{\theta}_{p_1} \geq \theta_p[l]$ holds for the obtained minimal level $\bar{\theta}_{p_1}$. We can show that the optimal values $\theta_j[l]$ for all $j \in [1 : J]$ are piecewise linear concave functions of $y_j^r[l]$, where, for $y_j^r[l] = 0$, we have $\theta_j[l] = 0$. Therefore, for each $j \in [1 : J]$, the graph of the function $\theta_j[l] = \theta_j(y_j^r[l])$ lies above the straight line that connects the point $(y_j^r[l]; \theta_j[l])$ and the origin $(0; 0)$.

We take the abscissa $\bar{y}_j^r$, $j \neq p$. It corresponds to the ordinate on the straight line equal to $\theta_p[l]$. The value of this ordinate is not greater than that computed by the analytic expression $\theta_j(\bar{y}_j^r[l])$. On the other hand, denote by $\hat{y}_j^r[l]$ the resource that corresponds to the minimal level $\theta_p[l]$ according to the analytical dependence $\theta_j[l] = \theta_j(y_j^r[l])$. Then, we have $\hat{y}_j^r[l] \leq \bar{y}_j^r[l]$, i.e., for each $j \in [1 : J]$, $j \neq p$, more resources than necessary are given at each iteration to attain the previous minimal level. Since the allocated resources will be redistributed among all subbranches, including $j = p$, we obtain the following strict inequality at the next iteration:

$$\min_{j \in [1:J]} \theta_j[l] < \min_{j \in [1:J]} \theta_j[l+1].$$

Here, the values $\theta_j[l+1] - \theta_p[l+1]$, $j \in [1:J]$ will decrease, and if the sequence $\min\limits_{j\in[1:J]} \theta_j[l]$ converges, then

$$\lim_{l\to\infty} (\bar{y}_j^s[l] - \hat{y}_j^s[l]) = 0, \qquad j \in [1:J], \qquad s \in [1:m].$$

The latter denotes that the error in the definition of $\hat{y}_j^s$, necessary to ensure the minimal level, decreases after the sufficiently large number of iterations.

We consider the determination of excess resources in subbranches. The excess is computed in the form

$$\Delta b_r[l] = b_r^0 - \sum_{j=1}^{J} \bar{y}_j^r[l].$$

The value $\Delta b_s[l]$ is distributed over subbranches inverse proportionally to the estimates $\bar{\lambda}_j^r[l]$, $j \in [1:J]$ that correspond to resources $y_j^r = \hat{y}_j^r[l]$, $y_j^s = y_j^s[l]$, $s \neq r$. This is made to align the levels of all these subbranches.

To avoid double counting, we propose interpolation for computing the variables $\bar{\lambda}_j^r[l]$ that correspond to the new values $\bar{y}_j^r[l]$. We have a piecewise linear dependence $\bar{\lambda}_j^r[l] = \bar{\lambda}_j^r(\bar{y}_j^r[l])$. For the values $y_j^r[l]$, the values λ_j^r are known, and, for w_j^r, they are equal to zero. According to the formula of linear interpolation over two points, we obtain

$$\bar{\lambda}_j^r[l] = \lambda_j^r[l](w_j^r - \bar{y}_j^r[l])/(w_j^r - y_j^r[l]). \tag{2.30}$$

Since the value $\max\limits_{j\in[1:J]} \theta_j[l] - \min\limits_{j\in[1:J]} \theta_j[l]$ decreases on the iterations, the differences $y_j^r[l] - \bar{y}_j^r[l]$ become less. If these differences tend to zero as $l \to \infty$, then the equality

$$\lim_{l\to\infty} \left(\bar{\lambda}_j^r[l] - \lambda_j^r[l]\right) = 0, \qquad j \in [1:J]$$

holds due to (2.29), (2.30). Upon determination of the estimates $\bar{\lambda}_j^r[l]$, the resources are redistributed over the subbranches:

$$\Delta y_j^r[l] = \Delta b_r[l](1/\bar{\lambda}_j^r[l]) \left(\sum_{j=1}^{J} 1/\lambda_j^r[l]\right).$$

The resources on the $l+1$-th step are determined in the form

$$y_j^r[l+1] = y_j^r[l] + \Delta y_j^r,$$

$$y_j^s[l+1] = y_j^s[l] \text{ for } s \neq r, \qquad j \in [1:J].$$

The algorithm forms a strictly monotone increasing sequence of values $\min_{j\in[1:J]} \theta_j[l]$. This sequence is bounded due to the technological constraints of problems (2.26). Therefore, it has the limit $\hat{\theta}$.

In [29], it was not proved that $\hat{\theta}$ is the optimal solution to the original problem (3.1). However, the series of experimental computations supports this conclusion.

The advantage of this method is essential simplification of the coordinating part. Some information about the levels θ_j, $j \in [1 : J]$ and the estimates of distributed resources λ_j^s, $j \in [1 : J]$, $s \in [1 : m]$ i.e., altogether $J(1+m)$ numbers, give the control system. The redistribution is accomplished using simple formulas. Here, we have the possibility of solving problems of practically any dimension.

In [18], by A. ten Kate, the drawbacks of the applying game procedure in the Kornai-Liptak decomposition method were indicated.

It is stressed that the use of sufficiently large bounds $\bar{\lambda}$, $\bar{y}$ in problem (2.11) results in a slow convergence of the algorithm. Moreover, this scheme gives an approximate solution. To obtain a precise solution of the parametric problem, we propose to apply the Dantzig-Wolfe principle for a dual problem, which has block structure with special binding constraints. We present the main constructions of this approach.

Return to problem (2.2) and, using the Kornai-Liptak conception, we write it in the following form:

$$
\begin{aligned}
&\sum_{j=1}^{J}(c_j, x_j) \to \max, \\
&A_j^0 x_j - y_j \le 0, \\
&\sum_{j=1}^{J} y_j \le b^0, \\
&x_j \ge 0, \qquad j \in [1 : J].
\end{aligned}
\tag{2.31}
$$

Consider the values y_j in (2.31) as unknowns. We formulate the dual for problem (2.31), where the vectors λ_j, $j \in [1 : J]$, as before, correspond to block conditions (2.31), and the vector μ is introduced for the second constraint in (2.31). We have

$$
\begin{aligned}
&(\mu, b^0) \to \min, \\
&\lambda_j A_j^0 \geq c_j, \\
&\lambda_j - \mu = 0, \\
&\lambda_j \geq 0, \qquad j \in [1:J], \\
&\mu \geq 0.
\end{aligned}
\tag{2.32}
$$

Problem (2.32) has block diagonal structure. We propose to solve problem (2.32) by the Dantzig-Wolfe decomposition method. Here, the constraints $\lambda_j A_j^0 \geq c_j$, $j \in [1:J]$ are considered as block constraints, and the equalities $\lambda_j - \mu = 0$, $j \in [1:J]$ give binding constraints. Due to the specificity of the binding constraints in problem (2.32), the corresponding transformations by the Dantzig-Wolfe scheme have some specificity and allow simplifications.

In paper [59] by S.G. Timokhin, upon the decomposition of problem (2.2) in form (2.4), i.e.,

$$
\begin{aligned}
&f_j = (c_j, x_j) \to \max, \\
&A_j^0 x_j \leq y_j, \\
&x_j \geq 0,
\end{aligned}
\tag{2.33}
$$

the following coordinating linear programming problem is introduced:

$$
\begin{aligned}
&\Psi = (f_j, z) \to \max, \\
&Yz \leq b^0,
\end{aligned}
\tag{2.34}
$$

where Y is a $m \times J$-matrix whose columns are vectors y_j, $j \in [1:J]$; $f = (f_1, \ldots, f_J)$ is the vector of optimal values of functionals of subproblems (2.33). Finally, z is a J-dimensional nonnegative vector of variables of problem (2.34).

We state the problem of finding a matrix Y^*, for which there exists solutions of subproblems (2.33) and the coordinating problem (2.34), where the value Ψ of the functional (2.34) is maximum.

Let $\hat{x}_j$, $j \in [1:J]$ be optimal solutions of subproblems (2.33) and $\hat{z} = (\hat{z}_1, \ldots, \hat{z}_J)$ be an optimal solution of problem (2.34). Then, the vector $\hat{x}_z = (\hat{z}_1 \hat{x}_1, \ldots, \hat{z}_J \hat{x}_J)$ is an admissible solution of the original problem (2.2).

Our goal is to find a matrix Y such that $\hat{x}_z$ is an optimal solution of problem (2.2). It should be noted that parameters y_j, $j \in [1 : J]$ play a role different from that in the traditional Kornai-Liptak decomposition scheme. We can say that, modulo coefficients, the vectors y_j, $j \in [1 : J]$ are distributed resources. The values z_j, $j \in [1 : J]$ play the role of these coefficients.

We formulate the dual problems for (2.33):

$$\begin{aligned} &(y_j, \lambda_j) \to \min, \\ &\lambda_j A_j^0 \geq c_j, \\ &\lambda_j \geq 0 \end{aligned} \tag{2.35}$$

and for (2.34):

$$\begin{aligned} &(b^0, \lambda^0) \to \min, \\ &\lambda^0 Y \geq f, \\ &\lambda^0 \geq 0. \end{aligned} \tag{2.36}$$

Let some initial values $y_j = y_j(0)$, $j \in [1 : J]$ be given. Further, we give a rule for their transformation. According to this rule, the value of the functional Ψ of problem (2.34) monotonically increases. For simplicity, we assume that the original tuple $y_j(0)$, $j \in [1 : J]$ corresponds to single optimal solutions $\hat{x}_j$, $\hat{\lambda}_j$ and $\hat{z}$, $\hat{\lambda}^0$ of conjugated problems (2.33), (2.35) and problems (2.34), (2.36), respectively, and that the nondegenerate bases correspond to the optimal solutions. Subsystem with number j, where y_j occurs in the optimal basis of problem (2.34), will be called *the basis subsystem.* We vary the constraint vector of this subsystem by the following rule:

$$y_j(\rho) = y_j(0) + \rho \delta b_j, \tag{2.37}$$

where the sought for direction δb_j is defined below, and ρ is a positive parameter. Due to the assumed nondegeneracy, we can indicate the neighborhood of the point $\rho = 0$, where the numbers of optimal bases of problems (2.33) and (2.34) are retained. In this neighborhood, we have the following dependence of the optimal value $\hat{f}_j$ on ρ:

$$\hat{f}_j(\rho) = f_j + \rho \delta f_j, \tag{2.38}$$

where

$$\delta f_j = (\delta b_j, \hat{\lambda}_j). \tag{2.39}$$

We study the behavior of the optimal value of the functional of problem (2.34) as the function of ρ. Denote by $\hat{Y}(0)$ the matrix of the optimal basis for (2.34), and let $\hat{Y}_j(\rho)$ be the same matrix with the j column transformed by rule (2.37). Due to the First Duality Theorem 1.3 from Chapter 1 and to the definition of the basis, we have

$$\Psi(\rho) = (b^0, \hat{\lambda}^0(\rho)), \tag{2.40}$$

where

$$\hat{\lambda}^0(\rho) = \hat{f}(\rho)\hat{Y}_j^{-1}(\rho). \tag{2.41}$$

We represent $\hat{Y}_j(\rho)$ in the form $\hat{Y}_j(\rho) = \hat{Y}(0) + \rho\delta Y_j$, where δY_j is a matrix with the j column δb_j, and the remaining columns are zero. We introduce the vector γ such that $\hat{Y}(0)\gamma = \delta b_j$. Then, we can introduce the equality

$$\hat{Y}_j^{-1}(\rho) = \hat{Y}^{-1}(0) - \frac{\rho}{1+\rho\gamma_j}\gamma\hat{\beta}_j, \tag{2.42}$$

where γ_j is the j-th coordinate of the vector γ, $\hat{\beta}_j$ is the jth row of matrix $\hat{Y}^{-1}(0)$. Substituting (2.42) into (2.41) with due regard to (2.38), (2.39), we obtain

$$\hat{\lambda}^0(\rho) = \hat{\lambda}^0 + \frac{\rho h_j(0)}{1+\gamma_j\rho}\bar{\beta}_j, \tag{2.43}$$

where

$$h_j(0) = \delta f_j - (\delta b_j, \hat{\lambda}^0). \tag{2.44}$$

Relations (2.40), (2.43) imply that

$$\Psi(\rho) = \Psi(0) + \frac{\rho h_j(0)}{1+\gamma_j\rho}(\hat{\beta}_j, b^0),$$

where $\Psi(0) = (\hat{\lambda}^0, b^0)$. Taking into account $(\hat{\beta}_j, b^0) = \hat{z}_j(0)$, we finally obtain

$$\Psi(\rho) = \Psi(0) + \frac{\rho h_j(0)}{1+\gamma_j\rho}\hat{z}_j(0).$$

By differentiating this equality with respect to ρ, we have

$$\left.\frac{d\Psi(\rho)}{d\rho}\right|_{\rho=0} = \frac{h_j(0)\hat{z}_j(0)}{(1+\gamma_j\rho)^2}.$$

The latter relation holds in a neighborhood of the point $\rho = 0$. The sign of the derivative depends on $h_j(0)$, since $\hat{z}_j(0) > 0$. Relations (2.44), (2.40) imply that

$$h_j(0) = (\delta b_j, \hat{\lambda}_j - \hat{\lambda}^0),$$

therefore, the best direction of the variation of δb_j is $(\hat{\lambda}_j - \lambda_j^0)$. Thus, a change of the constraint y_j in the direction of the difference of the dual estimates of the local and coordinating problems results in an increase of the functional of the latter. The case of the nonuniqueness of optimal dual variables is studied by scheme [59].

In [45], B. Schwartz and P. Tichatschke proposed a method also based on the Kornai-Liptak decomposition principle. Upon the partitioning into columns, we have the following problem:

$$\begin{aligned} &(c_1, x_1) + (c_2, x_2) \to \max, \\ &A_1 x_1 + A_2 x_2 \le b^0, \\ &x_1 \ge 0, \qquad x_2 \ge 0, \end{aligned} \tag{2.45}$$

where the matrices and vectors have compatible dimensions. We introduce the vectors y_1 and y_2 such that $y_1 + y_2 \le b^0$ and obtain subproblems

$$\begin{aligned} &(c_1, x_1) \to \max, \\ &A_1 x_1 \le y_1, \\ &x_1 \ge 0; \end{aligned} \tag{2.46}$$

$$\begin{aligned} &(c_2, x_2) \to \max, \\ &A_2 x_2 \le y_2, \\ &x_2 \ge 0. \end{aligned} \tag{2.47}$$

The proposed procedure is constructed in the following way. Let the vector $x_2^{(0)}$ be an optimal solution of subproblem (2.47) with the goal function $f_2^{(0)} = \left(c_2, x_2^{(0)}\right)$. Then, we assume $y_2^{(0)} = A_2 x_2^{(0)}$ and solve the following problem:

$$\begin{aligned} &f_1(x_1, \omega_2) = (c_1, x_1) + \omega_2 f_2^{(0)} \to \max, \\ &A_1 x_1 + \omega_2 y_2^{(0)} \le b^0, \\ &x_1 \ge 0, \qquad \omega_2 \ge 0. \end{aligned} \tag{2.48}$$

Let the optimal solution of this problem be $\left(x_1^{(1)}, \omega_2^{(1)}\right)$. We assume

$f_1^{(1)} = \left(c_1, x_1^{(1)}\right)$, $y_1^{(1)} = A_1 x_1^{(1)}$ and solve the problem

$$
\begin{aligned}
& f_2(x_2, \omega_1) = (c_2, x_2) + \omega_1 f_1^{(1)} \to \max, \\
& A_2 x_2 + \omega_1 y_1^{(1)} \le b^0, \qquad\qquad (2.49)\\
& x_2 \ge 0, \qquad \omega_1 \ge 0.
\end{aligned}
$$

If the optimal solution of problem (2.49) is $\left(x_2^{(2)}, \omega_1^{(2)}\right)$, then we introduce the following values: $f_2^{(2)} = \left(c_2, x_2^{(2)}\right)$, $y_2^{(2)} = A_2 x_2^{(2)}$. Then, we solve again the problem (2.48), where $f_2^{(0)} = f_2^{(2)}$ and $y_2^{(0)} = y_2^{(2)}$. Analogously, we come to problem (2.49) and so on.

We can see that the sequence of optimal values of the functionals of problems (2.48), (2.49) is monotonically nondecreasing. The criterion for termination of the iterations is equality of the goal functions for two sequential problems. The optimal solution of the original problem (2.45) is attained in a finite number of steps. The method allows us to construct the sequence of admissible solutions of problem (2.45). If we have the optimal solution $\left(x_1^{(l)}, \omega_2^{(l)}\right)$ of problem (2.48) at the l-th step and the same solution $\left(x_2^{(l-1)}, \omega_1^{(l-1)}\right)$ for (2.49) at the $l-1$-th step, then the vector $\left(x_1^{(l)}, \omega_2^{(l)}, x_2^{(l-1)}\right)$ is not an admissible solution to problem (2.45). The scheme obviously generalizes to the case of partitioning the initial matrix into an arbitrary number of submatrices. The method is efficient for the block diagonal structure of the constraint matrix.

§3. Parametric Decomposition

This approach proposed in papers [23, 24] by G.M. Levin and V.S. Tanaev is based on the introduction of special variables (parameters). It generalizes several well-known decomposition methods.

In this section, we consider decomposition methods based on the introduction of special variables, i.e., parameters. We study decomposition transformations that allow us, instead of solving the initial problem, to solve a number of interrelated problems.

Consider the problem of mathematical programming

$$\begin{aligned} &f(x) \to \min, \\ &g(x) \le 0, \\ &x \in M_x^0, \end{aligned} \tag{3.1}$$

where M_x^0 is a set in $\mathbf{R}^n$, the function $f(x)$ and the m-dimensional vector function $g(x)$ are assumed to be continuous.

We introduce the p-dimensional vector of parameters y and the set M_y^0 in $\mathbf{R}^p$ as well as the function $\tilde{f}(x, y)$ and the m-dimensional vector function $\tilde{g}(x, y)$ defined on the product $M_x^0 \times M_y^0$. Problem (3.1) is replaced by the following one:

$$\begin{aligned} &\tilde{f}(x, y) \to \min, \\ &\tilde{g}(x, y) \le 0, \\ &(x, y) \in M_x^0 \times M_y^0. \end{aligned} \tag{3.2}$$

Here, it is assumed that the introduced function has the following property: if (x^*, y^*) is the solution of problem (3.2), then x^* is the solution of the original problem (3.1), where $\tilde{f}(x^*, y^*) = f(x^*)$..

Problem (3.2) is solved by means of the two-level scheme. The coordinating problem has the following form:

$$\begin{aligned} &F(y) \to \min, \\ &y \in M_y, \end{aligned} \tag{3.3}$$

where M_y is a subset of M_y^0, and $F(y)$ is the optimal value of the goal function of the local problem

$$\begin{aligned} &\tilde{f}(x, y) \to \min, \\ &\tilde{g}(x, y) \le 0, \\ &(x, y) \in M_x^0 \times M_y^0 \end{aligned} \tag{3.4}$$

for a fixed value of y. We denote the solution of problem (3.4) by $x^*(y)$.

Procedures based on the Kornai-Liptak decomposition principle are particular cases of parametric decomposition. Let the functions $f(x)$, $g(x)$ be separable over the groups of variables $x_1, \ldots, x_J$, i.e.,

$$f(x) = \sum_{j=1}^{J} f_j(x_j), \qquad g(x) = \sum_{j=1}^{J} g_j(x_j),$$

$x_j \in M_x^j \subset \mathbf{R}_j^n$, $j \in [1:J]$. We put $M_x^0 = \prod M_x^j$, then the set M_y is defined in the form

$$M_y^0 = M_y = \left\{ y = (y_1, \ldots, y_J) \,\middle|\, y_j \in M_y^j,\ \sum_{j=1}^{J} y_j = 0 \right\},$$

where $M_y^j = \left\{ y_j \in \mathbf{R}^m \,\middle|\, (\exists \bar{x}_j \in M_x^j)(\tilde{g}_j(\bar{x}_j, y_j) \leq 0) \right\}$, $\tilde{g}_j(x_j, y_j) = g_j(x_j) - y_j$. The function $\tilde{f}(x,y)$ and the vector function $\tilde{g}(x,y)$ are defined as $f(x)$ and $(\tilde{g}_1(x_1, y_1), \ldots, \tilde{g}_J(x_J, y_J))$, respectively, and the parameters do not occur in the goal function. Here, local problem (3.4) is decomposed into J independent problems with dimensions $m \times n_j$, $j \in [1:J]$.

As a rule, the parametrization is realized in a way that problem (3.4) is simply decomposed into autonomous problems of lesser dimension. However, wherever possible, we need to have a simple goal function $F(y)$ of problem (3.3) and its domain. These two tendencies often trade-off against each other.

The approach of G.M. Levin, V.S. Tanaev proposes a formal scheme for the construction of methods of parametric decomposition. This formalism is based on the study of a number of particular problems of optimal design. Here, most attention is drawn to the question of the simplicity of the function $F(y)$ and its domain of application as compared to the same properties of the goal function of the original problem. We study this scheme in more detail.

First, we need to know the conditions, under which the solution of problem (3.1) can be obtained by means of the two-level optimization scheme. Let $\omega(x)$ be a one-to-one continuous mapping of the set M_x into the set M_y^0. By $M_x(y)$, we denote the set of admissible solutions of problem (4.4). By definition, $M_x(y) = \{x \in M_x^0 \,|\, \tilde{g}(x,y) \leq 0\}$. We require satisfaction of the following conditions:

Condition 1. $\tilde{f}(x, \omega(x)) = f(x)$ for all $x \in M_x$.

Condition 2. $M_x(y) \neq \varnothing$ for all $y \in M_y^0$, and if $x \in M + x(y_1)$, then $\tilde{f}(x, y_1) \leq f(x_1)$.

Condition 3. If $x_1 \in M_x$, $\omega(x_1) = y_1$, then there exists $x^*(y^*)$ such that $\tilde{f}(x, y_1) \leq f(x_1)$.

Then, we have the following statement.

Theorem 3.1. *If y^* is a solution of problem (3.3), then $x^*(y^*)$ is a solution of the original problem (3.1). If x^* is a solution of problem (3.1), then $\omega(x^*)$ is a solution of problem (3.3), where $f(x^*) = F(y^*)$.*

Thus, the solution of problem (3.1) can be obtained by solving problem (3.3).

The search for solution y^* of problem (3.3) is usually related to the computation of the values of function $F(y)$ at some points of the set M_y. This, in turn, requires the solution of problem (3.4) under fixed values of the vector parameter y. The solution complexity of problem (3.4) depends on the form of the function $\tilde{f}$ and $\tilde{g}$. In most well-known schemes of parametric decomposition, formation of these functions is determined by the choice of mapping $\omega(x)$. Under an appropriate choice of functions $\tilde{f}$, $\tilde{g}$, problem (3.4) can be essentially simpler than the original one.

The difficulty in solving problem (3.3) depends on the form of the set M_y. As a rule, M_y is chosen as a closed set, where each connected component contains one or several connected components of the set $\omega(M_x)$. In this case, the number of connected components of the set M_y does not exceed the same characteristic of the set $\omega(M_x)$ and, thus, the set M_x.

Behavior of the goal function is important for effective application of one or another search procedure for solutions of mathematical programming problems. The number of so-called stationary and local domains of problems (3.1) and (3.4) is essential.

The connected set $M_\Psi^a \subset M$ is called *a stationary domain (of level a)* of a real function $\Psi(u)$ with respect to a set $M \in \mathbf{R}^k$, if, for any $u \in M_\Psi^a$, we have equality $\Psi(u) = a$, and there exists a ε-neighborhood of point u in M such that $\Psi(u') \geq a$ for all u' from this neighborhood. The stationary domain M_Ψ^a is called *local* if it is either not a proper subset of another stationary domain and is closed (in the case of nonclosedness) or $\Psi(u) > a$ at all its boundary points.

The following statement shows that the number of local domains of the function $F(y)$ with respect to the set M_y does not exceed that corresponding to $f(x)$ and M_x. From this standpoint, solution of problem (4.4) is no more complex than solution of problem (3.1).

Theorem 3.2. *Let $y \in M_y$ and $y_1 = \omega(x^*(y))$. Then $F(y) \geq F(y_1)$. If*

$F(y) = F(y_1)$ and y_1 belongs to a stationary (local) domain M^a_{yF}, then there exists a stationary (local) domain M^a_{xf} such that $x^*(y) \in M^a_{xf}$.

Theorem 3.2 implies that the number of domains of local minima for problem (3.4) does not exceed the number of these domains for problem (3.1). Moreover, due to the inequality $F(y) \geq F(y_1)$ from the formulation of Theorem 3.2 in the solution of problem (3.4), we can transfer from the points $y \in M_y$ to the points $y' = \omega(x^*(y))$, which allows us to accelerate the search for y^*.

In the paper [63] by L.F. Verina, the proposed approach is applied for the special linear programming problem with a quasi-block constraint matrix. We have

$$f(x) = (c_1, x_1) + (c_2, x_2) + (c_3, x_3) \to \min, \tag{3.5}$$

$$\begin{aligned} A_{11}x_1 + A_{12}x_2 \qquad\qquad &\geq b_1, \\ A_{21}x_1 + A_{22}x_2 + A_{23}x_3 &\geq b_2, \\ + A_{32}x_2 + A_{33}x_3 &\geq b_3, \end{aligned} \tag{3.6}$$

$$x \geq 0, \tag{3.7}$$

where $x = (x_1, x_2, x_3)$, x_j is an n_j-dimensional vector column, A_{ij} is an $m_i \times n_j$-matrix, $i, j = 1, 2, 3$. It is assumed that constraints (3.6), (3.7) give a nonempty polyhedral domain $M_x \subset \mathbf{R}^{n_1+n_2+n_3}$.

A possible scheme of parametric decomposition looks as follows. The mapping $\omega(x)$ has the form

$$y_1 = x_2, \qquad y_2 = A_{23}x_3,$$

where $M_y = \omega(M_x) \subset \mathbf{R}^{n_2+m_2}$, and the function $\tilde{f}(x, y)$ is given as follows:

$$\tilde{f}(x, y) = (c_1, x_1) + (c_2, y_1) + (c_3, x_3).$$

The equality $M_x(y)$ holds for the set $M_x(y) = \omega^{-1}(y)$ and any $y \in M_y$. This set is given by the set of relations

$$A_{11}x_1 \geq b_1 - A_{12}y_1, \tag{3.8}$$

$$A_{21}x_1 \geq b_2 - A_{22}y_1 - y_2, \tag{3.9}$$

$$A_{33}x_3 \geq b_3 - A_{32}y_1, \tag{3.10}$$

$$A_{23}x_3 = y_2,$$
$$x_2 = y_1, \tag{3.11}$$
$$x \geq 0.$$

The reader can prove that conditions 1-3 hold for the introduced constructions.

In this case, local problem (3.4) is decomposed into two problems:

$$(c_1, x_1) \to \min \tag{3.12}$$

under constraints (3.8), (3.9), and $x_1 \geq 0$;

$$(c_3, x_3) \to \min \tag{3.13}$$

under constraints (3.10), (3.11) and $x_2 \geq 0$. The optimal values of the goal functions of problems (4.12) and (4.13) are denoted by $f_1(y)$ and $f_3(y)$, and the sets of admissible solutions are denoted by $M_x^1(y)$ and $M_x^3(y)$. The coordinating problem (4.4) has the form

$$F(y) = f_1(y) + (c_2, y_1) + f_3(y) \to \min, \qquad y \in M_y, \tag{3.14}$$

where the admissible set M_y is given as follows:

$$M_y = \left\{ y \in \mathbf{R}^{n_2+m_2} \,\middle|\, M_x^1(y) \neq \varnothing,\ M_x^3(y) \neq \varnothing,\ y_1 \geq 0 \right\}. \tag{3.15}$$

In order to determine M_y, we consider the duals of local problems (3.12) and (3.13):

$$\begin{aligned} &(d^1(y), \lambda^1) \to \max, \\ &A^1\lambda^1 \leq c_1, \qquad \lambda^1 \geq 0; \end{aligned} \tag{3.16}$$

$$\begin{aligned} &(d^3(y), \lambda^3) \to \max, \\ &A^3\lambda^3 \leq c_3, \qquad \lambda^3 \geq 0. \end{aligned} \tag{3.17}$$

Here,

$$d^1(y) = \begin{bmatrix} b_1 - A_{12}y_1 \\ b_2 - A_{22}y_1 - y_2 \end{bmatrix}, \qquad d^3(y) = \begin{bmatrix} b_3 - A_{32}y_1 \\ y_2 \end{bmatrix},$$

$$(A^1)^T = \begin{bmatrix} A_{11} \\ A_{21} \end{bmatrix}, \qquad (A^3)^T = \begin{bmatrix} A_{33} \\ A_{23} \end{bmatrix},$$

λ^1 and λ^3 are $(m_1 + m_2)$- and m_3-dimensional vectors of dual variables.

The sets of admissible solutions of problems (3.16) and (3.17) are denoted by M_λ^1 and M_λ^3. Relation (3.15) and the duality theory of linear programming imply that M_y is the set of $y \in \mathbf{R}^{n_2+m_2}$ such that $y_1 \geq 0$, and the goal functions of problems (3.16) and (3.17) are bounded from below on M_λ^1 and M_λ^3. Thus, M_y is described by the system of linear inequalities

$$\Lambda^1 d^1(y) \leq 0, \qquad \Lambda^3 d^3(y) \leq 0, \qquad y_1 \geq 0,$$

where Λ^1 and Λ^3 are matrices, whose columns are the directing vectors of unbounded edges of constraint polyhedrons of problems (3.16) and (3.17). If the polyhedrons of problems (3.16) and (3.17) are bounded, then $M_y = \{y \in \mathbf{R}^{n_2+m_2} \,|y_1 \geq 0\}$.

Consider the coordinating problem (3.14). The functions $f_1(y)$ and $f_3(y)$ are piecewise linear and convex. Generalized gradient descent is used for minimizing of $F(y)$. Without writing the corresponding formulas, we mention that the so-called generalized gradient of the function $F(y)$ is computed at each iteration. *The generalized gradient* of a convex function $\Phi(x)$ is, as usual, a row vector denoted by $\overline{\text{grad}}\Phi(z)$, such that the inequality

$$\Phi(x) \geq \Phi(z) + (\overline{\text{grad}}\Phi(z), (z - x))$$

holds for all $x \in \mathbf{R}^n$.

In this case, we have

$$\overline{\text{grad}}F(z) = \overline{\text{grad}}f_1(y) + c_2 + \overline{\text{grad}}f_3(y).$$

The formula

$$\overline{\text{grad}}f_i(y) = D_i^T \lambda^i(y), \qquad i = 1, 3,$$

holds, where

$$D_1 = \begin{bmatrix} -A_{12} & 0 \\ -A_{22} & -E \end{bmatrix}, \qquad D_3 = \begin{bmatrix} A_{32} & 0 \\ 0 & E \end{bmatrix},$$

$\mathring{\lambda}^1(y)$ and $\mathring{\lambda}^3(y)$ are optimal solutions of problems (3.16) and (3.17).

§4. Iterative Aggregation

The works by B.A. Shchennikov [46–48] provide the foundations of iterative aggregation. The following decomposition method of aggregation is proposed for the solution of systems of linear equations of the form

$$x = Ax + b, \tag{4.1}$$

where A is a square $n \times n$-matrix with components a_{ks}, $x = (x_1, \dots, x_n)$, and $b = (b_1, \dots, b_n)$ are n-dimensional vector columns. System (4.1) describes economical models of interbranch balance. Nonnegative solution (4.1) exists under conditions

$$a_{ks},\ k, s \in [1:n]; \qquad \sum_{k=1}^{n} a_{ks} \le 1,\ s \in [1:n]; \qquad b_k \ge 0,\ k \in [1:n].$$

We introduce the aggregated variable X_m for the first $m \le n$ components of the vector x: $X_m = \sum\limits_{k=1}^{m} x_k$, and the rest of components x will be renamed as follows: $X_{m+1} = x_{m+1}, \dots, X_n = x_n$. We have a vector of aggregated variables $X = (X_m, \dots, X_n)$. We introduce the aggregation weights $\alpha = (\alpha_1, \dots, \alpha_m)$, $\alpha_k = x_k / X_m$, $k \in [1:m]$. Consider the matrix $\bar{A}_\alpha$ and the vector $\bar{b}$:

$$\bar{A}_\alpha = \begin{bmatrix} \sum\limits_{k,s=1}^{m} a_{ks}a_s & \sum\limits_{k=1}^{m} a_{k,m+1} & \cdots & \sum\limits_{k=1}^{m} a_{kn} \\ \sum\limits_{s=1}^{m} a_{m+1,s}a_s & a_{m+1,m+1} & \cdots & a_{m+1,n} \\ \vdots & \vdots & \ddots & \vdots \\ \sum\limits_{s=1}^{m} a_{ns}a_s & a_{n,m+1} & \cdots & a_{nn} \end{bmatrix}, \qquad \bar{b} = \begin{bmatrix} \sum\limits_{k=1}^{m} b_k \\ b_{m+1} \\ \vdots \\ b_n \end{bmatrix}.$$

For (4.1), we compose the following system with aggregated variables:

$$X = \bar{A}_a X + \bar{b}. \tag{4.2}$$

This system corresponds to the constant weights $\alpha_1, \dots, a_m$. If they are known, we obtain a reduction of problem (4.1) to a problem of lesser dimension (4.2).

Further, we consider techniques for finding the vectors. Let vector X be known. According to (4.1), we define

$$x_k = \left(\sum_{s=1}^{m} a_{ks}a_s\right) X_m + \sum_{s=m+1}^{n} (a_{ks}X_s + b_s), \qquad k \in [1:m]. \tag{4.3}$$

for all $k \in [1:m]$.

Let X_m be strictly greater than zero, then relation (4.3) implies

$$a_k = \sum_{s=1}^{m} a_{ks} a_s + \left(\sum_{s=m+1}^{n} a_{ks} X_s + b_s \right) \Big/ X_m, \qquad k \in [1:m]. \tag{4.4}$$

It is easily seen that the equation

$$\alpha = C\alpha \tag{4.5}$$

holds, where C is a square $m \times m$-matrix with components $c_{ir} = a_{ir} + d_{ir}$, and the elements d_{ir} are defined in the following way. Let $\alpha(r)$ be an m-dimensional row with one in the r-th place and other zero components, $X(\alpha(r))$ be a solution of system (4.2), where $\bar{A}_{\alpha(r)}$ is substituted for $\bar{A}_\alpha$. Then, we have

$$d_{ir} = \left[\sum_{s=m+1}^{n} a_{is} X_s(\alpha(r)) + b_i \right] \Big/ X_m(\alpha(r)). \tag{4.6}$$

Thus, to define d_{ir}, we need to solve system (4.2) m multiplied by matrices $\bar{A}_{\alpha(r)}$, that differ from each other in the first column. Upon finding d_{ir}, system (4.5) is solved. Then, (4.2) and, therefore, (4.1) are solved. Eventually, the final algorithm for the solution of problem (4.1) is obtained.

The written relations allow us to construct the iterative process. At the l-th step, we have

$$\begin{aligned} X[l+1] &= \bar{A}_{\alpha[l]} X[l+1] + \bar{b}, \\ x_k[l+1] &= \left(\sum_{s=1}^{m} a_{ks} \alpha_s[l] \right) X_m[l+1] + \\ &\quad + \sum_{s=m+1}^{n} a_{ks} X_s[l+1] + b_k, \qquad k \in [1:m], \\ \alpha_s[l] &= x_s[l]/X_m[l], \qquad s \in [1:m]. \end{aligned} \tag{4.7}$$

In [48], convergence of the process (4.7) to the solution of problem (4.1) under $c_{ir} \geq 0$ with rate μ^l is established. Here, μ is the eigenvalue of matrix C with the greatest absolute value.

The iterative process (4.7) is especially simplified when all variables of vector x are aggregated, i.e., for $m = n$. In this case, we have one scalar equation for the aggregated variable at each step:

$$X[l+1] = \left(\sum_{k=1}^{n}\sum_{s=1}^{n} a_{ks}\alpha_s[l]\right)(X[l+1]+\hat{b}),$$

$$\hat{b} = \sum_{s=1}^{n} b_s, \qquad \alpha_s[l] = x_s[l]/X[l] = x_s[l]\left(\sum_{k=1}^{n} x_k[l]\right)^{-1},$$

$$x_k[l+1] = \sum_{s=1}^{n} a_{ks}\alpha_s[l]X[l+1] + b_k, \qquad k \in [1:n].$$

The components of the matrix C have the following form:

$$c_{ks} = a_{ks} + b_k\left(1 - \sum_{k=1}^{n} a_{ks}\right)\Big/\hat{b},$$

therefore, we have $c_{ks} \geq 0$ under $\sum\limits_{s}\sum\limits_{k=1}^{n} a_{ks} \leq 1$. In this case, it is expedient to use the iterative process when the norm of the matrix A is close to one.

Aggregation in problem (4.1) may be achieved by grouping variables. We introduce the set of indices I_m and aggregated variables X_m for $m \in [1:p]$ such that

$$\bigcup_{m=1}^{p} I_m = [1:n], \; I_r \bigcap I_m = \varnothing, \; r \neq m; \; X_m = \sum_{k \in I_m} x_k.$$

The aggregated problem has the form

$$X = \bar{A}X + \bar{b}, \tag{4.8}$$

where

$$\begin{gathered} \bar{a}_{ks} = \sum_{i \in I_k}\sum_{r \in I_s} a_{ir}\alpha_r, \qquad \alpha_r = \frac{x_r}{X_m}, \\ \sum_{r \in I_s} \alpha_r = 1, \; \bar{b} = \{(\bar{b}_s)^m\}, \; \bar{b}_s = \sum_{i \in I_s} b_i. \end{gathered} \tag{4.9}$$

In constructing the iterative process, we set an initial approximation $x_i[0]$, $i \in [1:n]$. At the first step, we solve the system

$$X[1] = \bar{A}[0]X[1] + \bar{b}, \tag{4.10}$$

where the elements of the matrix $\bar{A}[0]$ are defined according to (5.9) after α_r, x_r, X_m are replaced by $\alpha_r[0]$, $x_r[0]$, $X_m[0]$. Upon solution of (4.10), the values $x_i[1]$, $i \in [1:n]$ are determined by the formula

$$x_i[1] = \sum_{m=1}^{n} \sum_{r \in I_m} a_{ir} \alpha_r[0] X_m[1] + b_i.$$

The following iterations are constructed analogously.

Iterative aggregation is most effective when the matrix A in system (4.1) has the block triangular structure.

$$A = \begin{bmatrix} a_{11} & 0 & \dots & 0 \\ a_{21} & a_{22} & \dots & 0 \\ \vdots & \vdots & \ddots & \vdots \\ a_{m1} & a_{m2} & \dots & a_{mm} \end{bmatrix}$$

Here, all variables x_r, $r \in I_k$, $k \in [1:m]$ are aggregated in one value X_k. Finally, we obtain a system with aggregated variables, which is split into m separately solved equations at each step. A series of papers cited in [21] is concerned with the method of iterative aggregation. In these papers as well as in [19], problems of the method's convergence are studied.

In the paper [62] by I.Ya. Vakhutinskii, L.N. Dudkin, and B.A. Shchennikov, the method of aggregation under study is applied to special linear programming problems that describe particular economic models. In particular, the problem

$$\begin{aligned} &(c_1, x_1) + (c_2, x_2) \to \min, \\ &A^{11}x_1 + A^{12}x_2 + b_1 = x_1, \\ &A^{21}x_1 + A^{22}x_2 + b_2 = 0, \\ &x_1 \geq 0, \qquad x_2 \geq 0 \end{aligned} \tag{4.11}$$

is considered, where the vectors c_1, c_2, x_1, x_2, b_1, b_2 and matrices A^{11}, A^{12}, A^{21}, A^{22} have compatible dimensions. The vectors b_1 and $b_2 \geq 0$ are given.

The components of the vector x_1 are aggregated, where the weighted sum is taken with the weights equal to dual estimates that correspond to the first equality in (4.11). The latter fact corresponds to the particular economic meaning, where the total product is measured not in the natural, but in the value form. Here, since the estimates are not known in advance, an additional iterative process is constructed for their determination.

Let, on the l-th iteration step, the vectors $x_1[l]$, $x_2[l]$ and the vectors of dual variables $\lambda_1[l]$, $\alpha_2[l]$ that correspond to the first and second equalities in (4.11) be given. It is assumed that the scalar product $(\lambda_1[l-1], x_1[l])$ is not equal to zero. The vector $\alpha[l] = x_1[l]/(\lambda_1[l-1], x_1[l])$ is introduced. Here, the vector $\lambda_1[l-1]$ is obtained at the previous iteration. We consider the problem

$$\begin{aligned} &(c_1, \alpha[l]X) + (c_2, x_2) \to \min, \\ &\lambda_1[l-1]A^{11}\alpha[l]X + \lambda_1[l-1]A^{12}x_2 + \\ &\qquad +(\lambda_1[l-1], b_1) = X, \\ &A^{21}\alpha[l]X + A^{22}x_2 + b_2 = 0, \\ &X \geq 0, \qquad x_2 \geq 0, \end{aligned} \tag{4.12}$$

where the number X and vector x_2 are unknown variables. X is the sum of components of vector x_1 multiplied on the corresponding coordinates of vector $\lambda_1[l-1]$.

Let $X[l+1]$, $x_2[l+1]$ be optimal solution of problem (4.12). Let $R[l]$, denote the set of indices of r basis variables of the vector x_2 in the optimal solution. According to the first equality (5.11), we assume

$$x_1[l+1] = A^{11}\alpha[l]X[l+1] + A^{12}x_2[l+1] + b_1.$$

The next approximation for the dual variables is obtained in the following way. Here, $\lambda_2[l+1]$ is found from the system of linear equations

$$\lambda_1[l]A_r^{12} + \lambda_2[l+1]A_r^{22} = -c_{2r}, \qquad r \in R[l],$$

where A_r^{12}, A_r^{22}, c_{2r} are columns and components that correspond to basis indices. Then, we find $\lambda_1[l+1]$ by the formula

$$\lambda_1[l+1] = \lambda_1[l]A^{11} + \lambda_2[l+1]A^{21} + c_1.$$

The last equation corresponds to the dual conditions for the vector x_1 in the original problem (5.11).

In [62], convergence of the algorithm to optimal solutions of the primal and dual problem (5.11) was established. In that paper, the method of aggregation is applied to other economic models.

In [31], V.G. Mednitskii proposed a method of iterative aggregation for the block linear programming problem of the following form:

$$
\begin{aligned}
& f = (d_0, y_0) + \sum_{j=1}^{J}(c_j, x_j) + \sum_{j=1}^{J}(d_j, y_j) \to \max, \\
& D^0 y_0 + \sum_{j=1}^{J} A_j^0 x_j = b^0, \\
& A_j x_j + D_j y_j = b_j, \qquad j \in [1:J], \\
& y_0 \geq 0;\ x_j \geq 0,\ y_j \geq 0,\ j \in [1:J].
\end{aligned}
\tag{4.13}
$$

Here, as before, capital letters denote matrices and small letters denote vectors of corresponding dimensions, so that the products and sums in (4.13) have meaning. Note that the variables y_0, y_j, $j \in [1:J]$ are not binding, since y_0 occurs only in the binding condition, and y_j, $j \in [1:J]$ belongs to the corresponding blocks.

In what follows, we assume that the block conditions $A_j x_j + D_j y_j = b_j$, $x_j \geq 0$, $y_j \geq 0$ specify limited polyhedrons for all $j \in [1:J]$. Here, the aggregation means the following substitution of variables:

$$
x_j = \alpha_j X^j, \qquad j \in [1:J], \tag{4.14}
$$

where, for each $j \in [1:J]$, the vector of aggregation weights $\alpha = (\alpha_1, \ldots, \alpha_j)$ and the aggregated variable X^j are introduced. Note that, for the accepted notation, a subscript corresponds to a vector, and a superscript corresponds to a scalar. We will say that aggregation is *feasible* if, for all $j \in [1:J]$, the inequality $\alpha_j \geq 0$ holds.

We substitute (4.14) in (4.13) for fixed values α_j, $j \in [1:J]$ and obtain the problem with aggregated variables:

$$
\begin{aligned}
& \varphi = (d_0, y_0) + \sum_{j=1}^{J} c^j X^j + \sum_{j=1}^{J}(d_j, y_j) \to \max, \\
& D^0 y_0 + \sum_{j=1}^{J} a_j^0 X_j = b^0, \\
& a_j X^j + D_j y_j = b_j, \qquad j \in [1:J], \\
& y_0 \geq 0;\ X^j \geq 0,\ y_j \geq 0,\ j \in [1:J],
\end{aligned}
\tag{4.15}
$$

where $a_j^0 = Aj^0\alpha_j$, $a_j = A_j\alpha_j$, $c^j = (c_j, \alpha_j)$.

The aggregation weights can be chosen in different ways, thus obtaining problem (4.15), aggregated from the original problem (4.13). We establish some solvability conditions for these problems.

Theorem 4.1. *If an original problem is solvable, the aggregated problem has a feasible solution, and the aggregation is feasible, then the value of the functional of the aggregated problem is bounded from above on the whole set of its solutions.*

Let $\{\bar{y}_0, \bar{X}^i, \bar{y}_j\}$ be a feasible solution of the aggregated problem (4.15) that corresponds to the fixed weights α_j, $j \in [1 : J]$. Assuming $\bar{x}_j = \alpha_j \bar{X}^j$, we can easily see that the set $\{\bar{y}_0, \bar{x}_j, \bar{y}_j\}$ is feasible for the original problem (4.13), where the value $\bar{f}$ of the functional on it coincides with the values of the functional $\bar{\varphi}$ of problem (4.15) for the tuple $\{\bar{y}_0, \bar{X}^j, \bar{y}_j\}$. This gives us $\bar{\varphi} = \bar{f} \leq f^*$, where f^* is the optimal value of the functional of the original problem (5.13).

We introduce the so-called *adjoint* problem:

$$(d_0, y_0) + \sum_{j=1}^{J} (d_j, y_j) \to \max,$$

$$D^0 y_0 = b^0, \qquad y_0 \geq 0,$$

$$D_j y_j = b_j, \qquad y_j \geq 0, \qquad j \in [1 : J].$$

To ensure that any aggregated problem has a feasible solution, it is necessary and sufficient that the adjoint problem has an optimal solution. In fact, if $\{\tilde{y}_0, \tilde{y}_j\}$ is an optimal solution of the adjoint problem, by adding $X^j = 0$, $j \in [1 : J]$ to it, we have a feasible solution for any aggregated problem. Conversely, considering the aggregated problem for $\alpha_j = 0$, $j \in [1 : J]$ and taking a feasible solution for it, we establish that this solution is feasible for the adjoint problem, and, therefore, this problem is solvable.

Thus, if the original and the adjoint problems are solvable, and the aggregation is feasible, then every aggregated problem is solvable, which is assumed below.

It is easy to see that there exists an aggregated problem with the value of functional equal to the optimal value of the functional of the original problem. Let $\{y_0^*, x_j^*, y_j^*\}$ be an optimal solution of problem (4.13). We put $\alpha_j = x_j^*$, $X^j = 1$, $j \in [1 : J]$ and obtain a tuple $\{y_0^*, 1^j, y_j^*\}$, feasible for the corresponding aggregated problem with the value f^* of the functional. The vectors $\alpha_j = x_j^*$, $j \in [1:J]$ are called *vectors of optimal aggregation.*

The further goal is the construction of the iterative process that, starting from some feasible aggregation, results in the optimal one. Here, the criterion of optimality of feasible aggregation (the termination condition for the iteration process) is deduced.

Let the optimal solution $\{\overset{\circ}{y}_0, \overset{\circ}{X}^j, \overset{\circ}{y}_j\}$ of aggregation problem (4.15) be obtained for a vector α. We set $\bar{b}^0 = b^0 - \sum_{j=1}^{J} a_j \overset{\circ}{X}^j$ and introduce the so-called *bound* problem of parametric programming:

$$\begin{aligned} &(d_0, y_0) \to \max, \\ &D^0 y_0 = \bar{b}^0 + \rho\delta b, \\ &y_0 \geq 0, \end{aligned} \tag{4.16}$$

where δb is an arbitrary vector with dimension of the vector b^0, and ρ is a parameter. We can show that, for $\rho = 0$, the problem (5.16) has as a solution the vector $\overset{\circ}{y}_0$. We make the following assumption.

Condition A. *For any vector δb and some $\rho > 0$, a solution of problem (4.16) exists.*

We formulate the dual problem for $\rho = 0$ problem (4.16):

$$\begin{aligned} &(\lambda, \bar{b}^0) \to \min, \\ &\lambda D^0 \geq d_0. \end{aligned} \tag{4.17}$$

According to Theorem 1.12 of Chapter 1, Condition A holds if the set of optimal solutions of conjugated problems (4.16) (for $\rho = 0$), and (4.17) are bounded. For simplicity, we accept the latter assumption.

Let $\overset{\circ}{\lambda}$ be an optimal solution for problem (4.17). We introduce vectors $\overset{\circ}{\delta}_j$

$$\overset{\circ}{\delta}_j = c_j - \overset{\circ}{\lambda} A_j^0, \qquad j \in [1:J]$$

and formulate local problems for each $j \in [1:J]$:

$$\begin{aligned} &h_j = (\overset{\circ}{\delta}_j, x_j) + (d_j, y_j) \to \max, \\ &A_j x_j + D_j y_j = b_j, \\ &x_j \geq 0, \qquad y_j \geq 0. \end{aligned} \tag{4.18}$$

Let $\{\hat{x}_j, \hat{y}_j\}$, $j \in [1:J]$ be some optimal solutions of problem (4.18). We introduce the value $\hat{h}_j$, equal to the optimal value of the functional of the j-th problem: $\hat{h}_j = (\overset{\circ}{\delta}_j, \hat{x}_j) + (d_j, \hat{y}_j)$, $j \in [1:J]$. We also introduce the value $\overset{\circ}{h}$, which is equal to the value of the functional of the j-th problem on the disaggregated solution $\{\overset{\circ}{x}_j, \overset{\circ}{y}_j\}$: $\overset{\circ}{h}_j = (\overset{\circ}{\delta}_j, \overset{\circ}{x}_j) + (d_j, y_j)$ feasible for it. Since, generally speaking, disaggregated solutions are nonoptimal for local problems, the inequality $\hat{h} \geq \overset{\circ}{h}$ holds, where the notation is as follows:

$$\hat{h} = \sum_{j=1}^{J} \hat{h}_j, \qquad \overset{\circ}{h} = \sum_{j=1}^{J} \overset{\circ}{h}_j.$$

The following statement establishes the criterion of optimal aggregation:

Theorem 4.2. *For optimal aggregation, it is sufficient that equality*

$$\hat{h} = \overset{\circ}{h}. \tag{4.19}$$

holds for some optimal solution $\overset{\circ}{\lambda}$ *of the dual problem (4.17).*

Consider the dual of the aggregated problem (4.15):

$$\begin{aligned} &(\eta_0, b^0) + \sum_{j=1}^{J} (\eta_j, b_j) \to \min, \\ &\eta_0 D^0 \leq d_0, \\ &(\eta_0, a_j^0) + (\eta_j, a_j) \leq c^j, \\ &\eta_j D_j \leq d_j, \qquad j \in [1:J]. \end{aligned} \tag{4.20}$$

Let $\{\overset{\circ}{\eta}_0, \overset{\circ}{\eta}_j\}$, $j \in [1:J]$ be an optimal solution of problem (4.20). First, we establish equality (4.19) under the assumption that the vector $\overset{\circ}{\eta}_0$ is taken instead of $\overset{\circ}{\lambda}$ in the formation of local problems (5.18).

Applying the Second Duality Theorem 2.4 of Chapter 1 for the second inequality in (4.20), we have

$$\overset{\circ}{X}{}^j \left[(\overset{\circ}{\eta}_0, \overset{\circ}{a}_j) + (\overset{\circ}{\eta}_j, a_j) - c^j \right] = 0, \qquad j \in [1:J].$$

Recalling the definition of $\overset{\circ}{\delta}_j$, we rewrite the latter equalities in the form

$$\overset{\circ}{X}_j (\overset{\circ}{\eta}_j, a_j) - (\overset{\circ}{\delta}_j, \overset{\circ}{x}_j) = 0, \qquad j \in [1:J].$$

The expressions are transformed in accordance with the second equality in (4.15). We have

$$(\mathring{\eta}_j, b_j) = \mathring{\eta}_j D_j \mathring{y}_j + (\mathring{\delta}_j, \mathring{x}_j), \qquad j \in [1:J].$$

According to the Second Duality Theorem 1.4 of Chapter 1, the relations $\mathring{\eta}_j D_j \mathring{y}_j = (d_j, \mathring{y}_j)$, $j \in [1:J]$ hold for the third inequality in (4.20), hence we finally obtain

$$(\mathring{\eta}_j, b_j) = (d_j, \mathring{y}_j) + (\mathring{\delta}_j, \mathring{x}_j) = \mathring{h}_j, \qquad j \in [1:J]. \tag{4.21}$$

The value of the functional $\mathring{f}$ for the disaggregated solution $\{\mathring{y}_0, \mathring{x}_j, \mathring{y}_j\}$ feasible for problem (4.13) is equal to $\mathring{\varphi}$. From the First Duality Theorem 1.3 of Chapter 1, we have

$$\mathring{f} = \mathring{\varphi} = (\mathring{\eta}_0, b^0) + \sum_{j=1}^{J} (\mathring{\eta}_j, b_j).$$

Using (5.21), we obtain

$$\mathring{f} = (\mathring{\eta}_0, b^0) + \mathring{h}\,. \tag{4.22}$$

Now we consider the dual to the local problem, for every $j \in [1:J]$:

$$\begin{aligned} &\chi_j = (\zeta_j, b_j) \to \min, \\ &\zeta_j A_j \leq \mathring{\delta}_j, \\ &\zeta_j D_j \leq d_j. \end{aligned} \tag{4.23}$$

Let $\hat{\zeta}_j$, $j \in [1:J]$ be optimal solutions of problems (4.23). Then, the set $\{\mathring{\eta}_0, \hat{\zeta}_j\}$, where $j \in [1:J]$, is feasible for the dual to the initial problem. This follows if we write the first inequality in (4.23), using the definition of vector $\mathring{\delta}_j$, in the form $\hat{\zeta}_j A_j + \mathring{\eta}_0 A_j^0 \leq c_j$. The value of the functional $\hat{\Psi}$ of the dual to the original problem (4.13) for the feasible set $\{\mathring{\eta}_0, \hat{\zeta}_j\}$ is equal to

$$\Psi = (\mathring{\eta}_0, b^0) + \sum_{j=1}^{J} (\hat{\zeta}_j, b_j).$$

Using the First Duality Theorem 1.3 from Chapter 1 for the conjugated local problems (4.18), (4.23), we obtain

$$\hat{\Psi} = (\mathring{\eta}_0, b^0) + \hat{h}. \tag{4.24}$$

Comparing (4.22) and (4.24), we obtain feasible solutions of the conjugated problems, where the equality of functionals for these solutions hold if $\hat{h} = \overset{\circ}{h}$. Therefore, by Theorem 1.2 of Chapter 1, we conclude there is optimal aggregation when (4.19) holds.

Further, consider the formation of local problems by means of the optimal solution $\overset{\circ}{\lambda}$ of the dual (4.17) to the bound problem (4.16). First, note that any tuple $\{\lambda', \zeta_j'\}$, where λ' satisfies the condition $\lambda' D^0 \leq d_0$ and vectors δ_j', $j \in [1 : J]$ are formed by means of λ', is a feasible solution for the dual to problem (4.13). However, the latter inequality defines the feasible solution for the dual bound problem. Let, as before, $\overset{\circ}{\lambda}$ be an optimal solution of this problem for $\rho = 0$. Due to the First Duality Theorem 1.3 of Chapter 1 for the conjugated bound problems, we have

$$(d_0, \overset{\circ}{y}_0) = (\overset{\circ}{\lambda}, b^0) = (\overset{\circ}{\lambda}, b^0) - \left(\lambda^0 \sum_{j=1}^{J} a_j \overset{\circ}{X}^j \right) = (\overset{\circ}{\lambda}, b^0) - \sum_{j=1}^{J} (\overset{\circ}{\lambda}, A_j \overset{\circ}{x}_j).$$

Applying the Second Duality Theorem 1.4 of Chapter 1 for the first inequality in the dual aggregated problem (4.20), we obtain $(d_0, \overset{\circ}{y}_0) = (\overset{\circ}{\eta}_0, D^0 \overset{\circ}{y}_0)$. Because of the first equality (4.15), this relation becomes

$$(d_0, \overset{\circ}{y}_0) = (\overset{\circ}{\eta}_0, b^0) - \sum_{j=1}^{J} (\overset{\circ}{\eta}_0, A_j \overset{\circ}{x}_j).$$

Combining the latter relations, we come to the equality

$$(\overset{\circ}{\eta}_0, b^0) - \sum_{j=1}^{J} (\overset{\circ}{\eta}_0, A_j \overset{\circ}{x}_j) = (\overset{\circ}{\lambda}, b^0) - \sum_{j=1}^{J} (\overset{\circ}{\lambda}, A_j \overset{\circ}{x}_j).$$

We add the expression

$$\sum_{j=1}^{J} \left[(c_j, \overset{\circ}{x}_j) + (d_j, \overset{\circ}{y}_j) \right]$$

to both sides, recall the formula for the vectors $\overset{\circ}{\delta}_j$, $j \in [1 : J]$, and obtain $(\overset{\circ}{\eta}_0, b^0) + \overset{\circ}{h} = (\overset{\circ}{\lambda}, b^0) + \overset{\circ}{h}'$. In this case, the equality $\hat{h}' = \bar{h}'$ leads to the following one:

$$\overset{\circ}{f} = (\overset{\circ}{\eta}_0, b^0) + \overset{\circ}{h} = (\overset{\circ}{\lambda}, b^0) + \overset{\circ}{h}' = \hat{\Psi}',$$

Let, for some weight vector α, the optimal solution $\{\overset{\circ}{y}_0, \overset{\circ}{X}^j, \overset{\circ}{y}_j\}$ of the aggregated problem (4.15) be obtained. We compose the bound problem (4.16), and let, for $\rho = 0$, its optimal solution $\overset{\circ}{y}{}'_0$ be obtained, which corresponds to the nondegenerate basis and the vector of optimal dual estimates $\overset{\circ}{\lambda}{}'$. We form local problems by this vector $\overset{\circ}{\lambda}{}'$. The following statement shows the necessity of condition (4.19) for optimal aggregation.

Theorem 4.3. *If the strong inequality $\hat{h} > \overset{\circ}{h}$ holds, then the aggregation with vector α is nonoptimal.*

Proof. The inequality $\hat{h} > \overset{\circ}{h}$ implies $\hat{h}_j > \overset{\circ}{h}_j$ for some indices $j \in [1 : J_1]$. Let $\{\hat{x}_j, \hat{y}_j\}$ be a system of optimal solutions of local problems for $j \in [1 : J_1]$. We form a new vector of aggregation weights according to the rule

$$\alpha_j = \begin{cases} (1-\rho)\,\overset{\circ}{x}_j + \rho\hat{x}_j, j \in [1 : J_1], \\ \overset{\circ}{x}_j \qquad\qquad j \in [J_1 + 1 : J]. \end{cases} \tag{4.25}$$

This aggregation is feasible for $\rho \in [0, 1]$, since $\overset{\circ}{x}_j$, $\hat{x}_j \geq 0$.

Block constraints in the aggregated problem (4.15) with weights (4.25) are transformed to the form

$$A_j[(1-\rho)\,\overset{\circ}{x}_j + \rho\hat{x}_j]X^j + D_j y_j = b_j, \qquad j \in [1 : J_1],$$
$$A_j\,\overset{\circ}{x}_j X^j + D_j y_j = b_j, \qquad j \in [J_1 + 1 : J].$$

Therefore, the set

$$X^j = 1, \qquad j \in [1 : J],$$
$$y_j(\rho) = \begin{cases} (1-\rho)\,\overset{\circ}{y}_j + \rho\hat{y}_j, \; j \in [1 : J_1], \\ \overset{\circ}{y}_j, \qquad\qquad j \in [J_1 + 1 : J] \end{cases}$$

satisfies block constraints of the new aggregated problem. The binding constraints of the aggregated problem (4.15) have the form

$$D^0 y_0 = \bar{b} + \rho\Delta b,$$

where

$$\bar{b} = b^0 - \sum_{j=1}^{J} A_j^0\,\overset{\circ}{x}_j = b^0 - \sum_{j=1}^{J} a_j\,\overset{\circ}{X}^j,$$

$$\Delta b = \sum_{j\in[1:J_1]} A_j^0(\mathring{x}_j - \hat{x}_j).$$

Thus, let $\mathring{y}_0(\rho)$ be the optimal solution of the following parametric programming problem:

$$(d_0, y_0) \to \max, \qquad D^0 y_0 = \bar{b} + \rho\Delta b; \tag{4.26}$$

then the set

$$\{\mathring{y}_0(\rho), \hat{X}^j, y_j(\rho)\} \tag{4.27}$$

is feasible for the new aggregation problem for those $\rho \in [0,1]$, for which solution of (4.26) exists. However, for $\rho = 0$, this is our bound problem, which, by condition A, is solvable in the interval $[0, \bar{\rho}]$.

Since the basis of solution y_0' is nondegenerate, it is retained on the interval $[0, \rho']$, where $\rho' > 0$. Here, the optimal solution $\mathring{\lambda}'$ of the dual to the bound problem is also retained. By the First Duality Theorem 1.3 of Chapter 1, we have

$$(d_0, \mathring{y}_0(\rho)) = (\mathring{\lambda}', b^0) + \rho(\mathring{\lambda}', \Delta b) = (d_0, \mathring{y}_0) + \rho(\mathring{\lambda}', \Delta b) \tag{4.28}$$

for the conjugated bound problems.

The functional of the new aggregated problem has the following form on the solution (4.27):

$$\varphi(\rho) = (d_0, \mathring{y}_0(\rho)) + \sum_{j=1}^{J} c^j + \sum_{j=1}^{J}(d_j, y_j(\rho)).$$

Because of (4.28), the latter equality is transformed to the following one:

$$\varphi(\rho) = \mathring{\varphi} = \rho\left[(\mathring{\lambda}', \Delta b) + \sum_{j\in[1:J_1]} (d_j, \hat{y}_j - \mathring{y}_j) + \sum_{j\in[1:J_1]} (c_j, \hat{x}_j - \mathring{x}_j)\right],$$

and, according to the definition of Δb, we have

$$\varphi(\rho) = \mathring{\varphi} + \rho \sum_{j\in[1:J_1]} \left[(c_j - \mathring{\lambda}' A_j^0, \hat{x}_j - \mathring{x}_j) + (d_j, \hat{y}_j - \mathring{y}_j)\right] = \mathring{\varphi} + \rho(\hat{h} - \mathring{h}). \tag{4.29}$$

From the assumption, $\hat{h} - \overset{\circ}{h} > 0$, we have $\varphi(\rho) > \overset{\circ}{\varphi}$. In other words, we obtain an aggregated problem, whose functional value is greater than that for the original aggregated vector α. Theorem 4.3 is proved.

Formulas (4.25) and (4.29) dictate the construction of the iterative process with the growth of the functional for the disaggregated solutions feasible for the original problem (4.13). This is achieved as follows: if we have weights $\alpha_j[l]$ at the step l, then the aggregated problem is solved for these weights. By the optimal solution of this problem, we construct a bound problem and its dual. Dual estimates form the functionals of local problems, whose optimal solutions are used for determining the aggregation weights $\alpha_j[l+1]$, according to (4.25), for the next iteration step.

In [31], a case is studied, where the optimal solution of the bound problem corresponds to a nondegenerate basis, and the vector of estimates $\overset{\circ}{\lambda}{}'$ is defined nonuniquely.

In the method described above, the macrovariables correspond to each block. As we see, most of the book is devoted to the method, where variables entering different blocks are aggregated.

§5. The Use of Lagrange Functional in Block Dynamical Problems

As we have seen in Sections 5.1 and 5.2, for finite-dimensional block separable mathematical programming problems, the Lagrange function is considered, and its saddle point is sought by primal and dual variables. Optimization is realized by the two-level scheme. Lagrange multipliers of the binding constraints are fixed, and independent local problems for systems are solved at the low level. Then, by an appropriate rule, the upper level corrects the dual variables. J.D. Pearson extends this idea in [35] to block problems of optimal control of the following form:

$$f(u, z) = \sum_{j=1}^{J} \left[w_j(x_j(T)) + \int_0^T c_j(x_j(t), u_j(t), z_j(t), t)dt \right] \to \min, \tag{5.1}$$

$$\begin{aligned} dx_j(t)/dt = F_j(x_j(t), u_j(t), z_j(t)) = \\ = A_j(t)x_j(t) + B_j(t)u_j(t) + C_j(t)z_j(t),\ x_j(0) = \kappa_j, \end{aligned} \tag{5.2}$$

$$p_j(x_j(t), u_j(t), z_j(t), t) \geq 0, \tag{5.3}$$

$$z_j(t) = \sum_{i=1}^{J} L_{ji}(t) y_i(t), \tag{5.4}$$

$$y_j(t) = G_j(x_j(t), u_j(t), t) = M_j(t)x_j(t) + N_j(t)u_j(t), \qquad j \in [1:J], \tag{5.5}$$

where, as before, $j \in [1 : J]$ are numbers of subsystems: x_j is an n_j-dimensional control vector, u_j is an m_j-dimensional control vector, z_j and y_j are s_j- and r_j-dimensional vectors of input and output, respectively; $A_j(t)$, $B_j(t)$, $C_j(t)$, $L_j(t)$, $M_j(t)$, $N_j(t)$ are matrices of concerted dimensions; p_j is the vector function of its arguments of dimensions l_j.

Relations (5.4) define the dependence of the output of each subsystem on the inputs of other subsystems and define the feedback control. These equalities are binding conditions of problem (5.1)–(5.5).

We write the Lagrange functional as follows:

$$\begin{aligned}
&\mathcal{L}(x, u, z, \psi, \eta, \delta) = \\
&= \sum_{j=1}^{J} \Bigg\{ w_j(x_j(T)) + \int_0^T \Bigg[c_j(x_j(t), u_j(t), z_j(t), t)dt + \\
&+ \psi_j^T(t)(A_j(t)x_j(t) + B_j(t)u_j(t) + C_j(t)z_j(t) - dx_j(t)/dt) - \\
&- \eta_j^T(t) p_j(x_j(t), u_j(t), z_j(t), t) + \delta_j^T(t) \sum_{i=1}^{J} L_{ji}(t)(M_j(t)x_j(t) + \\
&+ N_j(t)u_j(t) - z_j(t)) \Bigg] dt \Bigg\}
\end{aligned} \tag{5.6}$$

Here, $\psi_j(t)$, $\eta_j(t)$; $\delta_j(t)$, $j \in [1 : J]$ are vectors of conjugated variables with dimensions η_j, l_j, s_j, respectively. The last but one term in (5.6) can easily be represented in the form

$$\sum_{j=1}^{J} \int_0^T \delta_j^T(t) \sum_{i=1}^{J} L_{ji}(t) G_j(t) dt = \sum_{j=1}^{J} \int_0^T G_j(t) k_j(t) dt,$$

$$\kappa_j(t) = \sum_{i=1}^{J} L_{ij}(t) \delta_i(t).$$

We can formulate the dual problem. This problem consists in finding a saddle point of the Lagrange functional

$$\mathcal{L}(x, u, z, \psi, \eta, \delta) \to \min_{x,u,z} \max_{\psi,\eta,\delta},$$

which is equivalent to

$$\mathcal{L} \to \max \tag{5.7}$$

under conditions

$$\begin{aligned} d\psi_j(t)/dt = -\partial H/\partial x_j = \\ = -(\partial G_j/\partial x_j)^T \kappa_j - A_j^T(t)\psi_j(t) + \eta_j^T(t)\partial p_j/\partial x_j - M_j^T(t)\kappa_j, \end{aligned} \tag{5.8}$$

$$\begin{aligned} 0 = \partial H_j/\partial u_j + (\partial G_j/\partial u_j)^T \kappa_j = \\ = \partial C_j/\partial u_j - \eta_j^T \partial p_j/\partial u_j + B_j^T(t)\psi_j(t) + M_j^T \kappa_j = 0, \end{aligned} \tag{5.9}$$

$$\partial H_j/\partial z_j = \partial c_j/\partial z_j + C_j^T(t)\psi_j(t) - \eta_j^T \partial p_j/\partial z_j + \delta_j = 0, \tag{5.10}$$

$$\psi_j(t) = \partial w_j/\partial(x_j(t)), \qquad \eta_j(t) \geq 0, \tag{5.11}$$

where the Hamiltonian function of the form

$$H(t) = \sum_{j=1}^{J} \left[c_j + \psi_j^T(t)F_j - \eta_j^T p_j\right].$$

is introduced.

If the functions entering in (5.1)–(5.5) are convex, then, under certain conditions (see, e.g., the book [58] by A.M. Ter-Krikorov), the solution of the dual problem (5.7)–(5.11) exists. Here, the optimal values of the functionals of conjugated problems coincide: the conditions of complementary slackness

$$\eta_j^*(t)p_j(x_j^*(t), u_j^*(t), u_j^*(t), z_j^*(t), t) = 0, \qquad j \in [1 : J]$$

hold. Moreover, the Pontryagin maximum principle also holds, i.e., for all admissible u, z the inequality

$$H(x^*, u^*, z^*, \psi^*, \eta^*, \delta^*) \leq H(x, u, z, \psi^*, \eta^*, \delta^*).$$

holds. (Here and in what follows, the sign $*$ denotes the optimal solution). Moreover, due to the convexity of the entering functions, the conditions (5.2)–(5.5) and (5.8)–(5.11) are sufficient conditions for the extremum of primal and dual problems. Below, we assume that, for the original problem and all intermediate local problems, the duality principles hold.

Methods of decomposition for problems (5.1)–(5.5) form the decomposition process, where, at each step, independent local problems for subsystems

$$f_j(u_j, z_j, \delta) = w_j(x_j(T)) + \int_0^T \Bigg[c_j(x_j(t), u_j(t), z_j(t), t) +$$

$$+ \sum_{i=1}^{J} \delta_i(t) L_{ij}(t) G_j(t) - \delta_j(t) z_j(t) \Bigg] dt \to \min \quad (5.12)$$

are solved for fixed variables $\delta(t) = \{\delta_1(t), \dots, \delta_J(t)\}$ under conditions (5.2), (5.3). By $f_j^*(\delta) = f_j^*(u_j^*(\delta), z_j^*(\delta), \delta)$, we denote the optimal value of the functional of the local problem j. According to a scheme (to be describe below), the coordinating problem

$$f^*(\delta) = \sum_{j=1}^{J} f_j^*(\delta) \underset{\delta}{\to} \max . \quad (5.13)$$

is solved. We can hope that the sequence of values of functionals (5.13) will converge to the optimal value of the functional of dual problem (5.7)–(5.11).

To justify the method and find out some its properties, we formulate a number of statements, following the work of J.D. Pearson [35]. We introduce the following value:

$$\xi_j(\delta, t) = \sum_{i=1}^{J} L_{ji}(t) G_j(x_j^*(\delta), u_j^*(\delta)) - z_j^*(\delta) \quad (5.14)$$

and call it *the error of interaction for the j-th subsystem.*

Theorem 5.1. *Satisfaction of equality $\delta(t) = \delta^*(t)$ is a necessary, and, given unique local optima, also a sufficient condition for $\xi_j(\delta, t)$, $j \in [1 : J]$, to be equal to zero.*

Proof. First, we establish the necessity. First, for some $\delta_j(t)$, $j \in [1 : J]$, we obtained optimal solutions of local problems, for which values (1.14) are equal to zero. According to the duality principle as applied to the local problems, the following necessary conditions hold:

$$\begin{aligned}
&d\psi_j^*(\delta,t)/dt = \\
&= -A_j^*(t)\psi_j^*(\delta,t) + (\eta_j^*(\delta,t))^T \partial p_j/\partial x_j - M_j^T \kappa_j(\delta^*,t), \\
&\partial c_j \partial u_j - (\eta_j^*(\delta,t))^T \partial p_j/\partial u_j + B_j^T(t)\psi_j^*(\delta,t) - N_j^T \kappa_j(\delta,t) = 0, \qquad (5.15) \\
&\partial c_j/\partial z_j + C_j^T(t)\psi_j^*(\delta,t) - (\eta_j(\delta,t))^T \partial p_j/\partial z_j = \delta_j(t), \\
&\psi_j^*(\delta,T) = \partial w_j/\partial x_j(T), \qquad \eta_j^*(\delta,t) \geq 0, \qquad j \in [1:J].
\end{aligned}$$

These conditions, together with (5.2), (5.3), (5.5), and the equality to zero of (5.14) guarantee the optimality of $\{x^*(\delta,t), u^*(\delta,t), z^*(\delta,t), \psi^*(\delta,t), \eta^*(\delta,t), \delta\}$ for the dual problem (5.7)–(5.11), from which it follows that $\delta(t) = \delta^*(t)$.

Now we establish sufficiency. Let $\delta(t) = \delta^*(t)$. Consider the tuple $\{x^*(t), u^*(t), z^*(t), \psi^*(t), \eta^*(t), \eta^*(t), \delta^*(t)\}$, which consists of the optimal solutions of the original problem (5.1)–(5.5) and the conjugated variables necessary for optimality (5.8)–(5.11). The components $x^*(t), u^*(t), z^*(t)$, $j \in [1:J]$ of this tuple are solutions of local problems for a given $\delta^*(t)$, since, in this case, conditions (5.15) and (5.2), (5.3) hold. However, since solutions of local problems are unique, then, for $\delta^*(t)$, they coincide with $x^*(t), u^*(t), z^*(t)$, for which equalities (5.4), (5.5) hold. The theorem is proved.

Theorem 5.1 implies that the error of interaction should tend to zero in approximating the optimum of the original problem. On the other hand, this value is the distance from the optimum.

Theorem 5.2. *If solutions of local problems are nonunique, then the strict inequality*

$$f^*(\delta) < f^*(\delta^*)$$

holds for $\delta(t) \neq \delta^*(t)$.

Proof. Due to the assumed uniqueness, the relation

$$f_j^*(u_j^*(\delta), z_j^*(\delta), \delta) < f_j(u_j^*(\delta^*), z_j^*(\delta^*), \delta)$$

holds. Summing these inequalities over $j \in [1:J]$, we obtain

$$f^*(\delta) < \sum_{j=1}^{J} \left\{ w_j(x_j^*(\delta^*,T)) + \int_0^T \left[c_j(x_j^*(\delta^*,t), u_j^*(\delta^*,t), z_j^*(\delta^*,t), t) + \right.\right.$$

$$+\sum_{i=1}^{J}\delta_j(t)L_{ij}(t)G_j(x_j^*(\delta^*),u_j^*(\delta^*),t)-\delta_j z_j^*(\delta^*))\Bigg]\,dt\Bigg\} \tag{5.16}$$

Changing the order of summation in the third summand of (5.16) and taking into account (5.4), (5.5), we obtain that the right-hand side of (5.16) does not depend on δ and is equal to $f^*(\delta^*)$. The theorem is proved.

Theorem 5.3. *If functions w_j, c_j, p_j are convex in variables x_j, u_j, z_j, and C_j are strictly convex in variables u_j, z_j, $j \in [1:J]$ for each $t \in [0,T]$, then the optimal solutions of the original problem and block problems are unique.*

Proof. Assume the converse. Let (u^*,z^*) and (u^0,z^0) be two solutions of the initial problem, therefore, $f(u^*,z^*) = f(u^0,z^0)$. In this case, the combination $(\alpha x^* + \beta x^0, \alpha z + \beta z^0)$, where $\alpha \geq 0$, $\beta \geq 0$, $\alpha+\beta = 1$, are solutions of problem (5.1)–(5.5). Conditions (5.2), (5.4), (5.5) hold, since they are linear. Constraints (5.3) are satisfied due to the convexity of functions in their arguments:

$$-p_j(\alpha x_j^*+\beta x_j^0,\alpha u_j^*+\beta u_j^0,\alpha z_j^*+\beta z_j^0) \leq -\alpha p_j(x_j^*,u_j^*,z_j^*)-\beta p_j(x_j^0,u_j^0,z_j^0) \leq 0.$$

Furthermore, using the definition of strict convexity, we establish

$$f(\alpha u^*+\beta u^0,\alpha z^*+\beta z^0) =$$

$$=\sum_{j=1}^{J}\Bigg[w_j(\alpha x_j^*(T)+\beta x_j^0(T)) + \int_0^T c_j(\alpha x_j^*(t)+\beta x_j^0(t),\alpha u_j^*(t)+$$

$$+\beta u_j^0(t),\alpha z_j^*(t)+\beta z_j^0(t))dt\Bigg] <$$

$$<\sum_{j=1}^{J}\Bigg[\alpha w_j(x_j^*(T))+\beta w_j(x_j^0(T))+\alpha\int_0^T c_j(x_j^*(t),u_j^*(t),z_j^*(t))dt+$$

$$+\beta x_j(x_j^0(t),u_j^0(t),z_j^0(t))dt\Bigg] = \alpha f(u^*,z^*)+\beta(u^0,z^0) = f(u^*,z^*).$$

The contradiction proves the theorem. Local problems are considered analogously.

Theorem 5.4. *The functional* $f^*(\delta) = \sum_{j=1}^{J} f_j^*(u_j^*(\delta), z_j^*(\delta), \delta)$ *is concave in* δ *on the convex subset* Ω' *of the set* Ω.

Proof. Take δ^1, $\delta^2 \in \Omega'$, then $\delta^3 = \alpha\delta^1 + \beta\delta^2 \in \Omega'$, where $\alpha \geq 0$, $\beta \geq 0$, $\alpha + \beta = 1$. We have the following chain of inequalities:

$$f^*(\delta^3) = \sum_{j=1}^{J} f_j^*(u_j^*(\delta^3), z_j^*(\delta^3), \delta^3) =$$

$$= \sum_{j=1}^{J} [(\alpha f_j^*(u_j^*(\delta^3), z_j^*(\delta^3), \delta^1) + \beta f_j^*(u_j^*(\delta^3), \delta^2)] \geq$$

$$\geq \sum_{j=1}^{J} [(\alpha f_j(u_j^*(\delta^1), z_j(\delta^1), \delta^1) + \beta f_j(u_j^*(\delta^2), z^*(\delta^2), \delta^2)] =$$

$$= \alpha f^*(\delta^1) + \beta f^*(\delta^2).$$

If we recall the definition of the concave function, this completes the theorem.

Let the values $\delta(\rho)$ depend on the parameter ρ in the form $\delta(\rho) = \delta + \rho\bar{\delta}\delta$. Consider local problems that depend parametrically on ρ in a right neighborhood of $\rho = 0$. Assume that, for $\rho = 0$, the local problems (5.12), (5.2), (5.3) have unique solutions $(x_j^*(\delta), u_j^*(\delta), z_j^*(\delta))$ and unique optimal conjugated variables $(\psi_j^*(\delta), \eta^*(\delta))$. We also consider, that, in a certain neighborhood of the point $\rho = 0$, there exist partial derivatives $\partial u_j^*(t,\rho)/\rho$, $\partial x_j^*(t,\rho)/\partial\rho$, $\partial z_j^*(t,\rho)/\partial\rho$. Assuming that the functions occurring in (5.1)–(5.5) are doubly continuously differentiable and, using the theorem on the marginal value [25], we compute the derivative $df(\rho)/d\rho$.

Theorem 5.5. *The formula*

$$df(0)/d\rho = \sum_{j=1}^{J} \int_0^T \bar{\delta}_j^T(t)\xi_j(x_j^*(\delta), u_j^*(\delta), z_j^*(\delta), t)dt \tag{5.17}$$

is valid.

Proof. The derivatives of the functionals of local problems are computed as follows:

$$dfj(0)/d\rho = [\partial w_j/\partial x_j(T,0)]\,(\partial x_j(T,0)/\partial\rho) +$$

$$+\int_0^T \Big\{ (\partial c_j/\partial x_j)(\partial x_j(t,0)/\partial \rho) + (\partial c_j/\partial u_j)(\partial u_j(t,0)/\partial \rho) +$$

$$(\partial c_j/\partial z_j)(\partial z_j(t,0)/\partial \rho) + (\psi_j^*(\delta))^T \Big[A_j(t)\partial x_j(t,0)/\partial \rho +$$

$$+B_j(t)\partial u_j(t,0)/\partial \rho + C_j(t)\partial z_j(t,0)/\partial \rho - d(\partial x_j(t,0)/\partial \rho)/dt\Big] -$$

$$-(\eta_j^*(\delta))^T \Big[(\partial p_j/\partial x_j)(\partial x_j(t,0)/\partial \rho) + (\partial p_j/\partial u_j)(\partial u_j(t,0)/\partial \rho) +$$

$$+(\partial p_j/\partial z_j)(\partial z_j(t,0)/\partial \rho)\Big] +$$

$$+(\partial \delta(t,0)/\partial \rho)^T \sum_{i=1}^{J} L_{ji}(t)(M_i(t)x_i(t) + N_i(t)u_i(t) - z_j(t)) +$$

$$+\delta t \sum_{i=1}^{J} L_{ji}(t)(M_i(t)\partial x_i(t,0)/\partial \rho + N_i(t)\partial u_i(t,0)/\partial \rho - \partial z_j(t,0)/\partial \rho)\Big\} dt. \qquad (5.18)$$

The seventh term in the right-hand side of (1.18) is integrated by parts as follows:

$$\int_0^T (\psi_j^*(\delta))^T (d/dt)\,[\partial x_j(t,0)/\partial \rho]\, dt = \psi_j^*(\delta)(\partial x_j(t,0)/\partial \rho\Big|_0^T -$$

$$-\int_0^T d(\psi_j^*(\delta))^T/dt(\partial x_j^*(t,0)/\partial \rho)dt. \qquad (5.19)$$

Since $x_j(0) = \kappa$, the value in the first term of the right-hand side of (5.19) is equal to zero for $t = 0$. According to (5.11), the value of this term for $t = T$ is cancelled with the first term in the right-hand side of (5.18). Due to (5.8), (5.9), (5.10), the terms that contain $\partial x_j(t,0)/\partial \rho$, $\partial u_j(t,0)/\partial \rho$, $\partial z_j(t,0)/\partial \rho$ are cancelled. Finally, we have

$$df_j(0)/d\rho = \int_0^T (\bar{\delta}_j^T(t) \left(\sum_{i=1}^{J} L_{ji}(t) G_j(x_j(\delta,t), u_j(\delta,t), t) - z_j(t) \right) dt.$$

If we recall the definition $\xi_j(\delta, t)$ by formula (5.14), then this completes the proof of the theorem.

From relation (5.17), we see that the derivative $\partial f/\partial \delta_j$ of the functional

f is equal to

$$\partial f/\partial \delta_j = \sum_{i=1}^{J} L_{ji}(M_j x_j^*(\delta, t) + N_j u_j^*(\delta, t) - z_j^*(\delta, t)).$$

Therefore, the iterative process can be organized, for example, by the method of steepest ascent $\delta_j^{(\nu+1)} = \delta_j^{(\nu)} + \rho^{(\nu)} \partial f(\delta^{(\nu)})/\partial \delta_j$.

Here, the value of the parameter $\rho^{(\nu)}$ is defined from the condition

$$f(\delta^{(\nu)} + \rho^{(\nu)} \nabla f(\delta^{(\nu)})) \underset{\rho^{(\nu)} \geq 0}{\longrightarrow} \max .$$

Uses of other gradient procedures within the framework of the two-level technique are represented in [15, 17, 26, 51, 53-56, 60].

J.D. Pearson [35] proposed another decomposition method for the problem (1.1)–(1.5). For (1.1)–(1.5), we formulate the following problem:

$$f(u, z, \hat{y}) = \sum_{j=1}^{J} \left\{ w_j(x_j(T)) + \int_0^T [c_j(x_j(t), u_j(t), z_j(t), t) + \right.$$
$$\left. + R(y_i - \hat{y}_j)^T (y_j - \hat{y}_j)] \, dt \right\} \to \min,$$
$$dx_j(t)/dt = b_j(x_j(t), u_j(t), z_j(t), t), \; x_j(0) = \kappa_j,$$
$$p_j(x_j(t), u_j(t), z_j(t), t) \geq 0, \; y_j(t) = G_j(x_j(t), z_j(t), t),$$
$$z_j(t) = \sum_{i=1}^{J} L_{ji} y_i(t),$$

where Q_j, $j \in [1 : J]$ are positive definite $S_j \times S_j$-matrices, and R is a sufficiently large positive number. If R is large, we can hope that $y_j^*(\hat{y}) \approx \hat{y}$.

Consider independent problems for fixed $\hat{y}(t)$:

$$f_j(u_j, \hat{y}) = w_j(x_j(T)) + \int_0^T [c_j(x_j(t), u_j(t), \hat{z}_j(t), dt) +$$
$$+ R(y_j(t) - \hat{y}_j(t))^T Q_j (y_j(t) - \hat{y}_j(t))] \, dt \to \min,$$
$$dx_j(t)/dt = b_j(x_j(t), u_j(t), \hat{z}_j(t), t), \qquad x_j(0) = \kappa_j,$$
$$p_j(x_j(t), u_j(t), \hat{z}_j(t), t) \geq 0, \qquad y_j(t) = G_j(x_j(t), z_j(t)),$$
$$\hat{z}_j(t) = \sum_{i=1}^{J} L_{ji} \hat{y}_i(t), \qquad j \in [1 : J]$$

and denote by $f_j(u_j^*(\hat{y}), \hat{y})$ the optimal value of the functional as well as

$$f(\hat{y}) = \sum_{j=1}^{J} f_j(u_j^*(\hat{y}), \hat{y}).$$

We can see that solution of the problem

$$f(\hat{y}) \underset{\hat{y}}{\longrightarrow} \min$$

gives the optimal value to the original problem (1.1)–(1.5). To this end, the condition $y_j^*(\hat{y}) = \hat{y}_j$, $j \in [1:J]$ is also needed.

We can avoid introducing the penalty functional $R(y_j(t) - \hat{y}_j(t))^T \times \times Q_j(y_j(t) - \hat{y}_j(t))$ by correcting the output variables $y_j(t)$, $j \in [1:J]$ at the upper level. Then, we have:

$$\bar{z}_j(t) = \sum_{i=1}^{J} L_{ji}\bar{y}_i(t),$$

and the local problems look as follows:

$$f_j(u_j, \bar{y}) = w_j(x_j(T)) + \int_0^T c_j(x_j(t), y_j(t), \bar{z}_j(t))dt \to \min,$$

$$dx_j(t)/dt = b_j(x_j(t), u_j(t), \bar{z}_j(t), t), \qquad x_j(0) = \kappa,$$

$$p_j(x_j(t), u_j(t), \bar{z}_j(t), t) \geq 0, \qquad \bar{y}_j(t) = G_j(x_j(t), u_j(t), t).$$

This is another approach of J.D. Pearson. It is applicable when the dimensions of outputs y_j do not exceed the dimensions of control vectors u_j.

Mixed coordination can be applied, when primal and conjugated variables are corrected simultaneously at the upper level. One of these schemes is represented in the paper [16] by T. Ishimatsu, A. Mori and M. Takata for a special quadratic problem with two subsystems. We have

$$f(u) = 0,5 \int_0^T (x(t), Zx(t) + (u(t), \bar{R}u(t)))dt \to \min,$$

$$dx(t)/dt = Ax(t) + Bu(t), \qquad x(0) = \kappa,$$

where Q is nonnegative and R is positive definite $n \times n$- and $m \times m$-matrices, respectively. It is assumed that

$$A = \begin{bmatrix} A_{11} & A_{12} \\ A_{21} & A_{22} \end{bmatrix}, \qquad B = \begin{bmatrix} B_1 & 0 \\ 0 & B_2 \end{bmatrix},$$

$$Q = \begin{bmatrix} Q_1 & 0 \\ 0 & Q_2 \end{bmatrix}, \qquad \bar{R} = \begin{bmatrix} R_1 & 0 \\ 0 & R_2 \end{bmatrix},$$

and the problem is rewritten in the form

$$dx_j(t)/dt = A_{jj}x_j(t) + B_j u_j(t) + C_j y_j, \qquad x_j(0) = \kappa,$$

$$y_j = G_j x_j,$$

$$x = (x_1^T, x_2^T)^T, \qquad u = (u_1^T, u_2^T), \qquad y = (y_1^T, y_2^T)^T,$$

where $C_j G_i = A_{ji}$.

The Lagrangian looks as follows:

$$\mathcal{L} = \sum_{j,i=1,\ j\neq i} \int_0^T \left[0,5(x_j(t), Q_j x_j(t)) + 0,5(u_j(t), R_j u_j(t)) + \right.$$

$$\left. +\delta_j^T(G_i x_i - y_i) + \psi_j^T(A_{jj}x_j(t) + B_j u_j(t) + C_j y_j - dx_j(t)/dt)\right] dt.$$

Necessary optimality conditions give

$$d\psi_j(t)/dt + A_{jj}^T\psi_j(t) + Q_j x_j(t) + G_j^T\delta_i = 0, \qquad \psi_j = 0,$$

$$C_j^T\psi_j(t) - \delta_j(t) = 0, \qquad u_j(t) = -R_j^{-1}B_j^T\psi_j(t),$$

from where we obtain the following two-point problem:

$$\begin{aligned} d\nu_j(t)/dt &= D_{jj}\nu_j(t) + D_{ji}q_i(t), \\ L_{j1}\nu_j(0) &+ L_{j2}\nu_j(T) = I_j, \end{aligned} \tag{5.20}$$

$$q_j(t) = H_j\nu_j(t), \qquad j,i = 1,2,\ j \neq i, \tag{5.21}$$

where

$$\nu_j(x_j^T, \psi_j^T), \qquad I_j = (x_j^T, 0^T)^T, \qquad q_j = (y_j^T, \delta_j^T)^T,$$

$$L_{j1} = \begin{bmatrix} E_{n_1} & 0 \\ 0 & 0 \end{bmatrix}, \qquad L_{j2} = \begin{bmatrix} 0 & 0 \\ 0 & E_{n_2} \end{bmatrix}, \qquad H_j = \begin{bmatrix} G_j & 0 \\ 0 & C_j^T \end{bmatrix}$$

$$D_{jj} = \begin{bmatrix} A_{jj} & -B_j R_j^{-1} B_j^T \\ -Q_j & -A_{jj}^T \end{bmatrix}, \qquad D_{ji} = \begin{bmatrix} C_j & 0 \\ 0 & G_j^T \end{bmatrix}.$$

Let $\Phi_j(t)$ be the fundamental solution for (5.20) that satisfies the relation $d\Phi_j(t)/dt = D_{jj}\Phi_j(t)$, $\Phi_j(0) = E_{wn_j}$. Assume that the matrix $L_j = L_{j1} + L_{j2}\Phi_j(T)$ is invertible. Then, we write down the solution

$$\nu_j(t) = \Phi_j(t) L_j^{-1} I_j + \int_0^t \Gamma_j(t,\tau) D_{jj} q_i(\tau) d\tau, \tag{5.22}$$

where

$$F_j(t,\tau) = \begin{cases} \Phi_j(t)(E_{2n_j} - L_j^{-1} L_{j2}\Phi_j(T))\Phi_j^{-1}(\tau), & 0 \le \tau \le t, \\ -\Phi_j(t) L_j^{-1} L_{j2}(T)\Phi_j^{-1}(\tau), & t \le \tau \le T. \end{cases}$$

The obtained solution does not satisfy (5.21). Therefore, we apply two-level scheme. At the lower level, independent two-point problems are solved:

$$d\nu_j(t)/dt = D_{jj}\nu_j(t) + D_{ji} q_i(t),$$

$$L_{j1}\nu_j(0) + L_{j2}\nu_j(T) = I_j,$$

where $q_j(t)$, $j = 1, 2$ are given from the upper level. The upper level corrects $q_j(t)$ according to equations

$$q_j^{(\nu+1)}(t) = q_j^{(\nu)}(t) + \rho(H_j \nu_j^{(\nu)}(t) - q_j^{(\nu)}(t)), \qquad 0 < \rho \le 1, \tag{5.23}$$

where ρ is the step of the gradient method. At the initial iteration, the functions $q_j(t)$ can be given as zeros, and the value

$$\Omega = \sum_{j=1}^{2} \int_0^T \left(H_j \nu_j^{(\nu)}(\tau) - q_j^{(\nu)}(\tau)\right)^T \left(H_j \nu_j^{(\nu)}(\tau) - q_j^{(\nu)}(\tau)\right) d\tau.$$

is the measure of convergency.

The following statement establishes the convergency of the iteration process:

Theorem 5.6. *The iterative method converges to the solution of problem (5.20), (5.21) if the inequality*

$$0 < a < 1, \qquad a = T \max_{j,i \in [1:2],\, j \ne i} \|H_j \Gamma_j D_{ji}\| \tag{5.24}$$

holds, where the norm is understood as $||A|| = \max_{t,\tau\in[0,T]} \left[\textit{trace } A(t,\tau))^T(A(t,\tau))\right]^{1/}$

Proof. We introduce the norm of the vector

$$||z|| = \max_{0\le t\le T}(z^T(t), z(t))^{1/2}.$$

From (5.22), (5.23), we obtain

$$\nu_j^{(\nu+1)}(t) - \nu_j^{(\nu)}(t) = \rho\int_0^T \Gamma_j(t,\tau)D_{ji}(H_i\nu_i^{(\nu)}(\tau) - q_i^{(\nu)}(\tau))d\tau.$$

Multiplying the latter inequality on H_j, we obtain

$$||H_j\nu_j^{(\nu+1)}(t) - q_j^{(\nu+1)}|| = ||\rho\int_0^T H_j\Gamma_j(t,\tau)D_{ji}(H_i\nu_i^{(\nu)}(\tau) - q_i^{(\nu)}(\tau))d\tau|| \le$$

$$\le \rho a||H_i\nu_i^{(\nu)}(t) - q_i^{(\nu)}(t)||.$$

Taking account of (5.23), the last inequality gives

$$\begin{aligned}||H_j\nu_j^{(\nu+1)}(t) - q_j^{(\nu+1)}(t)|| &= ||H_j\nu_j^{(\nu+1)}(t) - H_j\nu_j^{(\nu)}(t) + \\ &H_j\nu_j^{(\nu)}(t) - q_j^{(\nu)}(t) - \rho(H_j\nu_j^{(\nu)}(t) - q_j^{(\nu)}(t))|| \le \\ &\le ||H_j(\nu_j^{(v+1)}(t) - q_j^{(\nu)}(t))|| + (1-\rho)||H_j\nu_j^{(\nu)}(t) - q_j^{(\nu)}(t)|| \le \\ &\le \rho a||H_j\nu_j^{(\nu)}(t) - q_j^{(\nu)}(t)|| + (1-\rho)||H_j\nu_j^{(\nu)}(t) - q_j^{(\nu)}(t)||.\end{aligned}$$

Finally, we obtain

$$\sum_{j=1}^{2}||H_j\nu_j^{(\nu+1)}(t) - q_j^{(\nu+1)}(t)|| \le \left[1-\rho^{(1-a)}\right]\sum_{j=1}^{2}||H_j\nu_j^{(\nu)}(t) - q_j^{(\nu)}(t) - q_j^{(\nu)}(t)||,$$

where the strict inequality $[1-\rho(1-a)] < 1$ holds, since $0 < a < 1$, $0 \le \rho \le 1$ by the condition. Thus,

$$\lim_{\nu\to\infty}||H_j\nu^{(\nu+1)}(t) - q_j^{(\nu+1)}(t)|| = 0,$$

which proves Theorem 5.6.

It remains to note that condition (5.22) characterizes the weak connection between subsystems. We will encounter conditions of this type.

In [27], M.J. Mahmond considered a block optimal control problem with nonlinear functions in the right-hand sides of differential equations. We have

$$f(u) = 0,5 \int_0^T [((x(t) - \tilde{x}(t)), Q(x(t) - \bar{x}(t)))+$$

$$+(u(t) - \tilde{u}(t), R(u(t) - \tilde{u}(t)))]\, dt \to \min,$$

$$dx(t)/dt = Ax(t) + Bu(t) + c(x(t)), \qquad x(0) = \kappa,$$

where A, B, Q, R are block diagonal matrices, Q and R are symmetric, nonnegative and positive definite matrices, respectively; $c(x(t))$ is an n-dimensional vector function depending on all coordinates of the phase vector x. The functions $\tilde{x}(t)$, $\tilde{u}(t)$ are given. We partition the vector function $c(x(t))$ in accordance with the partitioning into subsystems: $c(x(t)) = \{c_1^T(x(t)), \dots, c_y^T(x(t))\}^T$ and introduce the notation

$$y_j(t) = c_j(x(t)).$$

We formulate problems for subsystems:

$$f_j(u_j) = 0,5 \int_0^T [(x_j(t) - \tilde{x}_j(t)), Q_j(x_j(t) - \tilde{x}_j(t))+$$

$$+(u_j(t) - \tilde{u}_j(t)), R_j(u_j(t) - \tilde{u}_j(t))]\, dt \to \min, \tag{5.25}$$

$$dx_j(t)/dt = A_j x_j(t) + B_j u_j(t) + y_j(t), \; x_j(0) = \kappa,$$

$$y_j(t) = c_j(x(t)).$$

The Lagrangian for (5.25) has the following form:

$$\mathcal{L}_j = \int_0^T \{0,5((x_j(t) - \tilde{x}_j(t), Q_j(x_j(t) - \tilde{x}_j(t)))+$$

$$+0,5((u_j(t) - \tilde{u}_j(t)), R_j(u_j(t) - \tilde{u}_j(t))) + \psi_j^T [A_j x_j(t) + B_j u_j(t)+$$

$$+ [y_j(t) - dx_j(t)/dt] + \delta_j^T [y_j - c_j(x(t))]\}\, dt =$$

$$= \int_0^T [H_j - \psi_j^T dx_j(t)/dt]\, dt,$$

where the following Hamiltonian

$$H_j = 0,5((x_j(t) - \tilde{x}_j(t)), Q_j(x_j(t) - \tilde{x}_j(t))) + 0,5((u_j(t) - \\ -\tilde{u}_j(t)), R_j(u_j(t) - \tilde{u}_j(t))) + \psi_j^T(t)(A_j x_j(t) + B_j u_j(t) + \\ + y_j(t)) + \delta_j^T(-g_j(t) + c_j(x(t))).$$

is introduced.

Necessary optimality conditions look as follows:

$$\partial H_j/\partial\psi_j = dx_j(t)/dt = A_j x_j(t) + B_j u_j(t) + y_j(t), \\ x_j(0) = \kappa_j, \tag{5.26}$$

$$-\partial H_j/\partial x_j = d\psi_j(t)/dt = -Q_j/x_j(t) - \tilde{x}_j(t)) - A_j^T\psi_j(t) - \\ - [\partial c_j(x(t))/\partial x_j]^T \delta_j(t), \qquad \psi)T) = 0, \tag{5.27}$$

$$\partial H_j/\partial u_j = 0 = R_j(u_j(t) - \tilde{u}_j(t)) + B_j^T\psi_j(t), \tag{5.28}$$

$$\partial H_j/\partial\delta_j = 0 = y_j(t) - c_j(x(t)), \tag{5.29}$$

$$\partial H_j/\partial y_j = 0 = \psi_j(t) - \delta_j(t), \tag{5.30}$$

Using (5.28), (5.30) in (5.26), we obtain

$$dx_j(t)/dt = A_j x_j(t) - B_j R_j^{-1} B_j^T \psi_j(t) + B_j\tilde{u}_j(t) + y_j(t), \; x_j(0) = \kappa_j, \tag{5.31}$$

$$d\psi_j(t)/dt = -Q_j(x_j(t) - \tilde{x}_j(t)) - \left\{A_j^T + [\partial c_j(x)/\partial x_j]^T\right\}\psi_j(t), \\ \psi_j(T) = 0. \tag{5.32}$$

Solution of the stated problem is obtained through the following three-level scheme:

Lower level. The trajectories $\psi_j^{(\nu)}(t)$, $y_j^{(\sigma)}(t)$ are given from the intermediary and upper level, and the differential equations (5.31) are integrated for $j \in [1 : J]$. We obtain $x_j(t)$, $j \in [1 : J]$, which is passed to the intermediate level.

Intermediate level. The differential equations (5.32) are back integrated for given $x(t)$. We obtain $\psi_j^{\delta)}(t)$, $j \in [1 : J]$ and test the condition

$$\left[\int_0^T ||\psi_j^{(\sigma)}(t) - \psi_j^{(\nu)}(t)||dt\right]^{1/2} \leq \varepsilon_\psi, \qquad j \in [1 : J], \tag{5.33}$$

where ε_ψ is a given precision. If condition (5.33) does not hold, then we come back to the lower level, where we pass $\psi_j^{(\sigma)}(t)$. In the opposite case, we go to the upper level with the values $x(t)$.

Upper level. The variables $y_j^{(\pi)}(t) = c_j(x(t))$ are computed, where $x(t)$ is obtained from the lower level. We test the condition

$$\left[\int_0^T ||c_j^{(\pi)}(t) - y_j^{(\sigma)}(t)||dt\right]^{1/2} \leq \varepsilon_y, \qquad j \in [1:J], \tag{5.34}$$

where ε_y is a given precision. If condition (5.34) does not hold, then we go to the lower level with $y^{(\pi)}(t)$ and $\psi^{(\sigma)}(t)$ from the intermediary level. The iterations are repeated. If inequality (5.34) holds, then the process terminates.

Iterative convergence to the optimum of the original problem is established under the assumption of smallness of partial derivatives $\partial c/\partial x$, which corresponds to weak cross-connections between subsystems.

M.G. Singh and A. Titli [53] attempted to extend the method of Lagrange functions to hierarchical problems of optimal control with block separable binding constraints

$$\begin{gathered} f = \sum_{j=1}^{J} \left[w_j(x_j(T)) + \int_0^T c_j(x_j(t), u_j(t), z_j(t), t)dt \right] \to \min, \\ dx_j(t)/dt = b_j(x_j(t), u_j(t), z_j(t), t), \; x_j(0) = \kappa_j, \; j \in [1:J], \\ \sum_{j=1}^{J} d_j(x_j(t), z_j(t)) = 0, \end{gathered} \tag{5.35}$$

where d_j, $j \in [1:J]$ are vector functions of dimension L.

The Lagrangian has the following form:

$$\mathcal{L}(x,u,z,\delta) = \sum_{j=1}^{J} \Bigg\{ w_j(x_j(T)) + \\ + \int_0^T \left[c_j(x_j(t), u_j(t), z_j(t), t) + \delta^T(t) d_j(x_j(t), u_j(t), z_j(t), t\right] \Bigg\}.$$

According to the duality principle, the saddle point

$$\mathcal{L}(x,u,z,\delta) \to \max_{\delta} \min_{x,u,z} \tag{5.36}$$

is sought, where the variables $x(t)$, $u(t)$, $z(t)$ satisfy differential equations in (5.35) with initial conditions. Problem (5.36) is solved by a two-level scheme. On the lower level, for fixed $\delta(t)$, subsystems solve independent problems of the form

$$f_j(\delta) = w_j(x_j(T)) + \int_0^T (c_j(x_j(t), u_j(t), z_j(t), t) +$$

$$+\delta^T(t) d_j(x_j(t), z_j(t)))dt \to \min,$$

$$dx_j(t)/dt = b_j(x_j(t), u_j(t), z_j(t), t), \qquad x_j(0) = \kappa.$$

The upper level changes dual variables by the formula

$$\delta^{(\nu+1)}(t) = \delta^{(\nu)}(t) + \rho \sum_{j=1}^{J} d_j \left(x_j(t), z_j(t), t^{(\nu)}\right),$$

where ν is the iteration number, ρ is the step of gradient method, and the functions d_j are computed from optimal solutions of the lower level problems. Convergence of the iterative algorithm is guaranteed by

$$\lim_{\nu \to \infty} \sum_{j=1}^{J} d_j \left(x_j(t), z_j(t), t^{(\nu)}\right) = 0.$$

Justification of this approach for problems of the type of (5.35) requires detailed analysis.

We enlarge on local problems. The Hamiltonian has the form

$$H_j = c_j/x_j(t), u_j(t), z_j(t), t) + \delta^T(t) d_j(x_j(t), z_j(t), t) + \psi_j^T(t) b_j(x_j(t), u_j(t), z_j(t), t)$$

Necessary optimality condition are written in the following form:

$$\begin{aligned} &dx_j(t)/dt = b_j(x_j(t), u_j(t), z_j(t), t), \; x_j(0) = \kappa_j, \\ &d\psi_j(t)/dt = \partial H_j/\partial x_j = \\ &= -\partial f_j/\partial x + (d_j/\partial x_j)^T \delta_j + (\partial b_j/\partial x_j)^T \psi_j(t), \\ &\psi_j(T) = - \; w_j/\partial x_j(T), \; \partial H_j/\partial u_j = 0, \; \partial H_j/\partial z_j = 0. \end{aligned} \tag{5.37}$$

Assume that the two latter equations in (5.37) determine the relation

$$u_j = u_j(x_j, \psi_j), \qquad z_j = z_j(x_j, \psi_j).$$

Using these relations in the differential equations in (5.35), we obtain a two-point problem with equations

$$d\nu_j(t)/dt = F_j(\nu_j(t), t),$$

where $\nu_j = [x_j^T, \psi_j^T]^T$, and the initial and boundary conditions are taken from (5.37). This two-point problem is solved by the quasilinearization method, see, e.g., the book by R. Bellman and R. Kalaba [6]. If an approximation $\nu_j^{(\sigma)}(t)$ is given, then the linear equations of the form

$$d\nu_j^{(\sigma+1)}(t)/dt = F_j(\nu_j^{(\sigma)}(t), t) + W_j(\nu_j^{(\sigma)}(t), t)(\nu_j^{(\sigma+1)}(t) - \nu^{(\sigma)}(t)),$$

are integrated, where $W_j(\nu_j^{(\sigma)}(t), t)$ is the Jacobi matrix for the function $F_j(\nu_j^{(\sigma)}(t), t)$ computed on the trajectory $\nu_j^{(\sigma)}(t)$.

The iterative decomposition of block optimal control problems proposed above assumes the solution of local problems for subsystems at each step. For dynamical subproblems one or other numerical methods of optimal control are used. Therefore, we always need to take care about the precision of computation in order to ensure the convergence of the iterative process. We can, nevertheless, do the opposite, namely, take a numerical optimal control method as an underlying method and construct a two-level optimization scheme in a way such that it is applied to the original block problem as well as to intermediary problems for subsystems. This approach is proposed in the work by G. Petriczek [37] based on the method of penalty functionals of A.B. Balakrishnanan [2,3], which will be our starting point.

Consider the problem of optimal control in the standard form:

$$\begin{aligned}
&f = w(x(T) + \int_0^T c(x(t), u(t), t)dt \to \min, \\
&dx(t)/dt = F(x(t), u(t), t), \qquad x(0) = \kappa, \qquad\qquad (5.38)\\
&p(x(t), u(t), t) = 0, \\
&u(t) \in U,
\end{aligned}$$

where, in particular, U is a compact set in Hilbert space $\mathcal{L}_2^m[0, T]$ of m-dimensional vector functions, and the components of n-dimensional phase vector $x(t)$ are assumed to be absolutely independent functions with derivatives in $\mathcal{L}_2^n[0, T]$.

Instead of (5.38), we solve a problem of unconstrained minimization of the form

$$h(\varepsilon, x(t), u(t)) = w(x(T)) + \int_0^T [(1/(2\varepsilon))||dx(t)/dt - F(x(t), u(t), t||+$$
$$+c(x(t), u(t), t) + (1/(2\varepsilon))||p(x(t), u(t), t)||_{R^n}]\, dt \to \min, \qquad (5.39)$$

where ε is a sufficiently small positive number.

In [2,3], it was established that, under some assumptions and in the case $\varepsilon \to 0$, the solution of problem (5.39) converges to solution (5.38) in the corresponding sense. We enlarge on a possible computation scheme for solution of (5.39). The procedure uses the Ritz method [28], and we do not need to solve dynamical equations. Let $C^n[0,T]$ be a Banach space of continuous vector functions with uniform norm. By $\{b_s(t)\}$ denote the sequence of base functions in $C^n[0,T]$. For given k, let S_k be a linear hull of $\{b_s(t)\}$ $s \in [1:k]$. We take the solution of problem (5.39) over phase coordinates in the form

$$x_1(t) = \sum_{s=1}^{k} a_s \int_0^t b_s(\xi) d\xi, \qquad (5.40)$$

where a_s, $s \in [1:k]$ are unknown coefficients. Obviously, the larger k the better is the accepted approximation.

First, we give some $x_1(t)$. For example, in (5.40), we put $a_s = 0$, $s \in [1:k]$ and find

$$\min_{u(t)\in U} [(1/(2\varepsilon))||dx(t)/dt-$$
$$- F(x, u, t)||^2_{R^n} + ||p(x, u, t)|| + c(x, u, t)]\,. \qquad (5.41)$$

We assume that this minimum is attained at a single point from U for every t, $dx(t)/dt$, x and ε Let $u_1(t)$ be an extremal element and the minimum value be equal to

$$r(\varepsilon, dx_1(t)/dt, x_1(t), u_1(t), t).$$

If $x_2(t)$ is chosen, then the problem

$$\inf_{x(t)\in S_k} \left[\int_0^T r(\varepsilon, x(t), dx(t)/dt, u_1(t), t)dt + w(x(T)) \right] \qquad (5.42)$$

is solved. The next control $u_2(t)$ is found from (5.41), where $x_2(t)$ is taken as $x(t)$. Thus, we obtain the sequence $x_\nu(t)$, $u_\nu(t)$, the convergence of which is analyzed in detail in [9]. It remains to note that coefficients $a = \{a_s\}$ in the solution of (5.42) are computed by iteration:

$$a_{\sigma+1} = a_\sigma - H_\sigma^{-1} G_\sigma,$$

where $G_0 = \partial/\partial a_s(h(\varepsilon, a_\sigma, u))$, and H_0 is a matrix of dimension $k \times k$-matrix with components

$$\frac{1}{\varepsilon}\int_0^T \left[\frac{\partial}{\partial a_i}(dx(t)/dt - F(x(t), u(t), t)),\ \frac{\partial}{\partial a_i}(dx(t)/dt - \right.$$
$$\left. - F(x(t), u(t), t))\right] dt + \frac{1}{\varepsilon}\int_0^T \left[\frac{\partial}{\partial a_i}(p(x(t), u(t), t)),\ \frac{\partial}{\partial a_i}(p(x(t), u(t), t))\right] dt +$$
$$+ \frac{\partial}{\partial a_i}\frac{\partial}{\partial a_i} w(x(T)).$$

Reformulation of problem (5.1)–(5.5) taking account of the described ε-technique gives $f(\varepsilon, x, u, z) \to \min$ under the condition

$$z(t) = LG(x(t), u(t), t), \tag{5.43}$$

where

$$f(\varepsilon, x, u, z) = w(x(T)) + \int_0^T c(x(t), u(t), z(t), t)dt +$$
$$+ (1/(2\varepsilon)) \int_0^T \|dx(t)/dt - F(x(t), u(t), z(t), t)\|_{R^n}^2 dt.$$

We have $c = \sum_{j=1}^J c_j$, $w = \sum_{j=1}^J w_j$ here, and the vector functions F and G have components F_j and G_j, $j \in [1 : J]$. The same holds for vectors x, u, z.

Solution of ε-problem (5.43) is obtained by a two-level scheme. The subsystems solve their problems of unconstrained minimization

$$f_j(\varepsilon, x_j, u_j, z_j, \delta) \to \min,$$

where

$$f_j(\varepsilon, x_j, u_j, z_j, \delta) =$$
$$= w_j(x_j(T)) + \int_0^T c_j(x_j(t), u_j(t), z_j(t), t)dt + \int_0^T (\delta_j^T z_j + \kappa_j^T G_j(x_j(t), u_j(t), t))dt +$$
$$+(1/(2\varepsilon)) \int_0^T ||dx_j(t)/dt - F_j(x_j(t), u_j(t), z_j(t), t)||_{R^{n_j}} dt.$$

On the upper level, the problem

$$f^*(\varepsilon, \delta) \underset{\delta \in \Omega}{\longrightarrow} \max,$$

$$f^*(\varepsilon, \delta) = \sum_{j=1}^{J} f_j^*(\varepsilon, \delta) = \sum_{j=1}^{J} f_j^*(\varepsilon, x_{\varepsilon j}^*(\delta), u_{\varepsilon j}^*(\delta), z_j^*(\delta))$$

is solved, where the index $*$ corresponds to the optimal solution of local ε-problems.

Comments and References to Chapter 5

In block programming, the method based on partition of variables should be mentioned. This approach is presented in detail by L.S. Lasdon [22]. It is applied, in particular, under binding constraints. It uses relaxation of constraints by the methods of M.S. Geoffrion [12]. This approach was started by J.C. Benders [7], by K. Ritter [40], and by J.B. Rosen [41]. In the papers of Yu.V. Mednitskii [32, 33], a block problem is considered, where the binding constraints use only subsystem resources. In this case, instead of laborious three-level algorithm, optimization at two levels, using primal and dual variables, is proposed. In the following papers, the simplex method is adapted to linear block problems: G.Sh.Rubinshtein [42], R.A. Zvyagina [73–75], M.A. Yakovleva [68, 69]. In this way, the partition into subproblems is done during the computation process.

Among numerous works in hierarchical optimal control systems, we should note the books by M.G. Singh and A.Titli [52], A.Titli [60].

J.Sanders in his work [44] proposes a decomposition approach, where first some of the variables are fixed, and the optimization is carried out with respect to the other variables. Then, the previously varied variables are fixed and the remaining variables are used for the optimization. This process is sequentially iterated. The proof of convergence is most important here. Y.Takahara [57] proves the convergence theorem in this process. Another fine result of this sort is found in G.Cohen [8].

Another approach is the artificial introduction of subsystems. D.A.Wismer [67] considers difference schemes over spatial coordinates in optimal control problems with partial differential equations. To obtain a more precise approximation, we need a great number of points in the net. Each point corresponds to a subsystem and approximation difference equations define the relations between subsystems. On the one hand they are weak, since they relate only to a neighboring point. On the other hand, there are no small coefficients in the cross-connections. In this case only computer experiments can verify the convergence of iterative procedures using the two-level technique. E.J.Bauman [4] partitions the time interval of optimization into subintervals. Here, each subsystem optimizes in its subinterval. Then, the upper level corrects the local solutions in order to find the extremum in the whole interval. Partition points occur, where smoothness conditions of the initial functions are violated or where additional conditions hold on the phase variables in intermediate instances.

References to Chapter 5

[1] Abadie I.M. and Williams A.C., Dual and Parametric Method in Decomposition, *Recent Advances Math. Programm.*, N.Y., Mc Graw Hill, 1963.

[2] Balakrishnan A.V., A Computational Approach to the Maximum Principle, J. Comp. and Syst. Sci., 1971, vol. 5, no. 2, pp. 163–191.

[3] Balakrishnan A.V., On a New Computing Technique in Optimal Control, SIAM J. Contr., 1968, vol. 6, no. 2, pp. 149–173.

[4] Bauman E.J., Trajectory Decompositions // Optimization Methods for Large-Scale Systems, New-York, Mc Graw-Hill, 1971, pp. 275-290.

[5] Bell E.J., Primal-Dual Decomposition Programming, *Ph.D. Thesis*, Operations Research Center, Univ. of California at Berkeley, 1965.

[6] Bellman R. and Kalaba R., *Kvazilinearizatsiya i linejnye kraevye zadachi* (Quasilinearization and Nonlinear Boundary Problems), Moscow: Mir, 1968.

[7] Benders J.C., Partitioning Procedures for Solving Mixed Variables Programming Problems, Numer. Math., vol. 4, no. 3, 1962, pp. 238–252.

[8] Cohen G., On an Algorithm of Decentralized Optimal Control, *J. Math. Anal. and Appl.*, 1977, vol. 59, no. 2, pp. 242-259.

[9] Dantzig G.B. and Wolfe P., Decomposition Algorithm for Linear Programs, *Econometrica*, vol. 29, no. 4, 1961, pp. 767–778.

[10] Dantzig G.B. and Wolfe P., Decomposition Principle for Linear Programs, *Oper. Res.*, vol. 8, no. 1, 1960, pp. 101–111.

[11] Dantzig G.B. and Wolfe P., Decomposition Algorithm for Problems of Linear Programs, *Matematika*, vol.8, no.1, 1964, pp. 151–160.

[12] Geoffrion A.M., Relaxation and the Dual Method in Mathematical Programming, Working paper 135, Western Management Science Institute, University of California at Los Angeles, 1968.

[13] Geoffrion A.M., Primal Resources-directive Approaches for Optimizing Nonlinear Decomposable Systems, Memorandum RM-5829-PR, The RAND Corporation, Santa Monica, California, 1958.

[14] Gol'shtein E.G. and Yudin D.B., *Novye napravleniya v linejnom programmirovanii* (New Trends in Linear Programming), Moscow: Sov. Radio, 1966.

[15] Grateloup G. and Titli A., Two-level Dynamic Optimization Methods, *J. Optimiz. Theory and Appl.*, 1975, vol. 15, no. 5, pp. 533–547.

[16] Ishimatsu T., Mohri A. and M. Takata, Optimization of Weakly Coupled Subsystems by a Two-level Method, *Int. J. Contr.*, 1975, vol. 22, no. 6, pp. 877-882.

[17] Jovdan M.R., A Unified Theory of Optimal Multilevel Control, *Int. J. Contr*, 1966, vol. 11. no. 1, pp. 93–101.

[18] Kate A., Decomposition of Linear Programms by Direct Distribution, *Econometrica*, vol. 40, no. 5, 1972, pp. 883–898.

[19] Khizder L.A., *Dokazatel'stvo chodimosti prochessa iterativnogo agregirovaniya v obshem sluchae* (Proof of Convergency of the Process of Iterative Aggregation in General Case) // Study in Mathematical Economics and Adjoint Problems, Moscow: Izd. Mosk. Gos. Univ., 1971.

[20] Kornai I. and Liptak T., *Planirovanie na dvykh urovnyakh*
(Planning on Two Levels) // Application of Mathematics in Economical Studies, vol. 3, Moscow: Mysl', 1965.

[21] Krasnosel'skii M.A., Ostrovskii A.Yu. and Sobolev A.V., *O shodimosti metode odnoparametricheskogo agregirovaniya* (On Convergence of a Method of One-parametric Aggregation), *Avtom. Telemekh.*, no. 9, 1978, pp. 102–109.

[22] Lasdon L.S., *Optimizatsiya bol'shikh sistem* (Optimization of Large Systems), Moscow: Nauka, 1975.

[23] Levin G.M. and Tanaev V.S., *Dekompozichionnye metody optimizatsii proektnykh reshenij* (Decomposition Methods of Optimization of Design Solutions), Minsk: Nauka Tekhn., 1978.

[24] Levin G.M. and Tanaev V.S., *O parametricheskoj dekompozichii ekstremak'nykh zadach* (On Parametric Decomposition of Extremal Problems), *Kibernetika*, no. 3, 1977.

[25] Levitin E.S., *O differentsiruemosti po parametru optimal'nogo znacheniya parametricheskikh zadach matematicheskogo programmirovaniya* (On Differentiability by the Parameter of Optimal Value of Parametric Mathematical Programming Problem), *Kibernetika*, 1976, no. 1, pp. 44-59.

[26] Luderer B., Losung Stkuktutierter Aufgaben der Optimalen Steuerung (mittels Zerlegungsverfahren), *Math. Operationsforsch. Stat., Series Optimization*, 1980, vol. 11, no. 4, pp. 593–603.

[27] Mahmoud M.S., Dynamic Multilevel Optimization for a Class of Non-linear Systems, *Int. J. Contr.*, 1979, vol. 30, no. 6, pp. 627-648.

[28] Marchuk G.I., *Metody vychislitel'noj matematiki* (Methods of Computational Mathematics), Moscow: Nauka, 1977.

[29] Marshak V.D., *Algoritm resheniya zadachi raspredeleniya resursov v otrasli* (An Algorithm for the Solution of a Distribution Problem in a Branch) // Proc. Inst. Matem. Sibirsk. Otdel. Akad. Nauk SSSR, vol. 10(27), Novosibirsk, 1973.

[30] Mauer I., *Shtrafnaya konstanta v blochnom programmirovanii* (Penalty Constant in Block Programming), *Izv. Akad. Nauk Estonsk. SSR, ser. Fizika Matem.*, vol. 20, no. 4, 1971, pp. 401–405.

[31] Mednitskii V.G., *Ob optimal'nosti agregirovaniya v blochnoj zadache linejnogo programmirovaniya* (On Optimality of Aggregation in the Block Linear Programming Problem) // Mathematical Methods for Solution of Economical Problems, vol. 3, 1972, pp.3–17.

[32] Mednitskii Yu.V., On Decomposition of the Linear Programming Problem with Binding Constraints and Variables, *Izv. Ross. Akad. Nauk*, Teor. Sist. Upr., 1998, no. 4, pp. 134-140.

[33] Mednitskii Yu.V., On Parallel Use of the Decomposition Method in a Pair of Dual Linear Programming Problems, *Izv. Ross. Akad. Nauk*, Teor. Sist. Upr., 1998, no. 1, pp. 107-112.

[34] Movshovich S.M., *Metod nevyazok dlya resheniya zadach blochnoj struktury* (Discrepancy Method for the Solution of Problems with Block Structure), *Ekonomika i Matem. Metody*, vol. II, no. 4, 1966, pp. 571–577.

[35] Pearson J.D., Dynamic Decomposition Techniques //Optimization Methods for Large-Scale Systems with Applications, New York, Mc Graw Hill, 1971, pp. 121-190.

[36] Pervozvanskaya T.N. and Pervozvanskii A.A., *Algoritm poiska optimal'nogo raspredeleniya chentralizovannykh resursov* (Algorithm for the Search of Optimal Distribution of Centralized Resources), *Izv. Akad. Nauk, Tekh. Kibern.*, no. 3, 1966, pp. 16–19.

[37] Petriczek G., On the Use of the *-technique in Two-level Dynamic Optimization, in Arch. Autom. i Telemec., 1978, vol. XXIII, no. 4, pp. 443-459.

[38] Polyak B.T. and Tret'yakov N.V., *Ob odnom oterachionnom metode linejnogo programmirovaniya i ego ekonomicheskoj interpretachii* (On an Iteration Method in Linear Programming and Its Economical Interpretation), *Ekonomika i Matem. Metody*, vol. VIII, no. 5, 1972.

[39] Razumikhin B.S., *Metod fizicheskogo modelirovaniya v matematicheskom programmirovanii i ekonomike. Metod iterativnoj dekompozichii i zadacha o raspredelenii resursov* (Method of Physical Modeling in Mathematical Programming and Economics. Method of Iterative Decomposition and the Problems of Resource Distribution), *Avtom. Telemekh.*, no. 11, 1972, pp. 111–123.

[40] Ritter K., A Decomposition Method for Linear Programming Problems with Coupling Constraints and Variables, Mathematical Research Center, University of Wisconsin, rept. 739, 1967.

[41] Rosen J.B., Primal Partition Programming for Block Diagonal Matrices, Numer. Math., vol. 6, 1964, pp. 250–260.

[42] Rubinshtein G.Sh., *O reshenii zadach linejnogo programmirovaniya bol'shogo ob'ema* (On Solution of Large Linear Programming Problems) //Optimal Planning, no. 2, Novosibirsk, 1964.

[43] Sanders J., A Nonlinear Decomposition Principle, *Oper. Res.*, vol. 13, no. 2, 1965, pp. 266-271.

[44] Sanders J., Multilevel Control, *IEEE Trans. Appl. and Ind.*, 1964, vol. 83, no. 75, pp. 473-479.

[45] Schwartz B. and Tichatschke P., Über eine Zerlegungsmethode zur Lösung grossdimensionierter linearer Optimierungsaufgaben, *Math. Operationsforsch. Stat.*, vol. 6, H. 1, 1975, pp. 123-128.

[46] Shchennikov B.A., *Blochnyj metod resheniya sistem linejnykh uravnenij bol'shoj razmernosti* (A Block Method for the Solution of Large Dimension Systems of Linear Equations), *Ekonomika i Mat. Metody*, vol. 1, no. 6, 1965, pp. 841–852.

[47] Shchennikov B.A., *Primenemie metode uterativnogo agregirovaniya dlja resheniya sistem linejnykh upavnenij* (Application of the Method of Iterative Aggregation for the Solution of Systems of Linear Equations), *Ekonomika i Mat. Metody*, vol. 11, no. 5, 1966, pp. 723–731.

[48] Shchennikov B.A., *Metod agregirovaniya dlja reshenija sistemy linejnykh uravnenij* (Aggregation Method for the Solution of a System of Linear Equations), *Dokl. Akad. Nauk SSSR*, vol. 173, no. 4, 1967, pp. 781–784.

[49] Shor N.Z., *Primenenie obobshennogo gradientnogo spuska v blochnom programmirovanii* (Application of the Generalized Gradient Descent in Block Descent), *Kibernetika*, no. 3, 1967, pp. 53–56.

[50] Silverman G., Primal Decomposition of Mathematical Programs by Resource Allocation, *Oper. Res.*, vol. 20, no. 1, 1972, pp. 58–74.

[51] Singh S.N., Decoupkling in Class of Nonlinear System by Output Feed-back, *Unoform. and Contr.*, 1974, vol. 26, no. 1, pp.61–81.

[52] Singh M.G. and Titli A., Systems: Decomposition, Optimization and Control, Oxford: Pergamon Press, 1978.

[53] Singh M.G. and Titli A., Closed-loop Hierarchical Control for Nonlinear Systems Using Quasilinearization, Automatica, 1975, vol. 11, no. 5, pp. 541-547.

[54] Singh M.G. and Hassan M.F., Local and Global Optinal Control for Nonlinear Systems Using Two-level Methods, *Int. J. Syst. Sci.*, 1976, vol. 8, no. 2, pp. 1375–1383.

[55] Singh M.G. and Hassan M.F., A Comparison of Two Hierarchical Optimization Methods, *Int. J. Syst. Sci.*, 1978, vol. 7, no. 6, pp. 603–611.

[56] Singh M.G. and Hassan M.F., Hierarchical Optimazation for Nonlinear Dynamical System with Non-separable cost function, *Automatica*, 1978, vol. 14, no. 1, pp. 99–101.

[57] Takahara Y., Multilevel Structure for a Class of Dynamic Optimization Problems, Optimaization Methods for Large-Scale Systems with Application, New-York: Mc Graw-Hill, 1971.

[58] Ter-Krikorov A.M., *Optimal'noe upravlenije i matematicheskaya ekonomika* (Optimal Control and Mathematical Economy), Moscow: Nauka, 1977.

[59] Timokhin S.G., *Dekompozichionnyj podhod k resheniju zadachi linejnogo programmirovaniya bol'shoj razmernosti* (Decomposition Approach to the Solution of a Linear Programming Problem of Higher Dimension), *Ekonomika i Matem. Metody*, vol. XIII, no. 2, 1977, pp. 330–341.

[60] Titli A., Commande hierarchisee et optimization des processus complexes, Dugon, 1975.

[61] Umnov A.E., *Metod shtrafnykh funkchij v zadachakh bol'shoj razmernosti* (The Method of Penalty Functions in Problems with Large Dimensions), *Zh. Vych. Mat. Mat. Fiz.*, vol. 15, no. 6, 1975, pp. 1399–1411.

[62] Vakhutinskii I.Ya., Dudkin L.N., and Shchennikov B.A., *Iterativnoe agregirovanie v nekotorykh optimal'nykh matematicheskikh modelyakh* (Iterative Aggregation in Some Optimal Economical Models), *Ekonomika i Matem. Metody*, vol. IX, no. 3, 1973, pp. 420–434.

[63] Verina L.F., *O dekompozichii zadach linejnogo programmirovaniya kvaziblochnoj struktury* (On Decomposition of Linear Programming Problems with Quasi-block Structure), *Izv. Akad. Nauk Belorus. SSR, Fiz.-Mat. Nauki*, no. 6, 1975, pp. 18-21.

[64] Verina L.F. and Tanaev V.S., *Dekompozichionnye podhody k resheniju zadach matematicheskogo programmirovanija* (Decompositional Approaches to the Solution of Mathematical Programming Problems), an Overview. *Ekonomika i Matem. Metody*, vol. XI, no. 6, 1975, pp. 1160–1172.

[65] Volkonskii V.A., *Optimal'noe planirovanie v uslovijakh bol'shoj razmernosti. Iterativnyje metody i princhip dekompozitsii* (Optimal Planning for Large Number of Dimensions. Iterative Methods and Decomposition Principle), *Ekonomika i Matem. Metody*, vol. 1, no. 2, 1965, pp. 195–219.

[66] Volkonskii V.A., Belen'kii V.Z., Ivankov S.A., Pomanskii A.B., Shapiro A.D., *Iterativnye metody v teorii igr i programmirovanii* (Iterative Methods in Game Theory and Programming), M.: Nauka, 1974.

[67] Wismer D.A., Distributed Multilevel Systems, Optimization Methods for Large-Scale Systems with Applications, New-York, 1971, pp. 233-260.

[68] Yakovleva M.A., *O passhirenii oblasti primenemija spechial'nykh algoritmov linejnogo programmirovaniya* (On Extending the Application Domain of Special Algorithms of Linear Programming), *Dokl. Akad. Nauk SSSR*, vol. 179, no. 5, 1968, pp. 1067–1069.

[69] Yakovleva M.A., *Dvuhkomponentnaya zadacha linejnogo programmirovaniya* (Two-Component Linear Programming Problem) // Optimal'noe Planirovanie, no.2, Novosibirsk, 1964.

[70] Yudin D.B. and Gol'shtein E.G., *Zadachi i metody linejnogo programmirovaniya* (Problems and Methods of Linear Programming), Moscow: Sov. Radio, 1964.

[71] Zschau E., A Primal Decomposition Algorithm for Linear Programming, *Graduate School of Buiseness, Stanford University*, no. 91, 1967.

[72] Zoutendijk G., *Metody vozmozhnykh napravlenij* (Methods of Feasible Directions), Moscow: Inostr. Lit., 1963.

[73] Zvyagina R.A., *Zadachi linejnogo programmirovaniya s matritsami proizvol'noj blochoj struktury* (Linear Programming Problem with Arbitrary Block Structure Matrices), *Dokl. Akad. Nauk SSSR*, vol. 196, no. 4, 1971.

[74] Zvyagina R.A., *Zadachi linejnogo programmirovaniya s blochno-diagonal'nymi matritsami* (Linear Programming Problem with Block Diagonal Matrices) // Optimal'noe Planirovanie, no. 2, Novosibirsk, 1964, pp. 755-758.

[75] Zvyagina R.A., *Ob obschem metode resheniya zadach linejnogo programmirovanija blochnoj struktury* (On a General Method of Solution of Linear Programming Problems with Block Structure) // Optimizatsiya, 1(18), Novosibirsk, 1971.

Index

Applied Optimization

1. D.-Z. Du and D.F. Hsu (eds.): *Combinatorial Network Theory.* 1996
ISBN 0-7923-3777-8
2. M.J. Panik: *Linear Programming: Mathematics, Theory and Algorithms.* 1996
ISBN 0-7923-3782-4
3. R.B. Kearfott and V. Kreinovich (eds.): *Applications of Interval Computations.* 1996 ISBN 0-7923-3847-2
4. N. Hritonenko and Y. Yatsenko: *Modeling and Optimimization of the Lifetime of Technology.* 1996 ISBN 0-7923-4014-0
5. T. Terlaky (ed.): *Interior Point Methods of Mathematical Programming.* 1996
ISBN 0-7923-4201-1
6. B. Jansen: *Interior Point Techniques in Optimization.* Complementarity, Sensitivity and Algorithms. 1997 ISBN 0-7923-4430-8
7. A. Migdalas, P.M. Pardalos and S. Story (eds.): *Parallel Computing in Optimization.* 1997 ISBN 0-7923-4583-5
8. F.A. Lootsma: *Fuzzy Logic for Planning and Decision Making.* 1997
ISBN 0-7923-4681-5
9. J.A. dos Santos Gromicho: *Quasiconvex Optimization and Location Theory.* 1998
ISBN 0-7923-4694-7
10. V. Kreinovich, A. Lakeyev, J. Rohn and P. Kahl: *Computational Complexity and Feasibility of Data Processing and Interval Computations.* 1998
ISBN 0-7923-4865-6
11. J. Gil-Aluja: *The Interactive Management of Human Resources in Uncertainty.* 1998
ISBN 0-7923-4886-9
12. C. Zopounidis and A.I. Dimitras: *Multicriteria Decision Aid Methods for the Prediction of Business Failure.* 1998 ISBN 0-7923-4900-8
13. F. Giannessi, S. Komlósi and T. Rapcsák (eds.): *New Trends in Mathematical Programming.* Homage to Steven Vajda. 1998 ISBN 0-7923-5036-7
14. Ya-xiang Yuan (ed.): *Advances in Nonlinear Programming.* Proceedings of the '96 International Conference on Nonlinear Programming. 1998 ISBN 0-7923-5053-7
15. W.W. Hager and P.M. Pardalos: *Optimal Control.* Theory, Algorithms, and Applications. 1998 ISBN 0-7923-5067-7
16. Gang Yu (ed.): *Industrial Applications of Combinatorial Optimization.* 1998
ISBN 0-7923-5073-1
17. D. Braha and O. Maimon (eds.): *A Mathematical Theory of Design: Foundations, Algorithms and Applications.* 1998 ISBN 0-7923-5079-0

Applied Optimization

18. O. Maimon, E. Khmelnitsky and K. Kogan: *Optimal Flow Control in Manufacturing.* Production Planning and Scheduling. 1998 ISBN 0-7923-5106-1
19. C. Zopounidis and P.M. Pardalos (eds.): *Managing in Uncertainty: Theory and Practice.* 1998 ISBN 0-7923-5110-X
20. A.S. Belenky: *Operations Research in Transportation Systems:* Ideas and Schemes of Optimization Methods for Strategic Planning and Operations Management. 1998 ISBN 0-7923-5157-6
21. J. Gil-Aluja: *Investment in Uncertainty.* 1999 ISBN 0-7923-5296-3
22. M. Fukushima and L. Qi (eds.): *Reformulation: Nonsmooth, Piecewise Smooth, Semismooth and Smooting Methods.* 1999 ISBN 0-7923-5320-X
23. M. Patriksson: *Nonlinear Programming and Variational Inequality Problems.* A Unified Approach. 1999 ISBN 0-7923-5455-9
24. R. De Leone, A. Murli, P.M. Pardalos and G. Toraldo (eds.): *High Performance Algorithms and Software in Nonlinear Optimization.* 1999 ISBN 0-7923-5483-4
25. A. Schöbel: *Locating Lines and Hyperplanes.* Theory and Algorithms. 1999 ISBN 0-7923-5559-8
26. R.B. Statnikov: *Multicriteria Design.* Optimization and Identification. 1999 ISBN 0-7923-5560-1
27. V. Tsurkov and A. Mironov: *Minimax under Transportation Constrains.* 1999 ISBN 0-7923-5609-8
28. V.I. Ivanov: *Model Development and Optimization.* 1999 ISBN 0-7923-5610-1
29. F.A. Lootsma: *Multi-Criteria Decision Analysis via Ratio and Difference Judgement.* 1999 ISBN 0-7923-5669-1
30. A. Eberhard, R. Hill, D. Ralph and B.M. Glover (eds.): *Progress in Optimization.* Contributions from Australasia. 1999 ISBN 0-7923-5733-7
31. T. Hürlimann: *Mathematical Modeling and Optimization.* An Essay for the Design of Computer-Based Modeling Tools. 1999 ISBN 0-7923-5927-5
32. J. Gil-Aluja: *Elements for a Theory of Decision in Uncertainty.* 1999 ISBN 0-7923-5987-9
33. H. Frenk, K. Roos, T. Terlaky and S. Zhang (eds.): *High Performance Optimization.* 1999 ISBN 0-7923-6013-3
34. N. Hritonenko and Y. Yatsenko: *Mathematical Modeling in Economics, Ecology and the Environment.* 1999 ISBN 0-7923-6015-X
35. J. Virant: *Design Considerations of Time in Fuzzy Systems.* 2000 ISBN 0-7923-6100-8

Applied Optimization

36. G. Di Pillo and F. Giannessi (eds.): *Nonlinear Optimization and Related Topics*. 2000
ISBN 0-7923-6109-1
37. V. Tsurkov: *Hierarchical Optimization and Mathematical Physics*. 2000
ISBN 0-7923-6175-X

KLUWER ACADEMIC PUBLISHERS – DORDRECHT / BOSTON / LONDON

Kolibri 360 GmbH
Ferdinand-Jühlke-Straße 7
99095 Erfurt, Deutschland
produktsicherheit@kolibri360.de